Cell Biology: Structure and Function of Cell

Cell Biology: Structure and Function of Cell

Natasha Rivera

www.larsen-keller.com

Cell Biology: Structure and Function of Cell
Natasha Rivera
ISBN: 978-1-64172-613-9 (Hardback)

Larsen & Keller

Published by Larsen and Keller Education,
5 Penn Plaza,
19th Floor,
New York, NY 10001, USA

Cataloging-in-Publication Data

Cell biology : structure and function of cell / Natasha Rivera.
    p. cm.
Includes bibliographical references and index.
ISBN 978-1-64172-613-9
1. Cytology. 2. Cells. 3. Biology. I. Rivera, Natasha.
QH581.2 .C45 2022
571.6--dc23

For more information regarding Larsen and Keller Education and its products, please visit the publisher's website www.larsen-keller.com

Table of Contents

Permissions

Index

Preface

This book aims to help a broader range of students by exploring a wide variety of significant topics related to this discipline. It will help students in achieving a higher level of understanding of the subject and excel in their respective fields. This book would not have been possible without the unwavered support of my senior professors who took out the time to provide me feedback and help me with the process. I would also like to thank my family for their patience and support.

The branch of biology that deals with the study of the structure and function of the cell is known as cell biology. It is involved in the study of various aspects of the cell such as its physiological properties, signaling pathways, metabolic processes and life cycle. It also studies the chemical composition and interactions of the cell with their environment. Research in this field is conducted at both microscopic and molecular levels. The cells which are studied in cell biology are broadly classified as either prokaryotic or eukaryotic. Prokaryotic cells do not have a membrane bound nucleus while eukaryotic cells have a membrane bound nucleus as well as membrane bound organelles. Cell biology plays an important role in the diagnosis and treatment of many diseases such as cancer. The study in cell biology is closely related to the fields of genetics, molecular biology, immunology, biochemistry and cytochemistry. The book aims to shed light on some of the unexplored aspects of cell biology. Different approaches, evaluations and concepts related to this field have been included herein. This textbook aims to serve as a resource guide for students and experts alike and contribute to the growth of the discipline.

A brief overview of the book contents is provided below:

Chapter – What is Cell Biology?

Cell biology primarily focuses on the study of the structure and functions of a cell as well as its physiological properties, life cycle, chemical composition, metabolic processes, signaling pathways and interactions with the environment. This is an introductory chapter which will introduce briefly all these aspects of cell biology.

Chapter – Fundamental Concepts in Cell Biology

Cell theory, cell division and symbiogenesis are some of the fundamental concepts in cell biology. The chapter closely examines these fundamental concepts of cell biology as well as important areas of study such as lipid bilayer to provide an extensive understanding of the subject.

Chapter – Different Types of Cell

There are two types of cells: eukaryotic and prokaryotic cells. Eukaryotic cells contain a nucleus and can be either single-celled or multicellular. Prokaryotic cells do not comprise of a nucleus and are single-celled organisms. The topics elaborated in this chapter will help in gaining a better perspective about these two categories of cells.

Chapter – Cell Structures

There are numerous structures which make up a cell. A few of the major structures are cell membrane, cell wall, cytoskeleton, cytoplasm, centriole, plasmid, cytosol and thylakoid. This chapter has been carefully written to provide an easy understanding of the varied aspects of these cell structures.

Chapter – Cell Organelles

Cell organelles are the specialized sub-units within a cell that have well-defined functions. They can be either membrane-bound or non-membrane bound. Various cell organelles include ribosome, lysosome, endosome, vacuole, cell nucleus, chloroplast, mitochondrion, etc. This chapter has been carefully written to provide an easy understanding of these cell organelles.

Chapter – Cellular Activities

Fundamental biological activities in cells include cellular reproduction, cell adhesion, cell signaling and cellular respiration. Some of the methods of cellular reproduction are meiosis, mitosis and fission. All these diverse biological activities of cells have been carefully analyzed in this chapter.

Natasha Rivera

1

What is Cell Biology?

Cell biology primarily focuses on the study of the structure and functions of a cell as well as its physiological properties, life cycle, chemical composition, metabolic processes, signalling pathways and interactions with the environment. This is an introductory chapter which will introduce briefly all these aspects of cell biology.

CELL

Cell is the basic membrane-bound unit that contains the fundamental molecules of life and of which all living things are composed. A single cell is often a complete organism in itself, such as a bacterium or yeast. Other cells acquire specialized functions as they mature. These cells cooperate with other specialized cells and become the building blocks of large multicellular organisms, such as humans and other animals. Although cells are much larger than atoms, they are still very small. The smallest known cells are a group of tiny bacteria called mycoplasmas; some of these single-celled organisms are spheres as small as 0.2 μm in diameter (1μm = about 0.000039 inch), with a total mass of 10^{-14} gram—equal to that of 8,000,000,000 hydrogen atoms. Cells of humans typically have a mass 400,000 times larger than the mass of a single mycoplasma bacterium, but even human cells are only about 20 μm across. It would require a sheet of about 10,000 human cells to cover the head of a pin, and each human organism is composed of more than 30,000,000,000,000 cells.

Nature and Function of Cells

A cell is enclosed by a plasma membrane, which forms a selective barrier that allows nutrients to enter and waste products to leave. The interior of the cell is organized into many specialized compartments, or organelles, each surrounded by a separate membrane. One major organelle, the nucleus, contains the genetic information necessary for cell growth and reproduction. Each cell contains only one nucleus, whereas other types of organelles are present in multiple copies in the cellular contents, or cytoplasm. Organelles include mitochondria, which are responsible for the energy transactions necessary for cell survival; lysosomes, which digest unwanted materials within the cell; and the endoplasmic reticulum and the Golgi apparatus, which play important roles in the internal organization of the cell by synthesizing selected molecules and then processing, sorting, and directing them to their proper locations. In addition, plant cells contain chloroplasts, which are responsible for photosynthesis, whereby the energy of sunlight is used to convert molecules of carbon dioxide (CO_2) and water (H_2O) into carbohydrates. Between all these organelles is the space in the cytoplasm called the cytosol. The cytosol contains an organized framework of fibrous molecules that constitute the cytoskeleton, which gives a cell its shape, enables organelles to move within the cell, and provides a mechanism by which the cell itself can move. The cytosol also contains more than 10,000 different kinds of molecules that are involved in cellular biosynthesis, the process of making large biological molecules from small ones.

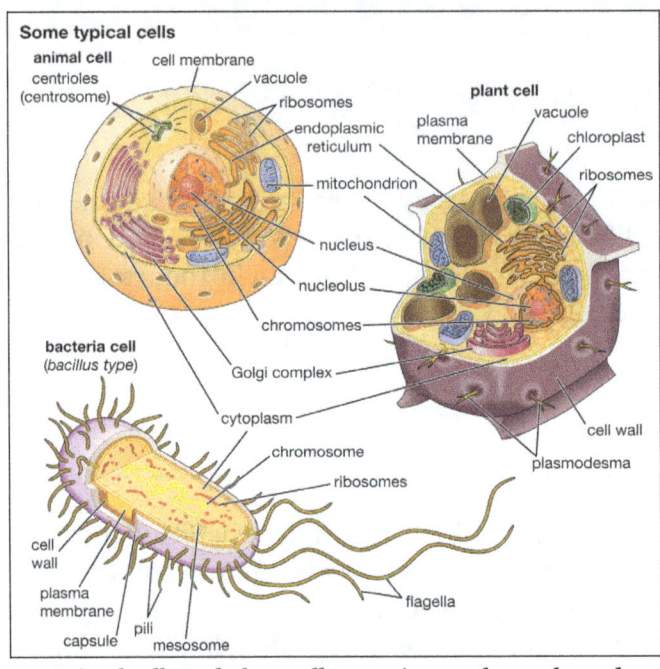

Animal cells and plant cells contain membrane-bound organelles, including a distinct nucleus. In contrast, bacterial cells do not contain organelles.

Specialized organelles are a characteristic of cells of organisms known as eukaryotes. In contrast, bcells of organisms known as prokaryotes do not contain organelles and are generally smaller than eukaryotic cells. However, all cells share strong similarities in biochemical function.

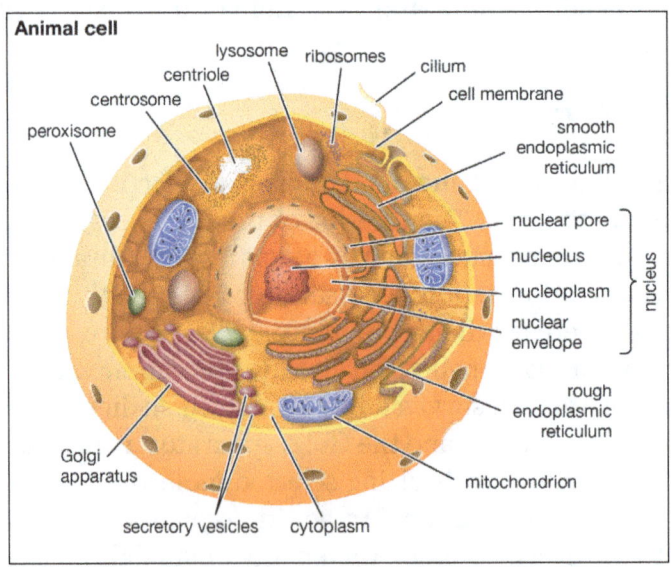

Cutaway drawing of a eukaryotic cell.

Molecules of Cells

Cells contain a special collection of molecules that are enclosed by a membrane. These molecules give cells the ability to grow and reproduce. The overall process of cellular reproduction occurs in two steps: cell growth and cell division. During cell growth, the cell ingests certain molecules from its surroundings by selectively carrying them through its cell membrane. Once inside the cell, these molecules are subjected to the action of highly specialized, large, elaborately folded molecules called enzymes. Enzymes act as catalysts by binding to ingested molecules and regulating the rate at which they are chemically altered. These chemical alterations make the molecules more useful to the cell. Unlike the ingested molecules, catalysts are not chemically altered themselves during the reaction, allowing one catalyst to regulate a specific chemical reaction in many molecules.

Biological catalysts create chains of reactions. In other words, a molecule chemically transformed by one catalyst serves as the starting material, or substrate, of a second catalyst and so on. In this way, catalysts use the small molecules brought into the cell from the outside environment to create increasingly complex reaction products. These products are used for cell growth and the replication of genetic material. Once the genetic material has been copied and there are sufficient molecules to support cell division, the cell divides to create two daughter cells. Through many such cycles of cell growth and division, each parent cell can give rise to millions of daughter cells, in the process converting large amounts of inanimate matter into biologically active molecules.

Structure of Biological Molecules

Cells are largely composed of compounds that contain carbon. The study of how carbon atoms interact with other atoms in molecular compounds forms the basis of the field of organic chemistry and plays a large role in understanding the basic functions of cells. Because carbon atoms can form stable bonds with four other atoms, they are uniquely suited for the construction of complex molecules. These complex molecules are typically made up of chains and rings that contain hydrogen, oxygen, and nitrogen atoms, as well as carbon atoms. These molecules may consist of anywhere from 10 to millions of atoms linked together in specific arrays. Most, but not all, of the carbon-containing molecules in cells are built up from members of one of four different families of small organic molecules: sugars, amino acids, nucleotides, and fatty acids. Each of these families contains a group of molecules that resemble one another in both structure and function. In addition to other important functions, these molecules are used to build large macromolecules. For example, the sugars can be linked to form polysaccharides such as starch and glycogen, the amino acids can be linked to form proteins, the nucleotides can be linked to form the DNA (deoxyribonucleic acid) and RNA (ribonucleic acid) of chromosomes, and the fatty acids can be linked to form the lipids of all cell membranes.

Approximate chemical composition of a typical mammalian cell	
component	percent of total cell weight
water	70
inorganic ions (sodium, potassium, magnesium, calcium, chloride, etc.)	1

miscellaneous small metabolites	3
proteins	18
RNA	1.1
DNA	0.25
phospholipids and other lipids	5
polysaccharides	2

Aside from water, which forms 70 percent of a cell's mass, a cell is composed mostly of macromolecules. By far the largest portion of macromolecules are the proteins. An average-sized protein macromolecule contains a string of about 400 amino acid molecules. Each amino acid has a different side chain of atoms that interact with the atoms of side chains of other amino acids. These interactions are very specific and cause the entire protein molecule to fold into a compact globular form. In theory, nearly an infinite variety of proteins can be formed, each with a different sequence of amino acids. However, nearly all these proteins would fail to fold in the unique ways required to form efficient functional surfaces and would therefore be useless to the cell. The proteins present in cells of modern animals and humans are products of a long evolutionary history, during which the ancestor proteins were naturally selected for their ability to fold into specific three-dimensional forms with unique functional surfaces useful for cell survival.

Most of the catalytic macromolecules in cells are enzymes. The majority of enzymes are proteins. Key to the catalytic property of an enzyme is its tendency to undergo a change in its shape when it binds to its substrate, thus bringing together reactive groups on substrate molecules. Some enzymes are macromolecules of RNA, called ribozymes. Ribozymes consist of linear chains of nucleotides that fold in specific ways to form unique surfaces, similar to the ways in which proteins fold. As with proteins, the specific sequence of nucleotide subunits in an RNA chain gives each macromolecule a unique character. RNA molecules are much less frequently used as catalysts in cells than are protein molecules, presumably because proteins, with the greater variety of amino acid side chains, are more diverse and capable of complex shape changes. However, RNA molecules are thought to have preceded protein molecules during evolution and to have catalyzed most of the chemical reactions required before cells could evolve.

Coupled Chemical Reactions

Cells must obey the laws of chemistry and thermodynamics. When two molecules react with each other inside a cell, their atoms are rearranged, forming different molecules as reaction products and releasing or consuming energy in the process. Overall, chemical reactions occur only in one direction; that is, the final reaction product molecules cannot spontaneously react, in a reversal of the original process, to reform the original molecules. This directionality of chemical reactions is explained by the fact that molecules only change from states of higher free energy to states of lower free energy. Free energy is the ability to perform work (in this case, the "work" is the rearrangement of atoms in the chemical reaction). When work is performed, some free energy is used and lost, with the result that the process ends at lower free energy. To use a familiar mechanical analogy, water at the top of a hill has the ability to perform the "work" of flowing downhill (i.e., it has high free energy), but, once it has flowed downhill, it cannot flow back up (i.e., it is in a state of low free energy). However, through another work process—that of a pump, for example—the water can be returned to the top of the hill, thereby recovering its ability to flow downhill. In thermodynamic terms, the free energy of the water has been increased by energy from an outside source (i.e., the pump). In the same way, the product molecules of a chemical reaction in a cell cannot reverse the reaction and return to their original state unless energy is supplied by coupling the process to another chemical reaction.

All catalysts, including enzymes, accelerate chemical reactions without affecting their direction. To return to the mechanical analogy, enzymes cannot make water flow uphill, although they can provide specific pathways for a downhill flow. Yet most of the chemical reactions that the cell needs to synthesize new molecules necessary for its growth require an uphill flow. In other words, the reactions require more energy than their starting molecules can provide.

Cells use a single strategy over and over again in order to get around the limitations of chemistry: they use the energy from an energy-releasing chemical reaction to drive an energy-absorbing reaction that would otherwise not occur. A useful mechanical analogy might be a mill wheel driven by the water in a stream. The water, in order to flow downhill, is forced to flow past the blades of the wheel, causing the wheel to turn. In this way, part of the energy from the moving stream is harnessed to move a mill wheel, which may be linked to a winch. As the winch turns, it can be used to pull a heavy load uphill. Thus, the energy-absorbing (but useful) uphill movement of a load can be driven by coupling it directly to the energy-releasing flow of water.

In cells, enzymes play the role of mill wheels by coupling energy-releasing reactions with energy-absorbing reactions. In cells the most important energy-releasing reaction serving a role similar to that of the flowing stream is the hydrolysis of adenosine triphosphate (ATP). In turn, the production of ATP molecules in the cells is an energy-absorbing reaction that is driven by being coupled to the energy-releasing breakdown of sugar molecules. In retracing this chain of reactions, it is necessary first to understand the source of the sugar molecules.

Photosynthesis: The Beginning of the Food Chain

Sugar molecules are produced by the process of photosynthesis in plants and certain bacteria. These organisms lie at the base of the food chain, in that animals and other nonphotosynthesizing organisms depend on them for a constant supply of life-supporting organic molecules. Humans, for example, obtain these molecules by eating plants or other organisms that have previously eaten food derived from photosynthesizing organisms.

Plants and photosynthetic bacteria are unique in their ability to convert the freely available electromagnetic energy in sunlight into chemical bond energy, the energy that holds atoms together in molecules and is transferred or released in chemical reactions. The process of photosynthesis can be summarized by the following equation:

$$\text{(solar) energy} + CO_2 + H_2O \rightarrow \text{sugar molecules} + O_2$$

The energy-absorbing photosynthetic reaction is the reverse of the energy-releasing oxidative decomposition of sugar molecules. During photosynthesis, chlorophyll molecules absorb energy from sunlight and use it to fuel the production of simple sugars and other carbohydrates. The resulting abundance of sugar molecules and related biological products makes possible the existence of nonphotosynthesizing life on Earth.

ATP: Fueling Chemical Reactions

Certain enzymes catalyze the breakdown of organic foodstuffs. Once sugars are transported into cells, they either serve as building blocks in the form of amino acids for proteins and fatty acids for lipids or are subjected to metabolic pathways to provide the cell with ATP. ATP, the common carrier of energy inside the cell, is made from adenosine diphosphate (ADP) and inorganic phosphate (P_i). Stored in the chemical bond holding the terminal phosphate compound onto the ATP molecule is the energy derived from the breakdown of sugars. The removal of the terminal phosphate, through the water-mediated reaction called hydrolysis, releases this energy, which in turn fuels a large number of crucial energy-absorbing reactions in the cell. Hydrolysis can be summarized as follows:

$$\text{ATP} + H_2O \rightarrow \text{ADP} + P_i + \text{energy}$$

The formation of ATP is the reverse of this equation, requiring the addition of energy. The central cellular pathway of ATP synthesis begins with glycolysis, a form of fermentation in which the sugar glucose is transformed into other sugars in a series of nine enzymatic reactions, each successive reaction involving an intermediate sugar containing phosphate. In the process, the six-carbon glucose is converted into two molecules of the three-carbon pyruvic acid. Some of the energy released through glycolysis of each glucose molecule is captured in the formation of two ATP molecules.

The second stage in the metabolism of sugars is a set of interrelated reactions called the tricarboxylic acid cycle. This cycle takes the three-carbon pyruvic acid produced in glycolysis and uses its carbon atoms to form carbon dioxide (CO_2) while transferring its hydrogen atoms to special carrier molecules, where they are held in high-energy linkage.

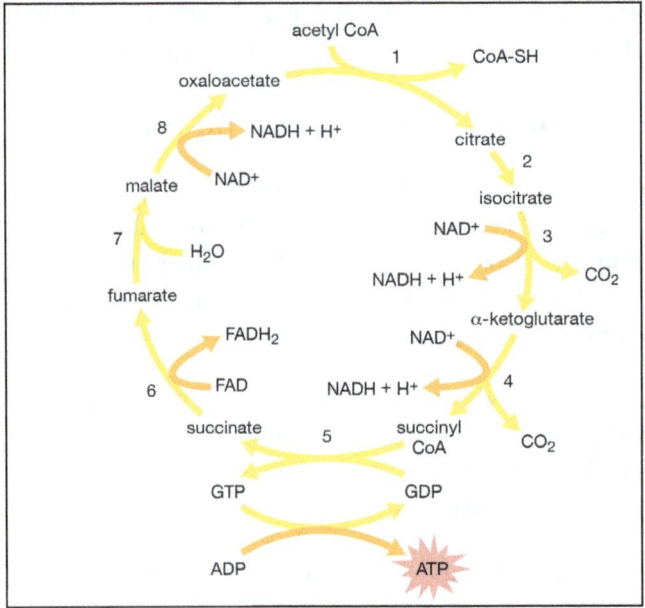

Tricarboxylic acid cycle.

In the third and last stage in the breakdown of sugars, oxidative phosphorylation, the high-energy hydrogen atoms are first separated into protons and high-energy electrons. The electrons are then passed from one electron carrier to another by means of an electron-transport chain. Each electron carrier in the chain has an increasing affinity for electrons, with the final electron acceptor being molecular oxygen (O_2). As separated electrons and protons, the hydrogen atoms are transferred to O_2 to form water. This reaction releases a large amount of energy, which drives the synthesis of a large number of ATP molecules from ADP and Pi.

Most of the cell's ATP is produced when the products of glycolysis are oxidized completely by a combination of the tricarboxylic acid cycle and oxidative phosphorylation. The process of glycolysis alone produces relatively small amounts of ATP. Glycolysis is an anaerobic reaction; that is, it can occur even in the absence of oxygen. The tricarboxylic acid cycle and oxidative phosphorylation, on the other hand, require oxygen. Glycolysis forms the basis of anaerobic fermentation, and it presumably was a major source of ATP for early life on Earth, when very little oxygen was available in the atmosphere. Eventually, however, bacteria evolved that were able to carry out photosynthesis. Photosynthesis liberated these bacteria from a dependence on the metabolism of organic materials that had accumulated from natural processes, and it also released oxygen into the atmosphere. Over a prolonged period of time, the concentration of molecular oxygen increased until it became freely available in the atmosphere. The aerobic tricarboxylic acid cycle and oxidative phosphorylation then evolved, and the resulting aerobic cells made much more efficient use of foodstuffs than their anaerobic ancestors, because they could convert much larger amounts of chemical bond energy into ATP.

Genetic Information of Cells

Cells can thus be seen as a self-replicating network of catalytic macromolecules engaged in a carefully balanced series of energy conversions that drive biosynthesis and cell movement. But energy alone is not enough to make self-reproduction possible; the cell must contain detailed instructions that dictate exactly how that energy is to be used. These instructions are analogous to the blueprints that a builder uses to construct a house; in the case of cells, however, the blueprints themselves must be duplicated along with the cell before it divides, so that each daughter cell can retain the instructions that it needs for its own replication. These instructions constitute the cell's heredity.

DNA: The Genetic Material

During the early 19th century, it became widely accepted that all living organisms are composed of cells arising only from the growth and division of other cells. The improvement of the microscope then led to an era during which many biologists made intensive observations of the microscopic structure of cells. By 1885 a substantial amount of indirect evidence indicated that chromosomes—dark-staining threads in the cell nucleus—carried the information for cell heredity. It was later shown that chromosomes are about half DNA and half protein by weight.

The revolutionary discovery suggesting that DNA molecules could provide the information for their own replication came in 1953, when American geneticist and biophysicist James Watson and British biophysicist Francis Crick proposed a model for the structure of the double-stranded DNA molecule (called the DNA double helix). In this model, each strand serves as a template in the synthesis of a complementary strand. Subsequent research confirmed the Watson and Crick model of DNA replication and showed that DNA carries the genetic information for reproduction of the entire cell.

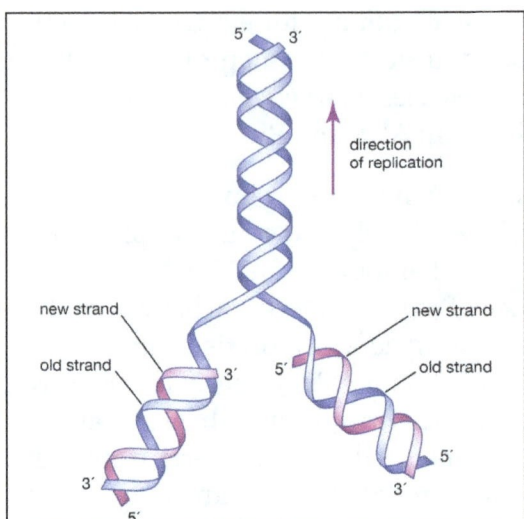

The initial proposal of the structure of DNA by James Watson and
Francis Crick was accompanied by a suggestion on the means of replication.

All of the genetic information in a cell was initially thought to be confined to the DNA in the chromosomes of the cell nucleus. Later discoveries identified small amounts of additional genetic information present in the DNA of much smaller chromosomes located in two types of organelles in the cytoplasm. These organelles are the mitochondria in animal cells and the mitochondria and chloroplasts in plant cells. The special chromosomes carry the information coding for a few of the many proteins and RNA molecules needed by the organelles. They also hint at the evolutionary origin of these organelles, which are thought to have originated as free-living bacteria that were taken up by other organisms in the process of symbiosis.

RNA: Replicated from DNA

It is possible for RNA to replicate itself by mechanisms related to those used by DNA, even though it has a single-stranded instead of a double-stranded structure. In early cells RNA is thought to have replicated itself in this way. However, all of the RNA in present-day cells is synthesized by special enzymes that construct a single-stranded RNA chain by using one strand of the DNA helix as a template. Although RNA molecules are synthesized in the cell nucleus, where the DNA is located, most of them are transported to the cytoplasm before they carry out their functions.

Molecular genetics emerged from the realization that DNA and RNA constitute the genetic material of all living organisms. (1) DNA, located in the cell nucleus, is made up of nucleotides that contain the bases adenine (A), thymine (T), guanine (G), and cytosine (C). (2) RNA, which contains uracil (U) instead of thymine, transports the genetic code to protein-synthesizing sites in the cell. (3) Messenger RNA (mRNA) then carries the genetic information to ribosomes in the cell cytoplasm that translate the genetic information into molecules of protein.

The RNA molecules in cells have two main roles. Some, the ribozymes, fold up in ways that allow them to serve as catalysts for specific chemical reactions. Others serve as "messenger RNA," which provides templates specifying the synthesis of proteins. Ribosomes, tiny protein-synthesizing machines located in the cytoplasm, "read" the messenger RNA molecules and "translate" them into proteins by using the genetic code. In this translation, the sequence of nucleotides in the messenger RNA chain is decoded three nucleotides at a time, and each nucleotide triplet (called a codon) specifies a particular amino acid. Thus, a nucleotide sequence in the DNA specifies a protein provided that a messenger RNA molecule is produced from that DNA sequence. Each region of the DNA sequence specifying a protein in this way is called a gene.

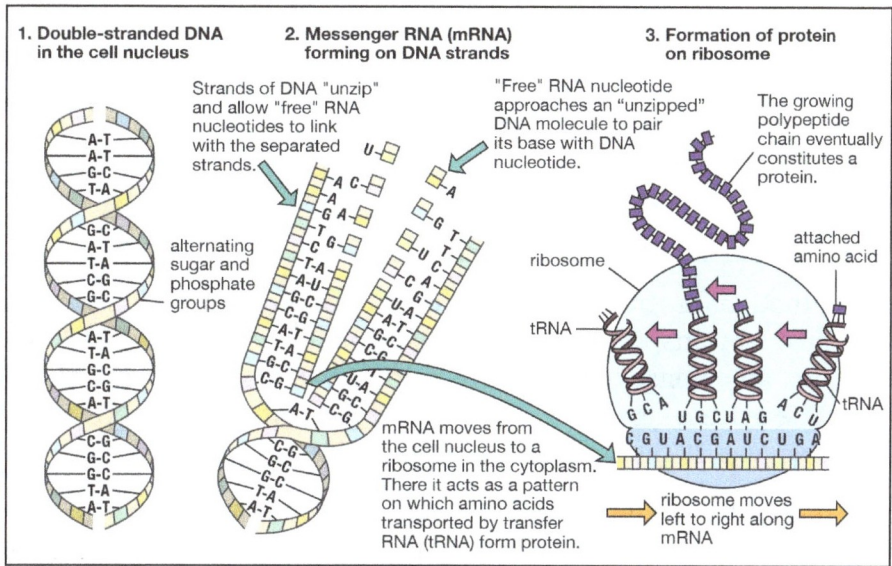

DNA directs protein synthesis.

By the above mechanisms, DNA molecules catalyze not only their own duplication but also dictate the structures of all protein molecules. A single human cell contains about 10,000 different proteins produced by the expression of 10,000 different genes. Actually, a set of human chromosomes is thought to contain DNA with enough information to express between 30,000 and 100,000 proteins, but most of these proteins seem to be made only in specialized types of cells and are therefore not present throughout the body.

Organization of Cells

Intracellular Communication

A cell with its many different DNA, RNA, and protein molecules is quite different from a test tube containing the same components. When a cell is dissolved in a test tube, thousands of different types of molecules randomly mix together. In the living cell, however, these components are kept in specific places, reflecting the high degree of organization essential for the growth and division of the cell. Maintaining this internal organization requires a continuous input of energy, because spontaneous chemical reactions always create disorganization. Thus, much of the energy released by ATP hydrolysis fuels processes that organize macromolecules inside the cell.

When a eukaryotic cell is examined at high magnification in an electron microscope, it becomes apparent that specific membrane-bound organelles divide the interior into a variety of subcompartments. Although not detectable in the electron microscope, it is clear from biochemical assays that each organelle contains a different set of macromolecules. This biochemical segregation reflects the functional specialization of each compartment. Thus, the mitochondria, which produce most of the cell's ATP, contain all of the enzymes needed to carry out the tricarboxylic acid cycle and oxidative phosphorylation. Similarly, the degradative enzymes needed for the intracellular digestion of unwanted macromolecules are confined to the lysosomes.

The relative volumes occupied by some cellular compartments in a typical liver cell		
cellular compartment	percent of total cell volume	approximate number per cell
cytosol	54	1
mitochondrion	22	1,700
endoplasmic reticulum plus Golgi apparatus	15	1
nucleus	6	1
lysosome	1	300

It is clear from this functional segregation that the many different proteins specified by the genes in the cell nucleus must be transported to the compartment where they will be used. Not surprisingly, the cell contains an extensive membrane-bound system devoted to maintaining just this intracellular order. The system serves as a post office, guaranteeing the proper routing of newly synthesized macromolecules to their proper destinations.

All proteins are synthesized on ribosomes located in the cytosol. As soon as the first portion of the amino acid sequence of a protein emerges from the ribosome, it is inspected for the presence of a short "endoplasmic reticulum (ER) signal sequence." Those ribosomes making proteins with such a sequence are transported to the surface of the ER membrane, where they complete their synthesis; the proteins made on these ribosomes are immediately transferred through the ER membrane to the inside of the ER compartment. Proteins lacking the ER signal sequence remain in the cytosol and are released from the ribosomes when their synthesis is completed. This chemical decision process places some newly completed protein chains in the cytosol and others within an extensive membrane-bounded compartment in the cytoplasm, representing the first step in intracellular protein sorting.

The newly made proteins in both cell compartments are then sorted further according to additional signal sequences that they contain. Some of the proteins in the cytosol remain there, while others go to the surface of mitochondria or (in plant cells) chloroplasts, where they are transferred through the membranes into the organelles. Subsignals on each of these proteins then designate exactly where in the organelle the protein belongs. The proteins initially sorted into the ER have an even wider range of destinations. Some of them remain in the ER, where they function as part of the organelle. Most enter transport vesicles and pass to the Golgi apparatus, separate membrane-bounded organelles that contain at least three subcompartments. Some of the proteins are retained in the subcompartments of the Golgi, where they are utilized for functions peculiar to that organelle. Most eventually enter vesicles that leave the Golgi for other cellular destinations such as the cell membrane, lysosomes, or special secretory vesicles.

Intercellular Communication

Formation of a multicellular organism starts with a small collection of similar cells in an embryo and proceeds by continuous cell division and specialization to produce an entire community of cooperating cells, each with its own role in the life of the organism. Through cell cooperation, the organism becomes much more than the sum of its component parts.

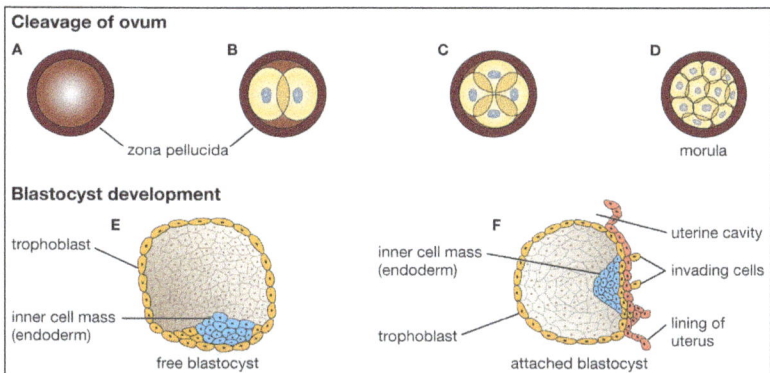

The ovum contains a small collection of cells in the early stages of human development. As cells divide (A–D), they are separated into different regions of the ovum. Each region of the ovum transmits a unique set of chemical signals to nearby cells. Thus, the signals detected by one cell differ from those detected by its neighbour cells. In this process, known as cell determination, cells are individually programmed to direct them toward development into different cell types.

A fertilized egg multiplies and produces a whole family of daughter cells, each of which adopts a structure and function according to its position in the entire assembly. All of the daughter cells contain the same chromosomes and therefore the same genetic information. Despite this common inheritance, different types of cells behave differently and have different structures. In order for this to be the case, they must express different sets of genes, so that they produce different proteins despite their identical embryological ancestors.

During the development of an embryo, it is not sufficient for all the cell types found in the fully developed individual simply to be created. Each cell type must form in the right place at the right time and in the correct proportion; otherwise, there would be a jumble of randomly assorted cells in no way resembling an organism. The orderly development of an organism depends on a process called cell determination, in which initially identical cells become committed to different pathways of development. A fundamental part of cell determination is the ability of cells to detect different chemicals within different regions of the embryo. The chemical signals detected by one cell may be different from the signals detected by its neighbour cells. The signals that a cell detects activate a set of genes that tell the cell to differentiate in ways appropriate for its position within the embryo. The set of genes activated in one cell differs from the set of genes activated in the cells around it. The process of cell determination requires an elaborate system of cell-to-cell communication in early embryos.

Cell Matrix and Cell-to-Cell Communication

The development of single cells into multicellular organisms involves a number of adaptations. The cells become specialized, acquiring distinct functions that contribute to the survival of the organism. The behaviour of individual cells is also integrated with that of similar cells, so that they act together in a regulated fashion. To achieve this integration, cells assemble into specialized tissues, each tissue being composed of cells and the spaces outside of the cells.

The surface of cells is important in coordinating their activities within tissues. Embedded in the plasma membrane of each cell are a number of proteins that interact with the surface or secretions of other cells. These proteins enable cells to "recognize" and adhere to the extracellular matrix and one another and to form populations distinct from surrounding cells. These interactions are key to the organizational behaviour of cell populations and contribute to the formation of embryonic tissues and the function of normal tissue in the adult organism.

Extracellular Matrix

A substantial part of tissues is the space outside of the cells, called the extracellular space. This is filled with a composite material, known as the extracellular matrix, composed of a gel in which a number of fibrous proteins are suspended. The gel consists of large polysaccharide (complex sugar) molecules in a water solution of inorganic salts, nutrients, and waste products known as the interstitial fluid. The major types of protein in the matrix are structural proteins and adhesive proteins.

Electron micrograph of a small area of dense fibrous connective tissue, illustrating the intimate association of cells and fibres. In the centre is a portion of a fibrocyte, and on either side are two collagen fibres. The collagen fibre on the left is cut transversely, showing round cross sections of the unit fibrils. The collagen fibre on the right has been cut nearly parallel to its long axis and shows extensive segments of the cross-striated fibrils.

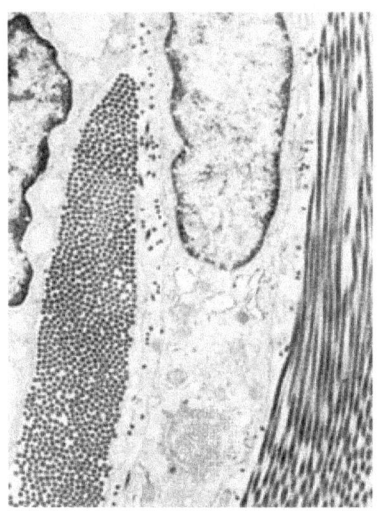

Electron micrograph of a small area of dense fibrous connective tissue, illustrating the intimate association of cells and fibres. In the centre is a portion of a fibrocyte, and on either side are two collagen fibres. The collagen fibre on the left is cut transversely, showing round cross sections of the unit fibrils. The collagen fibre on the right has been cut nearly parallel to its long axis and shows extensive segments of the cross-striated fibrils.

There are two general types of tissues distinct not only in their cellular organization but also in the composition of their extracellular matrix. The first type, mesenchymal tissue, is made up of clusters of cells grouped together but not closely adherent to one another. They synthesize a highly hydrated gel, rich in salts, fluid, and fibres, known as the interstitial matrix. Connective tissue is a mesenchyme that fastens together other more highly organized tissues. The solidity of various connective tissues varies according to the consistency of their extracellular matrix, which in turn depends on the water content of the gels, the amount and type of polysaccharides and structural proteins, and the presence of other salts. For example, bone is rich in calcium phosphate, giving that tissue its rigidity; tendons are mostly fibrous structural proteins, yielding a ropelike consistency; and joint spaces are filled with a lubricating fluid of mostly polysaccharide and interstitial fluid.

Epithelial tissues, the second type, are sheets of cells adhering at their side, or lateral, surfaces. They synthesize and deposit at their bottom, or basal, surfaces an organized complex of matrix materials known as the basal lamina or basement membrane. This thin layer serves as a boundary with connective tissue and as a substrate to which epithelial cells are attached.

Matrix Polysaccharides

The polysaccharides, or glycans, of the extracellular matrix are responsible for its gel-like quality and for organizing its components. These large acidic molecules exist alone (as glycosaminoglycans) or in combination with small proteins (as proteoglycans). They bind an extraordinarily large amount of water, thus forming massively swollen gels that fill the spaces between cells. Bound to proteins, they also organize other molecules in the extracellular matrix. The firmness and resiliency of cartilage, as at the surface of joints, is due to highly organized proteoglycans that bind water tightly.

Matrix Proteins

Matrix proteins are large molecules tightly bound to form extensive networks of insoluble fibres. These fibres may even exceed the size of the cells themselves. The proteins are of two general types, structural and adhesive.

The structural proteins, collagen and elastin, are the dominant matrix proteins. At least 10 different types of collagen are present in various tissues. The most common, type I collagen, is the most abundant protein in vertebrate animals, accounting for nearly 25 percent of the total protein in the body. The various collagen types share structural features, all being composed of three intertwined polypeptide chains. In some collagens the chains are linked together by covalent bonds, yielding a ropelike structure of great tensile strength. Indeed, the toughness of leather, chemically treated animal skin, is due to its content of collagen. Elastin is also a cross-linked protein, but, instead of forming rigid coils, it imparts elasticity to tissues. Only one type of elastin is known; it varies in elasticity according to variations in its cross-linking.

The adhesive proteins of the extracellular matrix bind matrix molecules to one another and to cell surfaces. These proteins are modular in that they contain several functional domains packaged together in a single molecule. Each domain binds to a specific matrix component or to a specific site on a cell. The major adhesive protein of the interstitial matrix is called fibronectin; that of the basal lamina is known as laminin.

Cell-matrix Interactions

Molecules intimately associated with the cell membrane link cells to the extracellular matrix. These molecules, called matrix receptors, bind selectively to specific matrix components and interact, directly or indirectly, with actin protein fibres that form the cytoskeleton inside the cell. This association of actin fibres with matrix components via receptors on the cell membrane can influence the organization of membrane molecules as well as matrix components and can modify the shape and function of the cytoskeleton. Changes in the cytoskeleton can lead to changes in cell shape, movement, metabolism, and development.

Intercellular Recognition and Cell Adhesion

The ability of cells to recognize and adhere to one another plays an important role in cell survival and reproduction. For example, when starved, several types of single-cell organisms band together to develop the specialized cells needed for reproduction. In this process, certain cells at the centre of the developing aggregate secrete chemicals that cause the other cells to adhere tightly into a group. In the case of slime mold amoebas, starvation causes the secretion of a compound, cyclic adenosine monophosphate (cyclic AMP, or CAMP), that induces the cells to stick together end to end. With further aggregation, the cells produce another cell-surface glycoprotein with which they stick to one another over their entire surfaces. The cellular aggregates then produce an extracellular matrix, which holds the cells together in a specific structural form.

Tissue and Species Recognition

Marine sponges are multicellular animals that can regenerate from single cells. The cells of a sponge rely on the processes of intercellular recognition and cellular adhesion to form aggregates of cells of the same species that eventually develop into an adult sponge.

Some multicellular animals or tissues can be dissociated into suspensions of single cells that show the same cellular recognition and adhesion as do aggregates of single-cell organisms. The marine sponge, for example, can be sieved through a mesh, yielding single cells and cells in clumps. When this cell suspension is rotated in culture, the cells reaggregate and in time reform a normal sponge. This reassociation shows selective cell recognition; that is, only cells of the same species reassociate. The ability of the cells to distinguish cells of their own species from those of others is mediated by proteoglycan molecules in the extracellular matrix. The proteoglycan binds to specific cell-surface receptor sites that are unique to a single species of sponge.

Cells from tissues of vertebrate animals can, like sponge cells, be dissociated and allowed to reaggregate. For example, when vertebrate embryonic cells from two different tissues are dissociated and then rotated together in culture, the cells form a multicellular aggregate within which they sort according to the type of tissue, a sorting that occurs regardless of whether the cells are from the same or different species. The specificity is due to a set of cell-surface glycoproteins called cell adhesion molecules (CAM). A portion of the CAM that extends from the surface of a cell adheres to identical molecules on the surface of adjacent cells. These CAM appear early in embryonic life, and their amounts in tissues change as the organs develop. The CAM, however, are not responsible for the stable adhesion of one cell to another; this more permanent adhesion is carried out by cell junctions.

Cell Junctions

There are three functional categories of cell junction: adhering junctions, often called desmosomes; tight, or occluding, junctions; and gap, or permeable, junctions. Adhering junctions hold cells together mechanically and are associated with intracellular fibres of the cytoskeleton. Tight junctions also hold cells together, but they form a nearly leakproof intercellular seal by fusion of adjacent cell membranes. Both adhering junctions and tight junctions are present primarily in epithelial cells. Many cell types also possess gap junctions, which allow small molecules to pass from one cell to the next through a channel.

Adhering Junctions

Cells subject to abrasion or other mechanical stress, such as those of the surface epithelia of the skin, have junctions that adhere cells to one another and to the extracellular matrix. These adhering junctions are called desmosomes when occurring between cells and hemidesmosomes (half-desmosomes) when linked to the matrix. Adhering junctions distribute mechanical shear force throughout the tissue and to the underlying matrix by virtue of their association with intermediate filaments crossing the interior of the cell. The linkage of these filaments, also called keratin filaments, to the desmosomes and, through these junctions, to adjacent cells provides a nearly continuous fibrous network throughout an epithelial sheet. Adhering junctions are also seen in other types of cells—for example, in the muscles of the heart and uterus—allowing these cells to remain anchored together despite the contractions of the muscles.

Tight Junctions

Sheets of cells separate fluids within the organs from fluids outside, as in the epithelial layer lining the intestine. This separation requires leakproof junctions between cells. Tight junctions form leakproof seals by fusing the plasma membranes of adjacent cells, creating a continuous barrier through which molecules cannot pass. The membranes are fused by tight associations of two types of specialized integral membrane proteins, in turn repelling large water-soluble molecules. In invertebrates this function is provided by septate junctions, in which the proteins of the membrane rather than the lipids form the seal.

Gap Junctions

These junctions allow communication between adjacent cells via the passage of small molecules directly from the cytoplasm of one cell to that of another. Molecules that can pass between cells coupled by gap junctions include inorganic salts, sugars, amino acids, nucleotides, and vitamins but not large molecules such as proteins or nucleic acids.

Gap junctions are crucial to the integration of certain cellular activities. For example, heart muscle cells generate electrical current by the movement of inorganic salts. If the cells are coupled, they will share this electrical current, allowing the synchronous contraction of all the cells in the tissue. This coupling function requires the regulation of molecular traffic through the gaps. The junctions are not open pores but dynamic channels, which change their permeability with changes in cellular activity. They consist of proteins completely crossing the cell membrane as six-sided columns with central pores. Under certain conditions the proteins are thought to change shape, causing the pores to become smaller or larger and thus changing the permeability of the junction.

Gap junctions are also found in tissues that are not electrically active. In these tissues, the junctions allow nutrients and waste products to travel throughout the tissue. Cells in such tissues are said to be metabolically coupled. During the formation of embryos, gap junctions are crucial to establishing differences between separate groups of cells, the coupled cells undergoing development together to become a specialized tissue.

Cell-to-cell Communication Via Chemical Signaling

In addition to cell-matrix and cell-cell interactions, cell behaviour in multicellular organisms is coordinated by the passage of chemical or electrical signals between cells. The most common form of chemical signaling is via molecules secreted from the cells and moving through the extracellular space. Signaling molecules may also remain on cell surfaces, influencing other cells only after the cells make physical contact. Finally, gap junctions allow small molecules to move between the cytoplasms of adjacent cells.

Types of Chemical Signaling

Chemical signals secreted by cells can act over varying distances. In the autocrine signaling process, molecules act on the same cells that produce them. In paracrine signaling, they act on nearby cells. Autocrine signals include extracellular matrix molecules and various factors that stimulate cell growth. An example of paracrine signals is the chemical transmitted from nerve to muscle that causes the muscle to contract. In this instance, the muscle cells have regions specialized to receive chemical signals from an adjacent nerve cell. In both autocrine and paracrine signaling, the chemical signal works in the immediate vicinity of the cell that produces it and is present at high concentrations. A chemical signal picked up by the bloodstream and taken to distant sites is called an endocrine signal. Most hormones produced in vertebrates are endocrine signals, such as the hormones produced in the pituitary gland at the base of the brain and carried by the bloodstream to act at low concentrations on the thyroid or adrenal glands.

The concentration at which a chemical signal acts has significance for its target cell. Chemical signals that act at high concentration act locally and rapidly. On the other hand, chemical signals that act at low concentrations act at distances and are generally slow.

Signal Receptors

The ability of a cell to respond to an extracellular signal depends on the presence of specific proteins called receptors, which are located on the cell surface or in the cytoplasm. Receptors bind chemical signals that ultimately trigger a mechanism to modify the behaviour of the target cell. Cells may contain an array of specific receptors that allow them to respond to a variety of chemical signals.

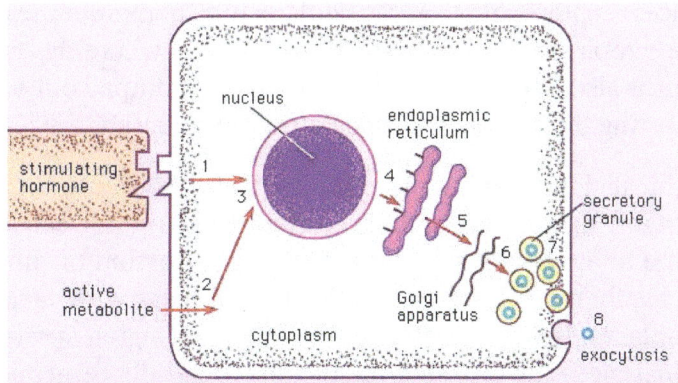

Hormones and active metabolites bind to different types of receptors. Water-soluble molecules (i.e., insulin) cannot pass through the lipid membrane of a cell and thus rely on cell surface receptors to transmit messages to the interior of the cell. In contrast, lipid-soluble molecules (i.e., certain active metabolites) are able to diffuse through the lipid membrane to communicate messages directly to the nucleus.

Signal molecules are either soluble or insoluble. Water-soluble molecules, such as the polypeptide hormone insulin, bind to receptors at cell surfaces. On the other hand, lipid-soluble molecules, such as the steroid hormones produced by the ovary or testis, pass through the lipid bilayer of the cell membrane to reach receptors within the cytoplasm. Extracellular matrix molecules are chemical signals, but, because of their size and insolubility, they act only on cell surface receptors and are neither taken up by the cells nor rapidly destroyed.

Cellular Response

The binding of chemical signals to their corresponding receptors induces events within the cell that ultimately change its behaviour. The nature of these intracellular events differs according to the type of receptor. Also, the same chemical signal can trigger different responses in different types of cell.

Cell surface receptors work in several ways when they are occupied. Some receptors enter the cell still bound to the chemical signal. Others activate membrane enzymes, which produce certain intracellular chemical mediators. Still other receptors open membrane channels, allowing a flow of ions that causes either a change in the electrical properties of the membrane or a change in the ion concentration in the cytoplasm. This regulation of enzymes or membrane channels produces changes in the concentration of intracellular signaling molecules, which are often called second messengers (the first messenger being the extracellular chemical signal bound to the receptor).

Two common intracellular signaling molecules are cyclic AMP and the calcium ion. Cyclic AMP is a derivative of adenosine triphosphate, the ubiquitous energy-carrying molecule of the cell. The intracellular concentrations of both cyclic AMP and calcium ions are normally very low. The binding of an extracellular chemical signal to a cell surface receptor stimulates an enzyme complex in the membrane to produce cyclic AMP. This second messenger then diffuses into the cytoplasm and acts on intracellular enzymes called kinases that modify the behaviour of the cell, culminating in the activation of target genes that increase the synthesis of certain proteins. The action of cyclic AMP is brief because it is rapidly degraded by specific enzymes.

Occupancy of other surface receptors causes a transient opening of membrane channels. This can allow calcium ions to enter the cytoplasm from the extracellular space, where their concentration is higher. The action of calcium ions is also brief because they are rapidly pumped out of the cell or bound to intracellular molecules, lowering the cytoplasmic concentration to the state existing before stimulation.

Some extracellular chemical signals enter the cell intact, still bound to the receptor, without generating a second messenger. In this mechanism, receptor occupancy causes individual receptors within the cell membrane to aggregate spontaneously. That portion of the membrane containing the aggregated receptors is then taken into the cell, where it fuses with various membrane-bounded organelles in the cytoplasm. In some instances the chemical signal is released within the organelles, and in almost all instances the ingested membrane is rapidly returned to the cell membrane along with the surface receptors.

Plant Cell Wall

The plant cell wall is a specialized form of extracellular matrix that surrounds every cell of a plant and is responsible for many of the characteristics distinguishing plant from animal cells. Although often perceived as an inactive product serving mainly mechanical and structural purposes, the cell wall actually has a multitude of functions upon which plant life depends. Such functions include: (1) providing the protoplast, or living cell, with mechanical protection and a chemically buffered environment, (2) providing a porous medium for the circulation and distribution of water, minerals, and other small nutrient molecules, (3) providing rigid building blocks from which stable structures of higher order, such as leaves and stems, can be produced, and (4) providing a storage site of regulatory molecules that sense the presence of pathogenic microbes and control the development of tissues.

Mechanical Properties of Wall Layers

All cell walls contain two layers, the middle lamella and the primary cell wall, and many cells produce an additional layer, called the secondary wall. The middle lamella serves as a cementing layer between the primary walls of adjacent cells. The primary wall is the cellulose-containing layer laid down by cells that are dividing and growing. To allow for cell wall expansion during growth, primary walls are thinner and less rigid than those of cells that have stopped growing. A fully grown plant cell may retain its primary cell wall (sometimes thickening it), or it may deposit an additional, rigidifying layer of different composition; this is the secondary wall. Secondary cell walls are responsible for most of the plant's mechanical support as well as the mechanical properties prized in wood. In contrast to the permanent stiffness and load-bearing capacity of thick secondary walls, the thin primary walls are capable of serving a structural, supportive role only when the vacuoles within the cell are filled with water to the point that they exert a turgor pressure against the cell wall. Turgor-induced stiffening of primary walls is analogous to the stiffening of the sides of a pneumatic tire by air pressure. The wilting of flowers and leaves is caused by a loss of turgor pressure, which results in turn from the loss of water from the plant cells.

Components of the Cell Wall

Although primary and secondary wall layers differ in detailed chemical composition and structural organization, their basic architecture is the same, consisting of cellulose fibres of great tensile strength embedded in a water-saturated matrix of polysaccharides and structural glycoproteins.

Cellulose

Cellulose consists of several thousand glucose molecules linked end to end. The chemical links between the individual glucose subunits give each cellulose molecule a flat ribbonlike structure that allows adjacent molecules to band laterally together into microfibrils with lengths ranging from two to seven micrometres. Cellulose fibrils are synthesized by enzymes floating in the cell membrane and are arranged in a rosette configuration. Each rosette appears capable of "spinning" a microfibril into the cell wall. During this process, as new glucose subunits are added to the growing end of the fibril, the rosette is pushed around the cell on the surface of the cell membrane, and its cellulose fibril becomes wrapped around the protoplast. Thus, each plant cell can be viewed as making its own cellulose fibril cocoon.

Matrix Polysaccharides

The two major classes of cell wall matrix polysaccharides are the hemicelluloses and the pectic polysaccharides, or pectins. Both are synthesized in the Golgi apparatus, brought to the cell surface in small vesicles, and secreted into the cell wall.

Hemicelluloses consist of glucose molecules arranged end to end as in cellulose, with short side chains of xylose and other uncharged sugars attached to one side of the ribbon. The other side of the ribbon binds tightly to the surface of cellulose fibrils, thereby coating the microfibrils with hemicellulose and preventing them from adhering together in an uncontrolled manner. Hemicellulose molecules have been shown to regulate the rate at which primary cell walls expand during growth.

The heterogeneous, branched, and highly hydrated pectic polysaccharides differ from hemicelluloses in important respects. Most notably, they are negatively charged because of galacturonic acid residues, which, together with rhamnose sugar molecules, form the linear backbone of all pectic polysaccharides. The backbone contains stretches of pure galacturonic acid residues interrupted by segments in which galacturonic acid and rhamnose residues alternate; attached to these latter segments are complex, branched sugar side chains. Because of their negative charge, pectic polysaccharides bind tightly to positively charged ions, or cations. In cell walls, calcium ions cross-link the stretches of pure galacturonic acid residues tightly, while leaving the rhamnose-containing segments in a more open, porous configuration. This cross-linking creates the semirigid gel properties characteristic of the cell wall matrix—a process exploited in the preparation of jellied preserves.

Proteins

Although plant cell walls contain only small amounts of protein, they serve a number of important functions. The most prominent group are the hydroxyproline-rich glycoproteins, shaped like rods with connector sites, of which extensin is a prominent example. Extensin contains 45 percent hydroxyproline and 14 percent serine residues distributed along its length. Every hydroxyproline residue carries a short side chain of arabinose sugars, and most serine residues carry a galactose sugar; this gives rise to long molecules, resembling bottle brushes, that are secreted into the cell wall toward the end of primary-wall formation and become covalently cross-linked into a mesh at the time that cell growth stops. Plant cells may control their ultimate size by regulating the time at which this cross-linking of extensin molecules occurs.

In addition to the structural proteins, cell walls contain a variety of enzymes. Most notable are those that cross-link extensin, lignin, cutin, and suberin molecules into networks. Other enzymes help protect plants against fungal pathogens by breaking fragments off of the cell walls of the fungi. The fragments in turn induce defense responses in underlying cells. The softening of ripe fruit and dropping of leaves in the autumn are brought about by cell wall-degrading enzymes.

Plastics

Cell wall plastics such as lignin, cutin, and suberin all contain a variety of organic compounds cross-linked into tight three-dimensional networks that strengthen cell walls and make them more resistant to fungal and bacterial attack. Lignin is the general name for a diverse group of polymers of aromatic alcohols. Deposited mostly in secondary cell walls and providing the rigidity of terrestrial vascular plants, it accounts for up to 30 percent of a plant's dry weight. The diversity of cross-links between the polymers—and the resulting tightness—makes lignin a formidable barrier to the penetration of most microbes.

Agave shawii (top) and Echeveria (bottom), two types of xerophytes (plants adapted to arid habitats). They develop highly cutinized fleshy leaves and stems for water storage with which they modulate the effects of strong sunlight, low humidity, and scant water.

Cutin and suberin are complex biopolyesters composed of fatty acids and aromatic compounds. Cutin is the major component of the cuticle, the waxy, water-repelling surface layer of cell walls exposed to the environment aboveground. By reducing the wetability of leaves and stems—and thereby affecting the ability of fungal spores to germinate—it plays an important part in the defense strategy of plants. Suberin serves with waxes as a surface barrier of underground parts. Its synthesis is also stimulated in cells close to wounds, thereby sealing off the wound surfaces and protecting underlying cells from dehydration.

Intercellular Communication

Plasmodesmata

Similar to the gap junction of animal cells is the plasmodesma, a channel passing through the cell wall and allowing direct molecular communication between adjacent plant cells. Plasmodesmata are lined with cell membrane, in effect uniting all connected cells with one continuous cell membrane. Running down the middle of each channel is a thin membranous tube that connects the endoplasmic reticula (ER) of the two cells. This structure is a remnant of the ER of the original parent cell, which, as the parent cell divided, was caught in the developing cell plate.

Although the precise mechanisms are not fully understood, the plasmodesma is thought to regulate the passage of small molecules such as salts, sugars, and amino acids by constricting or dilating the openings at each end of the channel.

Oligosaccharides with Regulatory Functions

The discovery of cell wall fragments with regulatory functions opened a new era in plant research. For years scientists had been puzzled by the chemical complexity of cell wall polysaccharides, which far exceeds the structural requirements of plant cell walls. The answer came when it was found that specific fragments of cell wall polysaccharides, called oligosaccharins, are able to induce specific responses in plant cells and tissues. One such fragment, released by enzymes used by fungi to break down plant cell walls, consists of a linear polymer of 10 to 12 galacturonic acid residues. Exposure of plant cells to such fragments induces them to produce antibiotics known as phytoalexins. In other experiments it has been shown that exposing strips of tobacco stem cells to a different type of cell wall fragment leads to the growth of roots; other fragments lead to the formation of stems, and yet others to the production of flowers. In all instances the concentration of oligosaccharins required to bring about the observed responses is equal to that of hormones in animal cells; indeed, oligosaccharins may be viewed as the oligosaccharide hormones of plants.

Cell Differentiation

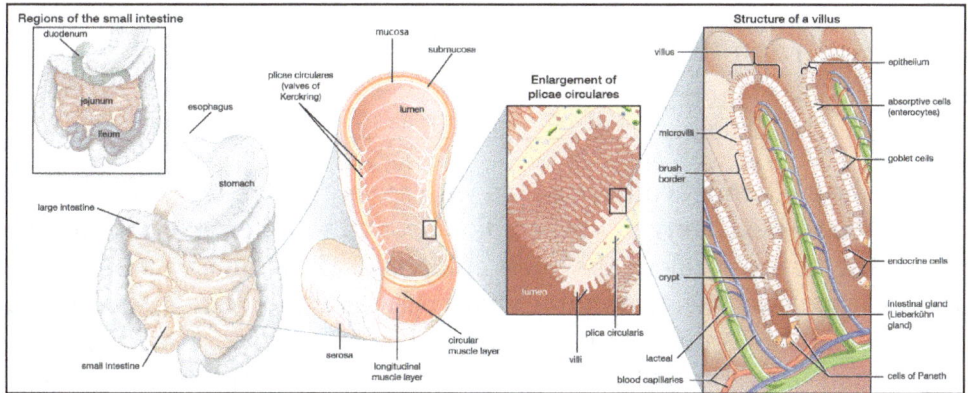

The small intestine contains many distinct types of cells,
each of which serves a specific function.

Adult organisms are composed of a number of distinct cell types. Cells are organized into tissues, each of which typically contains a small number of cell types and is devoted to a specific physiological function. For example, the epithelial tissue lining the small intestine contains columnar absorptive cells, mucus-secreting goblet cells, hormone-secreting endocrine cells, and enzyme-secreting Paneth cells. In addition, there exist undifferentiated dividing cells that lie in the crypts between the intestinal villi and serve to replace the other cell types when they become damaged or worn out. Another example of a differentiated tissue is the skeletal tissue of a long bone, which contains osteoblasts (large cells that synthesize bone) in the outer sheath and osteocytes (mature bone cells) and osteoclasts (multinucleate cells involved in bone remodeling) within the matrix.

In general, the simpler the overall organization of the animal, the fewer the number of distinct cell types that they possess. Mammals contain more than 200 different cell types, whereas simple invertebrate animals may have only a few different types. Plants are also made up of differentiated cells, but they are quite different from the cells of animals. For example, a leaf in a higher plant is covered with a cuticle layer of epidermal cells. Among these are pores composed of two specialized cells, which regulate gaseous exchange across the epidermis. Within the leaf is the mesophyll, a spongy tissue responsible for photosynthetic activity. There are also veins composed of xylem elements, which transport water up from the soil, and phloem elements, which transport products of photosynthesis to the storage organs.

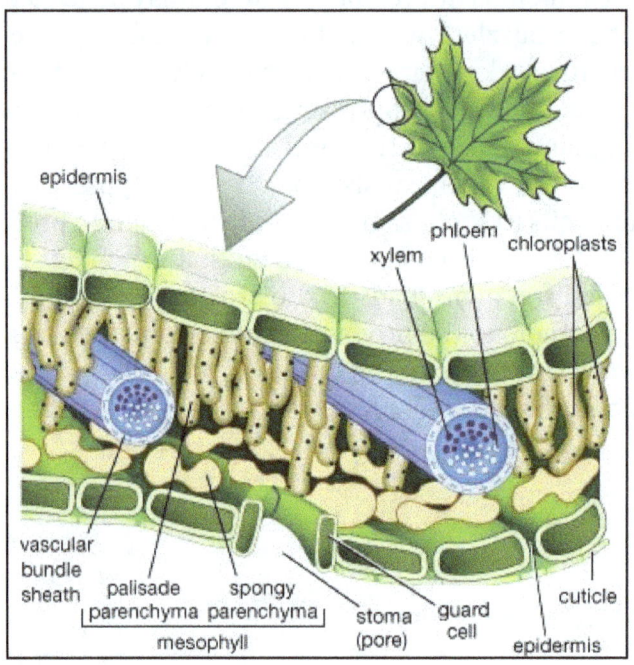

Structures of a leaf the epidermis is often covered with a waxy protective cuticle that helps prevent water loss from inside the leaf. Oxygen, carbon dioxide, and water enter and exit the leaf through pores (stomata) scattered mostly along the lower epidermis. The stomata are opened and closed by the contraction and expansion of surrounding guard cells. The vascular, or conducting, tissues are known as xylem and phloem; water and minerals travel up to the leaves from the roots through the xylem, and sugars made by photosynthesis are transported to other parts of the plant through the phloem. Photosynthesis occurs within the chloroplast-containing mesophyll layer.

The various cell types have traditionally been recognized and classified according to their appearance in the light microscope following the process of fixing, processing, sectioning, and staining tissues that is known as histology. Classical histology has been augmented by a variety of more discriminating techniques. Electron microscopy allows for higher magnifications. Histochemistry involves the use of coloured precipitating substrates to stain particular enzymes in situ. Immunohistochemistry uses specific antibodies to identify particular substances, usually proteins or carbohydrates, within cells. In situ hybridization involves the use of nucleic acid probes to visualize the location of specific messenger RNAs (mRNA). These modern methods have allowed the identification of more cell types than could be visualized by classical histology, particularly in the brain, the immune system, and among the hormone-secreting cells of the endocrine system.

Differentiated State

The biochemical basis of cell differentiation is the synthesis by the cell of a particular set of proteins, carbohydrates, and lipids. This synthesis is catalyzed by proteins called enzymes. Each enzyme in turn is synthesized in accordance with a particular gene, or sequence of nucleotides in the DNA of the cell nucleus. A particular state of differentiation, then, corresponds to the set of genes that is expressed and the level to which it is expressed.

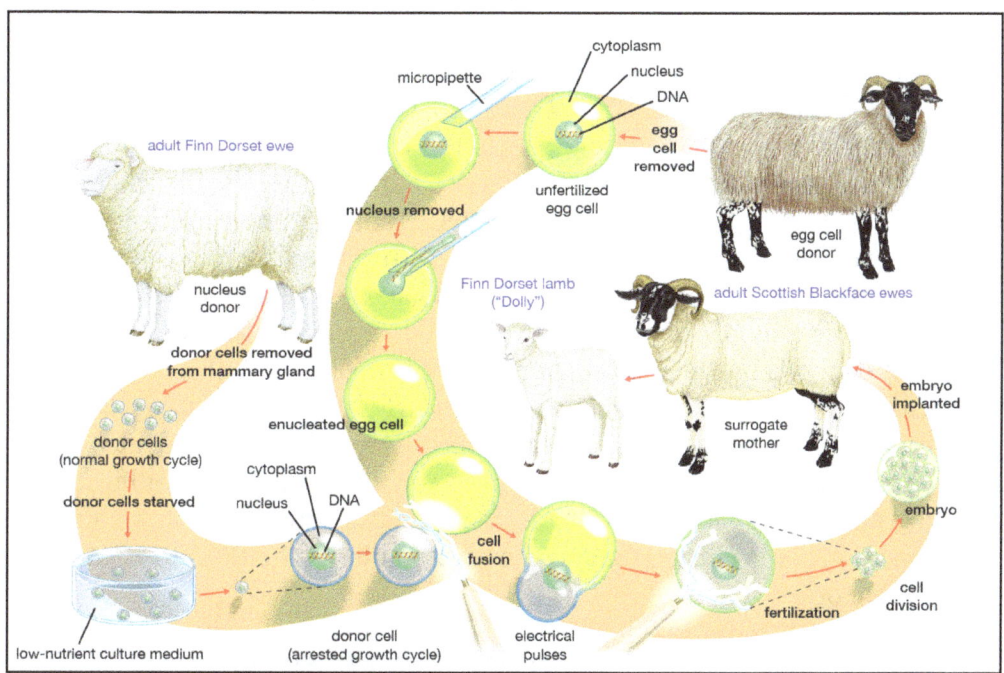

Dolly: The cloning of a Sheep, 1996.

Dolly the sheep was successfully cloned in 1996 by fusing the nucleus from a mammary-gland cell of a Finn Dorset ewe into an enucleated egg cell taken from a Scottish Blackface ewe. Carried to term in the womb of another Scottish Blackface ewe, Dolly was a genetic copy of the Finn Dorset ewe.

It is believed that all of an organism's genes are present in each cell nucleus, no matter what the cell type, and that differences between tissues are not due to the presence or absence of certain genes but are due to the expression of some and the repression of others. In animals the best evidence for retention of the entire set of genes comes from whole animal cloning experiments in which the nucleus of a differentiated cell is substituted for the nucleus of a fertilized egg. In many species this can result in the development of a normal embryo that contains the full range of body parts and cell types. Likewise, in plants it is often possible to grow complete embryos from individual cells in tissue culture. Such experiments show that any nucleus has the genetic information required for the growth of a developing organism, and they strongly suggest that, for most tissues, cell differentiation arises from the regulation of genetic activity rather than the removal or destruction of unwanted genes. The only known exception to this rule comes from the immune system, where segments of DNA in developing white blood cells are slightly rearranged, producing a wide variety of antibody and receptor molecules.

At the molecular level there are many ways in which the expression of a gene can be differentially regulated in different cell types. There may be differences in the copying, or transcription, of the gene into RNA; in the processing of the initial RNA transcript into mRNA; in the control of mRNA movement to the cytoplasm; in the translation of mRNA to protein; or in the stability of mRNA. However, the control of transcription has the most influence over gene expression and has received the most detailed analysis.

The DNA in the cell nucleus exists in the form of chromatin, which is made up of DNA bound to histones (simple alkaline proteins) and other nonhistone proteins. Most of the DNA is complexed into repeating structures called nucleosomes, each of which contains eight molecules of histone. Active genes are found in parts of the DNA where the chromatin has an "open" configuration, in which regulatory proteins are able to gain access to the DNA. The degree to which the chromatin opens depends on chemical modifications of the outer parts of the histone molecules and on the presence or absence of particular nonhistone proteins. Transcriptional control is exerted with the help of regulatory sequences that are found associated with a gene, such as the promoter sequence, a region near the start of the gene, and enhancer sequences, regions that lie elsewhere within the DNA that augment the activity of enzymes involved in the process of transcription. Whether or not transcription occurs depends on the binding of transcription factors to these regulatory sequences. Transcription factors are proteins that usually possess a DNA-binding region, which recognizes the specific regulatory sequence in the DNA, and an effector region, which activates or inhibits transcription. Transcription factors often work by recruiting enzymes that add modifications (e.g., acetyl groups or methyl groups) to or remove modifications from the outer parts of the histone molecules. This controls the folding of the chromatin and the accessibility of the DNA to RNA polymerase and other transcription factors.

In general, it requires several transcription factors working in combination to activate a gene. For example, the chicken delta 1 crystallin gene, normally expressed only in the lens of the eye, has a promoter that contains binding sites for two activating transcription factors and an enhancer that contains binding sites for two other activating transcription factors. There is also an additional enhancer site that can bind either an activator (deltaEF3) or a repressor (deltaEF1). Successful transcription requires that all these sites are occupied by the correct transcription factors.

Fully differentiated cells are qualitatively different from one another. States of terminal differentiation are stable and persistent, both in the lifetime of the cell and in successive cell generations (in the case of differentiated types that are capable of continued cell division). The inherent stability of the differentiated state is maintained by various processes, including feedback activation of genes by their own products and repression of inactive genes. Chromatin structure may be important in maintaining states of differentiation, although it is still unclear whether this can be maintained during DNA replication, which involves temporary removal of chromosomal proteins and unwinding of the DNA double helix.

A type of differentiation control that is maintained during DNA replication is the methylation of DNA, which tends to recruit histone deacetylases and hence close up the structure of the chromatin.

DNA methylation occurs when a methyl group is attached to the exterior, or sugar-phosphate side, of a cytosine (C) residue. Cytosine methylation occurs only on a C nucleotide when it is connected to a G (guanine) nucleotide on the same strand of DNA. These nucleotide pairings are called CG dinucleotides. One class of DNA methylase enzyme can introduce new methylations when required, whereas another class, called maintenance methylases, methylates CG dinucleotides in the DNA double helix only when the CG of the complementary strand is already methylated. Each time the methylated DNA is replicated, the old strand has the methyl groups and the new strand does not. The maintenance methylase will then add methyl groups to all the CGs opposite the existing methyl groups to restore a fully methylated double helix. This mechanism guarantees stability of the DNA methylation pattern, and hence the differentiated state, during the processes of DNA replication and cell division.

Process of Differentiation

Differentiation from visibly undifferentiated precursor cells occurs during embryonic development, during metamorphosis of larval forms, and following the separation of parts in asexual reproduction. It also takes place in adult organisms during the renewal of tissues and the regeneration of missing parts. Thus, cell differentiation is an essential and ongoing process at all stages of life.

The visible differentiation of cells is only the last of a progressive sequence of states. In each state, the cell becomes increasingly committed toward one type of cell into which it can develop. States of commitment are sometimes described as "specification" to represent a reversible type of commitment and as "determination" to represent an irreversible commitment. Although states of specification and determination both represent differential gene activity, the properties of embryonic cells are not necessarily the same as those of fully differentiated cells. In particular, cells in specification states are usually not stable over prolonged periods of time.

Two mechanisms bring about altered commitments in the different regions of the early embryo: cytoplasmic localization and induction. Cytoplasmic localization is evident in the earliest stages of development of the embryo. During this time, the embryo divides without growth, undergoing cleavage divisions that produce separate cells called blastomeres. Each blastomere inherits a certain region of the original egg cytoplasm, which may contain one or more regulatory substances called cytoplasmic determinants. When the embryo has become a solid mass of blastomeres (called a morula), it generally consists of two or more differently committed cell populations—a result of the blastomeres having incorporated different cytoplasmic determinants. Cytoplasmic determinants may consist of mRNA or protein in a particular state of activation. An example of the influence of a cytoplasmic determinant is a receptor called Toll, located in the membranes of Drosophila (fruit fly) eggs. Activation of Toll ensures that the blastomeres will develop into ventral (underside) structures, while blastomeres containing inactive Toll will produce cells that will develop into dorsal (back) structures.

In induction, the second mechanism of commitment, a substance secreted by one group of cells alters the development of another group. In early development, induction is usually instructive; that is, the tissue assumes a different state of commitment in the presence of the signal than it would in the absence of the signal. Inductive signals often take the form of concentration gradients of substances that evoke a number of different responses at different concentrations. This leads to the formation of a sequence of groups of cells, each in a different state of specification. For example, in Xenopus (clawed frog) the early embryo contains a signaling centre called the organizer that secretes inhibitors of bone morphogenetic proteins (BMPs), leading to a ventral-to-dorsal (belly-to-back) gradient of BMP activity. The activity of BMP in the ventral region of the embryo suppresses the expression of transcription factors involved in the formation of the central nervous system and segmented muscles. Suppression ensures that these structures are formed only on the dorsal side, where there is decreased activity of BMP.

The final stage of differentiation often involves the formation of several types of differentiated cells from one precursor or stem cell population. Terminal differentiation occurs not only in embryonic development but also in many tissues in postnatal life. Control of this process depends on a system of lateral inhibition in which cells that are differentiating along a particular pathway send out signals that repress similar differentiation by their neighbours. For example, in the developing central nervous system of vertebrates, neurons arise from a simple tube of neuroepithelium, the cells of which possess a surface receptor called Notch. These cells also possess another cell surface molecule called Delta that can bind to and activate Notch on adjacent cells. Activation of Notch initiates a cascade of intracellular events that results in suppression of Delta production and suppression of neuronal differentiation. This means that the neuroepithelium generates only a few cells with high expression of Delta surrounded by a larger number of cells with low expression of Delta. High Delta production and low Notch activation makes the cells develop into neurons. Low Delta production and high Notch activation makes the cells remain as precursor cells or become glial (supporting) cells. A similar mechanism is known to produce the endocrine cells of the pancreas and the goblet cells of the intestinal epithelium. Such lateral inhibition systems work because cells in a population are never quite identical to begin with. There are always small differences, such as in the number of Delta molecules displayed on the cell surface. The mechanism of lateral inhibition amplifies these small differences, using them to bring about differential gene expression that leads to stable and persistent states of cell differentiation.

Errors in Differentiation

Three classes of abnormal cell differentiation are dysplasia, metaplasia, and anaplasia. Dysplasia indicates an abnormal arrangement of cells, usually arising from a disturbance in their normal growth behaviour. Some dysplasias are precursor lesions to cancer, whereas others are harmless and regress spontaneously. For example, dysplasia of the uterine cervix, called cervical intraepithelial neoplasia (CIN), may progress to cervical cancer. It can be detected by cervical smear cytology tests (Pap smears).

Metaplasia is the conversion of one cell type into another. In fact, it is not usually the differentiated cells themselves that change but rather the stem cell population from which they are derived. Metaplasia commonly occurs where chronic tissue damage is followed by extensive regeneration. For example, squamous metaplasia of the bronchi occurs when the ciliated respiratory epithelial cells of people who smoke develop into squamous, or flattened, cells. In intestinal metaplasia of the stomach, patches resembling intestinal tissue arise in the gastric mucosa, often in association with gastric ulcers. Both of these types of metaplasia may progress to cancer.

Anaplasia is a loss of visible differentiation that can occur in advanced cancer. In general, early cancers resemble their tissue of origin and are described and classified by their pattern of differentiation. However, as they develop, they produce variants of more abnormal appearance and increased malignancy. Finally, a highly anaplastic growth can occur, in which the cancerous cells bear no visible relation to the parent tissue.

The Evolution of Cells

Development of Genetic Information

Life on Earth could not exist until a collection of catalysts appeared that could promote the synthesis of more catalysts of the same kind. Early stages in the evolutionary pathway of cells presumably centred on RNA molecules, which not only present specific catalytic surfaces but also contain the potential for their own duplication through the formation of a complementary RNA molecule. It is assumed that a small RNA molecule eventually appeared that was able to catalyze its own duplication.

Imperfections in primitive RNA replication likely gave rise to many variant autocatalytic RNA molecules. Molecules of RNA that acquired variations that increased the speed or the fidelity of self-replication would have out multiplied other, less-competent RNA molecules. In addition, other small RNA molecules that existed in symbiosis with autocatalytic RNA molecules underwent natural selection for their ability to catalyze useful secondary reactions such as the production of better precursor molecules. In this way, sophisticated families of RNA catalysts could have evolved together, since cooperation between different molecules produced a system that was much more effective at self-replication than a collection of individual RNA catalysts.

Another major step in the evolution of the cell would have been the development, in one family of self-replicating RNA, of a primitive mechanism of protein synthesis. Protein molecules cannot provide the information for the synthesis of other protein molecules like themselves. This information must ultimately be derived from a nucleic acid sequence. Protein synthesis is much more complex than RNA synthesis, and it could not have arisen before a group of powerful RNA catalysts evolved. Each of these catalysts presumably has its counterpart among the RNA molecules that function in the current cell: (1) there was an information RNA molecule, much like messenger RNA (mRNA), whose nucleotide sequence was read to create an amino acid sequence; (2) there was a group of adaptor RNA molecules, much like transfer RNA (tRNA), that could bind to both mRNA and a specific activated amino acid; and (3) finally, there was an RNA catalyst, much like ribosomal RNA (rRNA), that facilitated the joining together of the amino acids aligned on the mRNA by the adaptor RNA.

At some point in the evolution of biological catalysts, the first cell was formed. This would have required the partitioning of the primitive biological catalysts into individual units, each surrounded by a membrane. Membrane formation might have occurred quite simply, since many amphiphilic molecules—half hydrophobic (water-repelling) and half hydrophilic (water-loving)—aggregate to form bilayer sheets in which the hydrophobic portions of the molecules line up in rows to form the interior of the sheet and leave the hydrophilic portions to face the water. Such bilayer sheets can spontaneously close up to form the walls of small, spherical vesicles, as can the phospholipid bilayer membranes of present-day cells.

Structure and properties of two representative lipidsBoth stearic acid (a fatty acid) and phosphatidylcholine (a phospholipid) are composed of chemical groups that form polar "heads" and nonpolar "tails." The polar heads are hydrophilic, or soluble in water, whereas the nonpolar tails are hydrophobic, or insoluble in water. Lipid molecules of this composition spontaneously form aggregate structures such as micelles and lipid bilayers, with their hydrophilic ends oriented toward the watery medium and their hydrophobic ends shielded from the water.

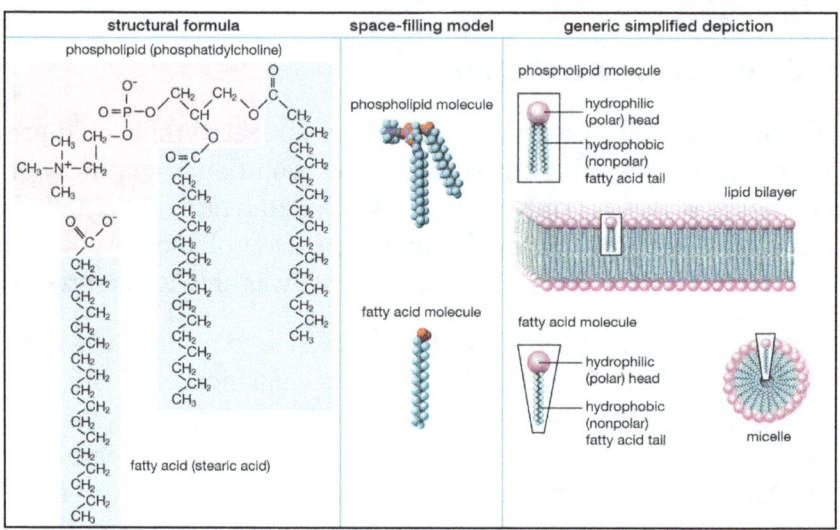

As soon as the biological catalysts became compartmentalized into small individual units, or cells, the units would have begun to compete with one another for the same resources. The active competition that ensued must have greatly accelerated evolutionary change, serving as a powerful force for the development of more efficient cells. In this way, cells eventually arose that contained new catalysts, enabling them to use simpler, more abundant precursor molecules for their growth. Because these cells were no longer dependent on preformed ingredients for their survival, they were able to spread far beyond the limited environments where the first primitive cells arose.

It is often assumed that the first cells appeared only after the development of a primitive form of protein synthesis. However, it is by no means certain that cells cannot exist without proteins, and it has been suggested that the first cells contained only RNA catalysts. In either case, protein molecules, with their chemically varied side chains, are more powerful catalysts than RNA molecules; therefore, as time passed, cells arose in which RNA served primarily as genetic material, being directly replicated in each generation and inherited by all progeny cells in order to specify proteins.

As cells became more complex, a need would have arisen for a stabler form of genetic information storage than that provided by RNA. DNA, related to RNA yet chemically stabler, probably appeared rather late in the evolutionary history of cells. Over a period of time, the genetic information in RNA sequences was transferred to DNA sequences, and the ability of RNA molecules to replicate directly was lost. It was only at this point that the central process of biology—the synthesis, one after the other, of DNA, RNA, and protein—appeared.

Development of Metabolism

The first cells presumably resembled prokaryotic cells in lacking nuclei and functional internal compartments, or organelles. These early cells were also anaerobic (not requiring oxygen), deriving their energy from the fermentation of organic molecules that had previously accumulated on the Earth over long periods of time. Eventually, more sophisticated cells evolved that could carry out primitive forms of photosynthesis, in which light energy was harnessed by membrane-bound proteins to form organic molecules with energy-rich chemical bonds. A major turning point in the evolution of life was the development of photosynthesizing prokaryotes requiring only water as an electron donor and capable of producing molecular oxygen. The descendants of these prokaryotes, the blue-green algae (cyanobacteria), still exist as viable life-forms. Their ancestors prospered to such an extent that the atmosphere became rich in the oxygen they produced. The free availability of this oxygen in turn enabled other prokaryotes to evolve aerobic forms of metabolism that were much more efficient in the use of organic molecules as a source of food.

The switch to predominantly aerobic metabolism is thought to have occurred in bacteria approximately 2 billion years ago, about 1.5 billion years after the first cells had formed. Aerobic eukaryotic cells (cells containing nuclei and all the other organelles) probably appeared about 1.5 billion years ago, their lineage having branched off much earlier from that of the prokaryotes. Eukaryotic cells almost certainly became aerobic by engulfing aerobic prokaryotes, with which they lived in a symbiotic relationship. The mitochondria found in both animals and plants are the descendants of such prokaryotes. Later, in branches of the eukaryotic lineage leading to plants and algae, a blue-green algae like organism was engulfed to perform photosynthesis. It is likely that over a long period of time these organisms became the chloroplasts.

The eukaryotic cell thus apparently arose as an amalgam of different cells, in the process becoming an efficient aerobic cell whose plasma membrane was freed from energy metabolism—one of the major functions of the cell membrane of prokaryotes. The eukaryotic cell membrane was therefore able to become specialized for cell-to-cell communication and cell signaling. It may be partly for this reason that eukaryotic cells were eventually more successful at forming complex multicellular organisms than their simpler prokaryotic relatives.

CELL BIOLOGY

Cell biology is the academic discipline that studies the basic unit of living things, cells. Cells are the smallest independently functioning unit in the structure of an organism and usually consist of a nucleus surrounded by cytoplasm and enclosed by a membrane. Cell biology examines, on microscopic and molecular levels, the physiological properties, structure, organelles (such as nuclei and mitochondria), interactions, life cycle, division and death of these basic units of organisms. Cell biology research extends to both the great diversity of single-celled organisms, such as bacteria and the many specialized cells in multicellular organisms, such as animals and plants.

The field of cell biology traditionally has focused on questions concerning how the various organelles work and work together, how these cellular processes are regulated and how the various cells within the organism communicate with each other. Understanding the composition of cells and how cells work is fundamental to all the biological and medical sciences. Examining the similarities and differences between cell types is particularly important to the fields of cell and molecular biology, because the principles learned from studying one cell type can be generalized to other cell types. Research in cell biology closely is related to genetics, biochemistry, molecular biology and developmental biology.

The structures and functions within a cell often are compared to similar activities in a typical city. Mitochondria are the cell's energy plants, plant chloroplasts are solar energy plants, chromosomes contain the original blueprints of the city, the endoplasmic reticulum represents the road system, Golgi apparatus is the post office and the nucleus is city hall. Proteins, however, participate in virtually every function of the cell city — in other words, they are the bricks and lumber, messengers, copy machines, waste recyclers and more. Every cell typically contains hundreds of different kinds of proteins that function together to generate the behavior of the cell. An important part of cell biology is investigation of molecular mechanisms by which proteins are moved to different places inside cells or secreted from cells.

Most proteins are synthesized by ribosomes in the cytoplasm. This process also is known as protein biosynthesis or protein translation. Some proteins, such as those to be incorporated in membranes (membrane proteins) are transported into the endoplasmic reticulum (ER) during synthesis and further processed in the Golgi apparatus. From the Golgi, membrane proteins can move to the plasma membrane or to other subcellular compartments, or they can be secreted from the cell. There is regular movement of proteins through these compartments. ER- and Golgi-resident proteins associate with other proteins but remain in their respective compartments. Other proteins pass through the ER and Golgi to the plasma membrane.

Most of the genetic information in the cell resides in the nucleus and is contained within the chromosomes (mitochondria also carry some DNA of their own). The study of the microscopically visible stages of cell division during mitosis and meiosis generally is considered part of cell biology, while the actual submicroscopic activity of DNA replication and protein synthesis is considered part of molecular biology.

2

Fundamental Concepts in Cell Biology

Cell theory, cell division and symbiogenesis are some of the fundamental concepts in cell biology. The chapter closely examines these fundamental concepts of cell biology as well as important areas of study such as lipid bilayer to provide an extensive understanding of the subject.

CELL THEORY

Cell theory is a proposed and widely accepted view of how most life on Earth functions. According to the theory, all organisms are made of cells. Groups of cells create tissues, organs, and organisms. Further, cells can only arise from other cells. These are the main tenants of cell theory.

Before the invention of advanced microscopes, microorganisms were unknown, and it was assumed that individuals were the basic units of life. However, in the 1800s, this view began to change thanks to the microscope. Microscopes allowed early scientists to view and postulate about the cells they could see. Even with a microscope, it is not always possible to see the exact functioning of a cell. Scientists formulated a general theory of how cells work that is very simple.

This theory revolves around the fact that no matter what type of organism we view under a microscope, the organisms are clearly divided into a number of different cells. Some cells are very large, such as a frog egg. Other cells, such as some bacterial cells, are so small we can barely see them with a normal light microscope. Viruses, which may or may not be living, are the only forms of reproducing DNA or RNA which are not always contained within a cell.

Parts of Cell Theory

Cell theory has three major hypotheses:

- First, all organisms are made of cells.
- Second, cells are the fundamental building blocks used to create tissues, organs, and entire functioning organisms.
- The third, and probably most important part of the theory is that cells can only arise from other cells.

Thus, all organisms start as single cells. These cells grow, divide through mitosis, and develop into multi-celled organisms. Mitosis is a form of cell division that produces identical cells. These cells can then differentiate when given different signals to produce different types of tissues and organs. This is how large and complex organisms are made. Single-celled organisms divide as well, but when they divide, the cells separate into two new individuals. This is known as asexual reproduction.

Cell Theory Examples

Single-Celled Organisms

Single-celled organisms are a great way to study cell theory. With modern microscopes, the processes behind cell theory can easily be viewed and studied. A great example of watching cell theory in action can be accomplished by putting a drop of pond water under a microscope. Below is a picture of two Euglena organisms, seen just after reproduction.

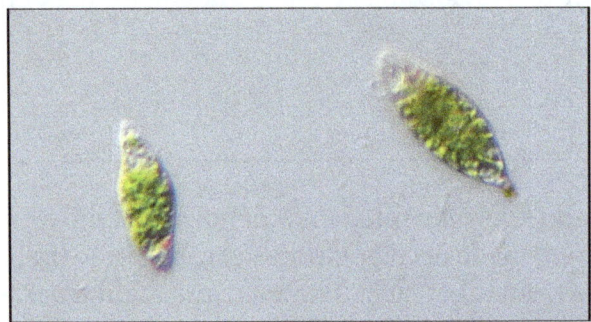

Euglena

Minutes before, these two cells were one. Euglena reproduces through simple cellular division. The DNA in the parent organism is duplicated, as are the internal organelles. Then, the large cell divides into two equally-sized smaller cells, as seen in the picture. These two cells are now independent organisms. Each will try to survive, grow, and eventually reproduce again.

In Plants

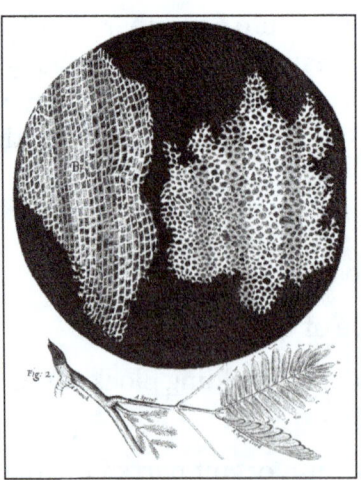

A thin slice of cork

Cells were first discovered in plants. Plants, unlike the other examples here, have large structures called cell walls, which enable the plant to remain rigid. These cell walls are easily visible, even with the first microscope invented in 1665. Robert Hooke, the man who first identified cells, did so using a simple microscope aimed at a thin slice of cork.

Hooke was clearly looking at cells. In fact, with a better microscope, he likely could have seen the cells in action and the many organelles inside. Instead, Hooke did not come to the immediate conclusion that all organisms were made of cells. He assumed the structures were limited to the tissues of plants. It was not until the 1840s that cell theory would be largely accepted by science.

In Animals

In 1839, scientist Theodor Schwann presented evidence that animals, like plants, were also fundamentally composed of different types of cells. Modern microscopy techniques allow scientists a much more comprehensive and accurate view of cells compared to early scientists. Below is a scanning electron micrograph of red blood cells. It distinctly shows how our red blood cells are separate, functional units of the human body.

Blood cell

Like red blood cells, every part of the body is composed of different types of cells. According to cell theory, all of these cells are derived from the zygote, which is a single cell that results from the fertilization of an egg with a sperm. This cell then divides, replicates, and begins to differentiate into the many different cell types of the body. Eventually, a fully-functional organism is formed.

Other Organisms

Cells are the basic building blocks of all life on Earth. This is true of fungi, the only kingdom not yet covered. In fact, fungi are a sort of intermediate between plants and animals. While they lack the sun-harvesting chloroplasts of plants, they do have cell walls. However, there is one form of life which does not strictly adhere to cell theory.

Viruses are small DNA or RNA particles, surrounded by a protective protein coating. Many scientists do not consider viruses a living organism, and thus it is okay that they do not conform to the typical cell theory. Other scientists consider them living, but suggest they are an exception to cell theory. For viruses to reproduce they must infect a host cell. Only by using the host cell's machinery can a virus replicate its genetic code and the proteins needed to create new virus particles.

Contributions to Cell Theory

Besides Robert Hooke and Theodor Schwann, a number of scientists have made significant contributions to cell theory. In fact, cell theory has been growing and changing since the first cells were observed, and many fantastic experiments have been devised to show various parts of cell theory.

CELL DIVISION

Cell division is the process by which a parent cell divides into two or more daughter cells. Cell division usually occurs as part of a larger cell cycle. In eukaryotes, there are two distinct types of cell division: a vegetative division, whereby each daughter cell is genetically identical to the parent cell (mitosis), and a reproductive cell division, whereby the number of chromosomes in the daughter cells is reduced by half to produce haploid gametes (meiosis). Meiosis results in four haploid daughter cells by undergoing one round of DNA replication followed by two divisions. Homologous chromosomes are separated in the first division, and sister chromatids are separated in the second division. Both of these cell division cycles are used in the process of sexual reproduction at some point in their life cycle. Both are believed to be present in the last eukaryotic common ancestor.

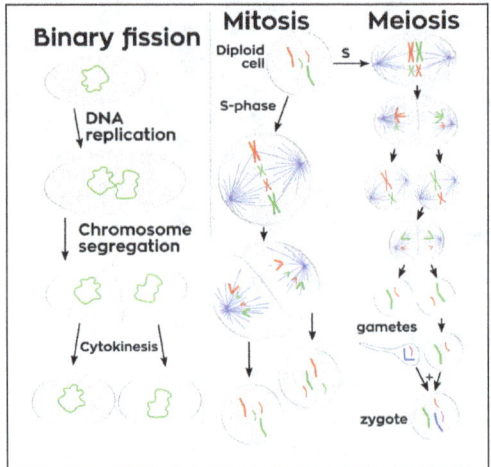

Three types of cell division

Prokaryotes (bacteria) undergo a vegetative cell division known as binary fission, where their genetic material is segregated equally into two daughter cells. While binary fission may be the means of division by most prokaryotes, there are alternative manners of division, such as budding, that have been observed. All cell divisions, regardless of organism, are preceded by a single round of DNA replication.

For simple unicellular microorganisms such as the amoeba, one cell division is equivalent to reproduction – an entire new organism is created. On a larger scale, mitotic cell division can create progeny from multicellular organisms, such as plants that grow from cuttings. Mitotic cell division enables sexually reproducing organisms to develop from the one-celled zygote, which itself was produced by meiotic cell division from gametes. After growth, cell division by mitosis allows for continual construction and repair of the organism. The human body experiences about 10 quadrillion cell divisions in a lifetime.

The primary concern of cell division is the maintenance of the original cell's genome. Before division can occur, the genomic information that is stored in chromosomes must be replicated, and the duplicated genome must be separated cleanly between cells. A great deal of cellular infrastructure is involved in keeping genomic information consistent between generations.

Phases of Eukaryotic Cell Division

Interphase

Interphase is the process a cell must go through before mitosis, meiosis, and cytokinesis. Interphase consists of three main phases: G_1, S, and G_2. G_1 is a time of growth for the cell where specialized cellular functions occur in order to prepare the cell for DNA Replication. There are checkpoints during interphase that allow the cell to be either advance or halt further development. In S phase, the chromosomes are replicated in order for the genetic content to be maintained. During G_2, the cell undergoes the final stages of growth before it enters the M phase, where spindles are synthesized. The M phase, can be either mitosis or meiosis depending on the type of cell. Germ cells, or gametes, undergo meiosis, while somatic cells will undergo mitosis. After the cell proceeds successfully through the M phase, it may then undergo cell division through cytokinesis. The control of each checkpoint is controlled by cyclin and cyclin-dependent kinases. The progression of interphase is the result of the increased amount of cyclin. As the amount of cyclin increases, more and more cyclin dependent kinases attach to cyclin signaling the cell further into interphase. At the peak of the cyclin attached to the cyclin dependent kinases this system pushes the cell out of interphase and into the M phase, where mitosis, meiosis, and cytokinesis occur. There are three transition checkpoints the cell goes through before entering the M phase. The most important being the G_1-S transition checkpoint. If the cell does not pass this checkpoint, then the cell will exit the cell cycle.

Prophase

Prophase is the first stage of division. The nuclear envelope is broken down, long strands of chromatin condense to form shorter more visible strands called chromosomes, the nucleolus disappears, and microtubules attach to the chromosomes at the kinetochores present in the centromere. Microtubules associated with the alignment and separation of chromosomes are referred to as the spindle and spindle fibers. Chromosomes will also be visible under a microscope and will be connected at the centromere. During this condensation and alignment period in meiosis, the homologous chromosomes undergo a break in their double-stranded DNA at the same locations followed by a recombination of the now fragmented parental DNA strands into non-parental combinations, known as crossing over. This process is evidenced to be caused in a large part by the highly conserved Spo11 protein through a mechanism similar to that seen with toposomerase in DNA replication and transcription.

Metaphase

In metaphase, the centromeres of the chromosomes convene themselves on the metaphase plate (or equatorial plate), an imaginary line that is equidistant from the two centrosome poles and held together by complex complexes known as cohesins. Chromosomes line up in the middle of the cell by microtubule organizing centers (MTOCs) pushing and pulling on centromeres of both chromatids thereby causing the chromosome to move to the center. At this point the chromosomes are still condensing and are currently one step away from being the most coiled and condensed they will be, and the spindle fibers have already connected to the kinetochores. During this phase all the microtubules, with the exception of the kinetochores, are in a state of instability promoting their progression towards anaphase. At this point, the chromosomes are ready to split into opposite poles of the cell towards the spindle to which they are connected.

Anaphase

Anaphase is a very short stage of the cell cycle and occurs after the chromosomes align at the mitotic plate. Kinetochores emit anaphase-inhibition signals until their attachment to the mitotic spindle. Once the final chromosome is properly aligned and attached the final signal dissipates and triggers the abrupt shift to anaphase. This abrupt shift is caused by the activation of the anaphase-promoting complex and its function of tagging degradation of proteins important towards the metaphase-anaphase transition. One of these proteins that is broken down is securin which through its breakdown releases the enzyme separase that cleaves the cohesin rings holding together the sister chromatids thereby leading to the chromosomes separating. After the chromosomes line up in the middle of the cell, the spindle fibers will pull them apart. The chromosomes are split apart as the sister chromatids move to opposite sides of the cell. While the sister chromatids are being pulled apart, the cell and plasma are elongated by non-kinetochore microtubules.

Telophase

Telophase is the last stage of the cell cycle in which a cleavage furrow splits the cells cytoplasm (cytokinesis) and chromatin. This occurs through the synthesis of a new nuclear envelopes that forms around the chromatin gathered at each pole and the reformation of the nucleolus as the chromosomes decondense their chromatin back to the loose state it possessed during interphase. The division of the cellular contents is not always equal and can vary by cell type as seen with oocyte formation where one of the four daughter cells possess the majority of the cytoplasm.

Variants

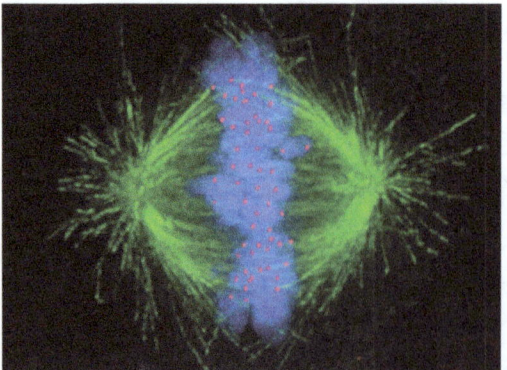

Image of the mitotic spindle in a human cell showing microtubules
in green, chromosomes (DNA) in blue, and kinetochores in red.

Cells are broadly classified into two main categories: simple, non-nucleated prokaryotic cells, and complex, nucleated eukaryotic cells. Due to their structural differences, eukaryotic and prokaryotic cells do not divide in the same way. Also, the pattern of cell division that transforms eukaryotic stem cells into gametes (sperm cells in males or egg cells in females), termed meiosis, is different from that of the division of somatic cells in the body. Image of the mitotic spindle in a human cell showing microtubules in green, chromosomes (DNA) in blue, and kinetochores in red.

The cells were directly imaged in the cell culture vessel,
using non-invasive quantitative phase contrast time-lapse microscopy.

Degradation

Multicellular organisms replace worn-out cells through cell division. In some animals, however, cell division eventually halts. In humans this occurs, on average, after 52 divisions, known as the Hayflick limit. The cell is then referred to as senescent. With each division the cells telomeres, protective sequences of DNA on the end of a chromosome that prevent degradation of the chromosomal DNA, shorten. This shortening has been correlated to negative effects such as age related diseases and shortened lifespans in humans. Cancer cells, on the other hand, are not thought to degrade in this way, if at all. An enzyme complex called telomerase, present in large quantities in cancerous cells, rebuilds the telomeres through synthesis of telomeric DNA repeats, allowing division to continue indefinitely.

SYMBIOGENESIS

Symbiogenesis, or endosymbiotic theory, is an evolutionary theory of the origin of eukaryotic cells from prokaryotic organisms, first articulated in 1905 and 1910 by the Russian botanist Konstantin Mereschkowski, and advanced and substantiated with microbiological evidence by Lynn Margulis in 1967. It holds that the organelles distinguishing eukaryote cells evolved through symbiosis of individual single-celled prokaryotes (bacteria and archaea). The theory holds that mitochondria, plastids such as chloroplasts, and possibly other organelles of eukaryotic cells represent formerly free-living prokaryotes taken one inside the other in endosymbiosis. In more detail, mitochondria appear to be related to Rickettsiales proteobacteria, and chloroplasts to nitrogen-fixing filamentous cyanobacteria. Among the many lines of evidence supporting symbiogenesis are that new mitochondria and plastids are formed only through binary fission, and that cells cannot create new ones otherwise; that the transport proteins called porins are found in the outer membranes of mitochondria, chloroplasts and bacterial cell membranes; that cardiolipin is found only in the inner mitochondrial membrane and bacterial cell membranes; and that some mitochondria and plastids contain single circular DNA molecules similar to the chromosomes of bacteria.

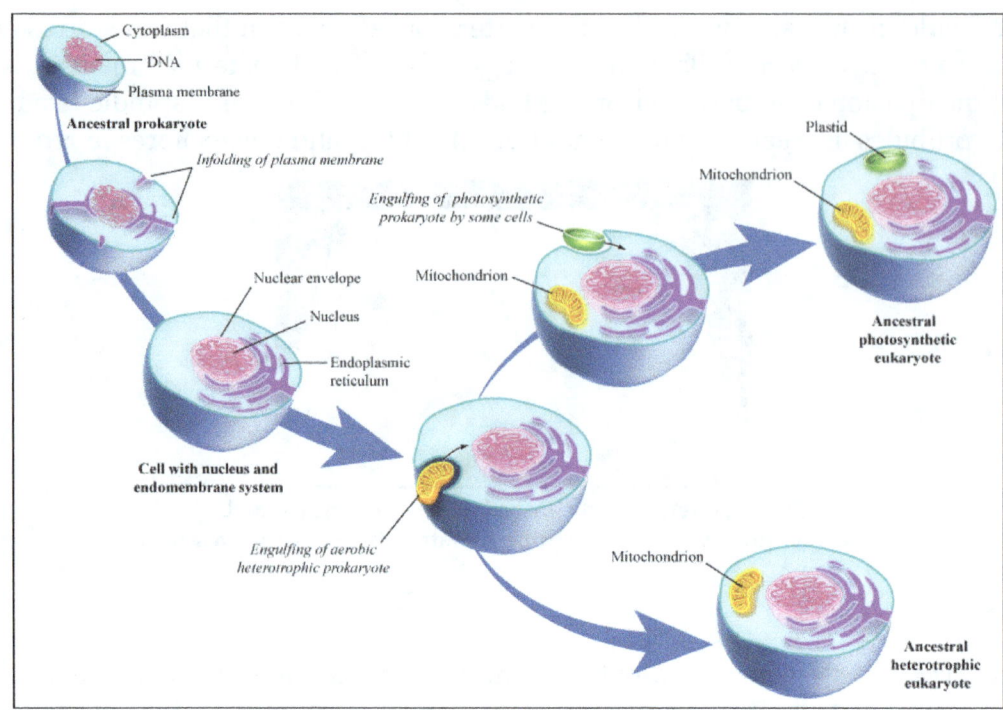

One model for the origin of mitochondria and plastids

Endosymbionts to Organelles

According to Keeling and Archibald, the usual way to distinguish organelles from endosymbionts is by their reduced genome sizes. As an endosymbiont evolves into an organelle, most of their genes are transferred to the host cell genome. The host cell and organelle need to develop a transport mechanism that enables the return of the protein products needed by the organelle but now manufactured by the cell. Cyanobacteria and α-proteobacteria are the most closely related free-living organisms to plastids and mitochondria respectively. Both cyanobacteria and α-proteobacteria maintain a large (>6Mb) genome encoding thousands of proteins. Plastids and mitochondria exhibit a dramatic reduction in genome size when compared to their bacterial relatives. Chloroplast genomes in photosynthetic organisms are normally 120-200kb encoding 20-200 proteins and mitochondrial genomes in humans are approximately 16kb and encode 37 genes, 13 of which are proteins. Using the example of the freshwater amoeboid, however, *Paulinella chromatophora*, which contains chromatophores found to be evolved from cyanobacteria, Keeling and Archibald argue that this is not the only possible criterion; another is that the host cell has assumed control of the regulation of the former endosymbiont's division, thereby synchronizing it with the cell's own division. Nowack and her colleagues performed gene sequencing on the chromatophore (1.02 Mb) and found that only 867 proteins were encoded by these photosynthetic cells. Comparisons with their closest free living cyanobacteria of the genus *Synechococcus* (having a genome size 3 Mb, with 3300 genes) revealed that chromatophores underwent a drastic genome shrinkage. Chromatophores contained genes that were accountable for photosynthesis but were deficient in genes that could carry out other biosynthetic functions; this observation suggests that these endosymbiotic cells are highly dependent on their hosts for their survival and growth mechanisms. Thus, these chromatophores were found to be non-functional for organelle-specific purposes when compared to mitochondria and plastids. This distinction could have promoted the early evolution of photosynthetic organelles.

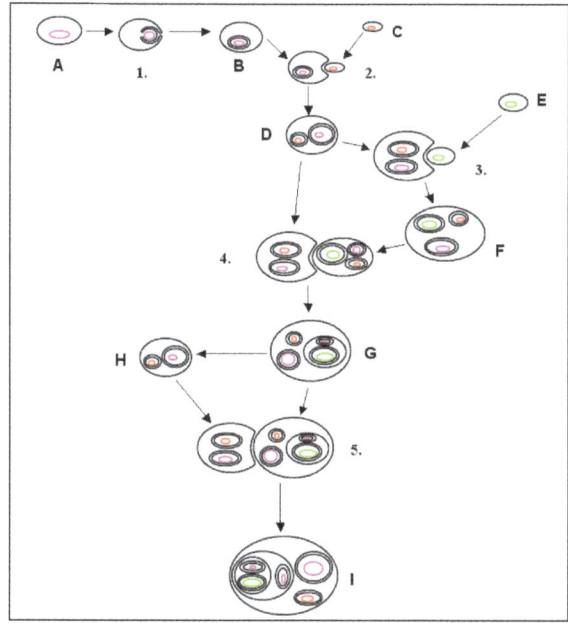

Modern endosymbiotic theory posits simply life forms
merged, forming cell organelles, like mitochondria.

The loss of genetic autonomy, that is, the loss of many genes from endosymbionts, occurred very early in evolutionary time. Taking into account the entire original endosymbiont genome, there are three main possible fates for genes over evolutionary time. The first fate involves the loss of functionally redundant genes, in which genes that are already represented in the nucleus are eventually lost. The second fate involves the transfer of genes to the nucleus. The loss of autonomy and integration of the endosymbiont with its host can be primarily attributed to nuclear gene transfer. As organelle genomes have been greatly reduced over evolutionary time, nuclear genes have expanded and become more complex. As a result, many plastid and mitochondrial processes are driven by nuclear encoded gene products. In addition, many nuclear genes originating from endosymbionts have acquired novel functions unrelated to their organelles. The mechanisms of gene transfer are not fully known; however, multiple hypotheses exist to explain this phenomenon. The cDNA hypothesis involves the use of messenger RNA (mRNAs) to transport genes from organelles to the nucleus where they are converted to cDNA and incorporated into the genome. The cDNA hypothesis is based on studies of the genomes of flowering plants. Protein coding RNAs in mitochondria are spliced and edited using organelle-specific splice and editing sites. Nuclear copies of some mitochondrial genes, however, do not contain organelle-specific splice sites, suggesting a processed mRNA intermediate.

The cDNA hypothesis has since been revised as edited mitochondrial cDNAs are unlikely to recombine with the nuclear genome and are more likely to recombine with their native mitochondrial genome. If the edited mitochondrial sequence recombines with the mitochondrial genome, mitochondrial splice sites would no longer exist in the mitochondrial genome. Any subsequent nuclear gene transfer would therefore also lack mitochondrial splice sites. The bulk flow hypothesis is the alternative to the cDNA hypothesis, stating that escaped DNA, rather than mRNA, is the mechanism of gene transfer. According to this hypothesis, disturbances to organelles, including autophagy (normal cell destruction), gametogenesis (the formation of gametes), and cell stress, release DNA which is imported into the nucleus and incorporated into the nuclear DNA using non-homologous end joining (repair of double stranded breaks). For example, in the initial stages of endosymbiosis, due to a lack of major gene transfer, the host cell had little to no control over the endosymbiont. The endosymbiont underwent cell division independently of the host cell, resulting in many "copies" of the endosymbiont within the host cell. Some of the endosymbionts lysed (burst), and high levels of DNA were incorporated into the nucleus. A similar mechanism is thought to occur in tobacco plants, which show a high rate of gene transfer and whose cells contain multiple chloroplasts. In addition, the bulk flow hypothesis is also supported by the presence of non-random clusters of organelle genes, suggesting the simultaneous movement of multiple genes. In 2015, the biologist Roberto Cazzolla Gatti provided evidence for a variant theory, endogenosymbiosis, in which not only are organelles endosymbiotic, but that pieces of genetic material from symbiotic parasites ("gene carriers" such as viruses, retroviruses and bacteriophages), are included in the host's nuclear DNA, changing the host's gene expression and contributing to the process of speciation. Molecular and biochemical evidence suggests that mitochondria are related to Rickettsiales proteobacteria (in particular, the SAR11 clade, or close relatives), and that chloroplasts are related to nitrogen-fixing filamentous cyanobacteria.

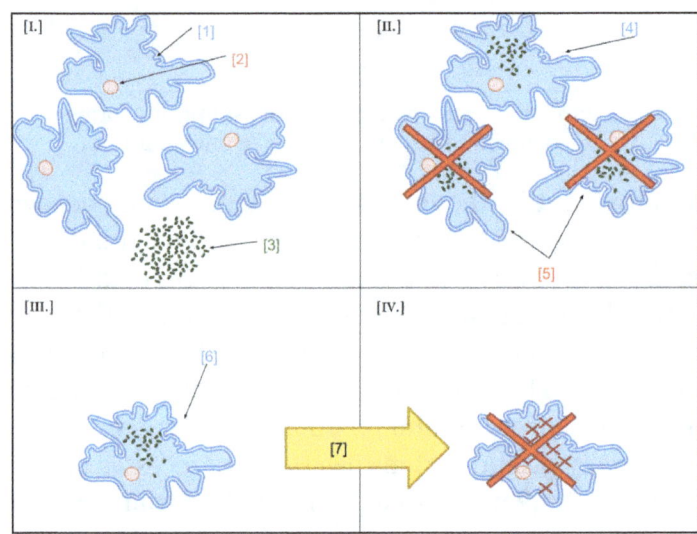

Kwang Jeon's experiment: [I] Amoebae infected by x-bacteria [II] Many amoebae become sick and die [III] Survivors have x-bacteria living in their cytoplasm [IV] Antibiotics kill x-bacteria: host amoebae die as now dependent on x-bacteria.

Endosymbiosis of Protomitochondria

Endosymbiotic theory for the origin of mitochondria suggests that the proto-eukaryote engulfed a protomitochondria, and this endosymbiont became an organelle.

Mitochondria

Mitochondria are organelles that synthesize ATP for the cell by metabolizing carbon-based macromolecules. The presence of deoxyribonucleic acid (DNA) in mitochondria and proteins, derived from mtDNA, suggest that this organelle may have been a prokaryote prior to its integration into the proto-eukaryote. Mitochondria are regarded as organelles rather than endosymbionts because mitochondria and the host cells share some parts of their genome, undergo mitosis simultaneously, and provide each other means to produce energy. Endomembrane system and nuclear membrane were derived from the protomitochondria.

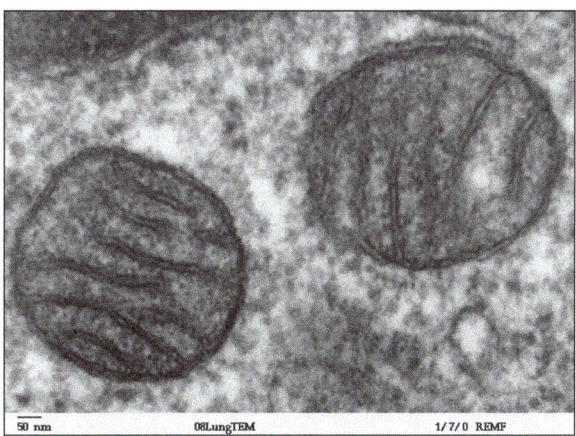

Mitochondria of a mammal lung cell visualized
using Transmission Electron Microscopy

Nuclear Membrane

The presence of a nucleus is one major difference between eukaryotes and prokaryotes. Some conserved nuclear proteins between eukaryotes and prokaryotes suggest that these two types had a common ancestor. Another theory behind nucleation is that early nuclear membrane proteins caused the cell membrane to fold inwardly and form a sphere with pores like the nuclear envelope. Strictly regarding energy expenditure, endosymbiosis would save the cell more energy to develop a nuclear membrane than if the cell was to fold its cell membrane to develop this structure since the interactions between proteins are usually enabled by ATP. Digesting engulfed cells without a complex metabolic system that produces massive amounts of energy like mitochondria would have been challenging for the host cell. This theory suggests that the vesicles leaving the protomitochondria may have formed the nuclear envelope.

Endomembrane System

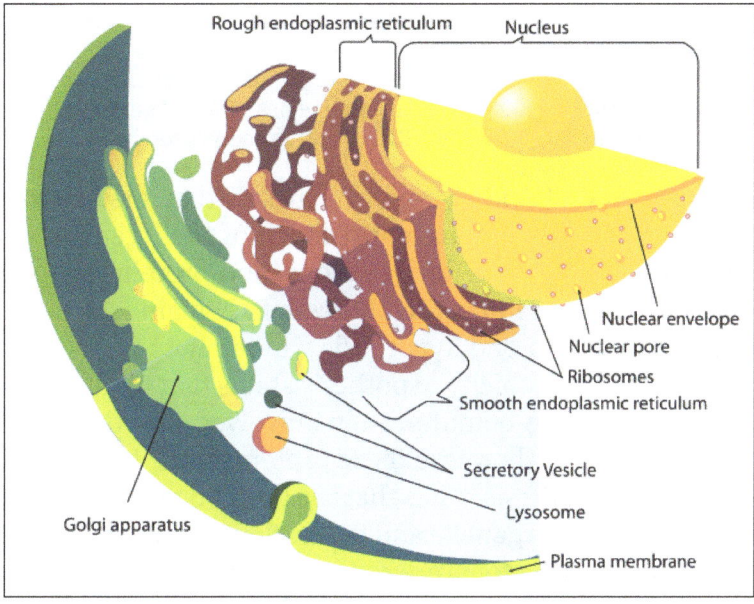

Diagram of endomembrane system in eukaryotic cell

Modern eukaryotic cells use the endomembrane system to transport products and wastes in, within, and out of cells. The membrane of nuclear envelope and endomembrane vesicles are composed of similar membrane proteins. These vesicles also share similar membrane proteins with the organelle they originated from or are traveling towards. This suggests that what formed the nuclear membrane also formed the endomembrane system. Prokaryotes do not have a complex internal membrane network like the modern eukaryotes, but the prokaryotes could produce extracellular vesicles from their outer membrane. After the early prokaryote was consumed by a proto-eukaryote, the prokaryote would have continued to produce vesicles that accumulated within the cell. Interaction of internal components of vesicles may have led to formation of the endoplasmic reticulum and contributed to the formation of Golgi apparatus.

Organellar Genomes

Plastomes and Mitogenomes

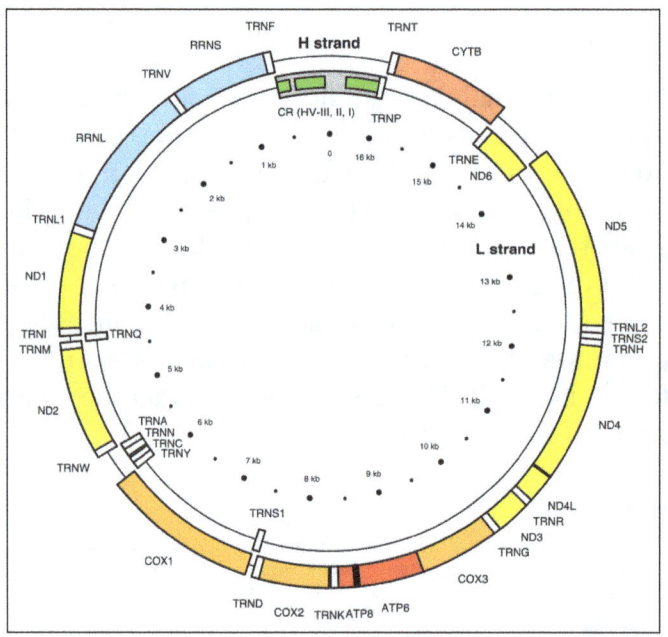

The human mitochondrial genome has retained genes
encoding 2 rRNAs, 22 tRNAs, and 13 redox proteins.

The third and final possible fate of endosymbiont genes is that they remain in the organelles. Plastids and mitochondria, although they have lost much of their genomes, retain genes encoding rRNAs, tRNAs, proteins involved in redox reactions, and proteins required for transcription, translation, and replication. There are many hypotheses to explain why organelles retain a small portion of their genome; however no one hypothesis will apply to all organisms and the topic is still quite controversial. The hydrophobicity hypothesis states that highly hydrophobic (water hating) proteins (such as the membrane bound proteins involved in redox reactions) are not easily transported through the cytosol and therefore these proteins must be encoded in their respective organelles. The code disparity hypothesis states that the limit on transfer is due to differing genetic codes and RNA editing between the organelle and the nucleus. The redox control hypothesis states that genes encoding redox reaction proteins are retained in order to effectively couple the need for repair and the synthesis of these proteins. For example, if one of the photosystems is lost from the plastid, the intermediate electron carriers may lose or gain too many electrons, signalling the need for repair of a photosystem. The time delay involved in signalling the nucleus and transporting a cytosolic protein to the organelle results in the production of damaging reactive oxygen species. The final hypothesis states that the assembly of membrane proteins, particularly those involved in redox reactions, requires coordinated synthesis and assembly of subunits; however, translation and protein transport coordination is more difficult to control in the cytoplasm.

Non-photosynthetic Plastid Genomes

The majority of the genes in the mitochondria and plastids are related to the expression (transcription, translation and replication) of genes encoding proteins involved in either photosynthesis (in plastids) or cellular respiration (in mitochondria). One might predict that the loss of photosynthesis or cellular respiration would allow for the complete loss of the plastid genome or the mitochondrial genome respectively. While there are numerous examples of mitochondrial descendants (mitosomes and hydrogenosomes) that have lost their entire organellar genome, non-photosynthetic plastids tend to retain a small genome. There are two main hypotheses to explain this occurrence: The essential tRNA hypothesis notes that there have been no documented functional plastid-to-nucleus gene transfers of genes encoding RNA products (tRNAs and rRNAs). As a result, plastids must make their own functional RNAs or import nuclear counterparts. The genes encoding tRNA-Glu and tRNA-fmet, however, appear to be indispensable. The plastid is responsible for haem biosynthesis, which requires plastid encoded tRNA-Glu (from the gene trnE) as a precursor molecule. Like other genes encoding RNAs, trnE cannot be transferred to the nucleus. In addition, it is unlikely trnE could be replaced by a cytosolic tRNA-Glu as trnE is highly conserved; single base changes in trnE have resulted in the loss of haem synthesis. The gene for tRNA-formylmethionine (tRNA-fmet) is also encoded in the plastid genome and is required for translation initiation in both plastids and mitochondria. A plastid is required to continue expressing the gene for tRNA-fmet so long as the mitochondrion is translating proteins. The limited window hypothesis offers a more general explanation for the retention of genes in non-photosynthetic plastids. According to the bulk flow hypothesis, genes are transferred to the nucleus following the disturbance of organelles. Disturbance was common in the early stages of endosymbiosis, however, once the host cell gained control of organelle division, eukaryotes could evolve to have only one plastid per cell. Having only one plastid severely limits gene transfer as the lysis of the single plastid would likely result in cell death. Consistent with this hypothesis, organisms with multiple plastids show an 80-fold increase in plastid-to-nucleus gene transfer compared to organisms with single plastids.

Evidence

There are many lines of evidence that mitochondria and plastids including chloroplasts arose from bacteria.

- New mitochondria and plastids are formed only through binary fission, the form of cell division used by bacteria and archaea.

- If a cell's mitochondria or chloroplasts are removed, the cell does not have the means to create new ones. For example, in some algae, such as Euglena, the plastids can be destroyed by certain chemicals or prolonged absence of light without otherwise affecting the cell. In such a case, the plastids will not regenerate.

- Transport proteins called porins are found in the outer membranes of mitochondria and chloroplasts and are also found in bacterial cell membranes.

- A membrane lipid cardiolipin is exclusively found in the inner mitochondrial membrane and bacterial cell membranes.

- Some mitochondria and some plastids contain single circular DNA molecules that are similar to the DNA of bacteria both in size and structure.

- Genome comparisons suggest a close relationship between mitochondria and Rickettsial bacteria.

- Genome comparisons suggest a close relationship between plastids and cyanobacteria.

- Many genes in the genomes of mitochondria and chloroplasts have been lost or transferred to the nucleus of the host cell. Consequently, the chromosomes of many eukaryotes contain genes that originated from the genomes of mitochondria and plastids.

- Mitochondrial and plastid ribosomes are more similar to those of bacteria (70S) than those of eukaryotes.

- Proteins created by mitochondria and chloroplasts use N-formylmethionine as the initiating amino acid, as do proteins created by bacteria but not proteins created by eukaryotic nuclear genes or archaea.

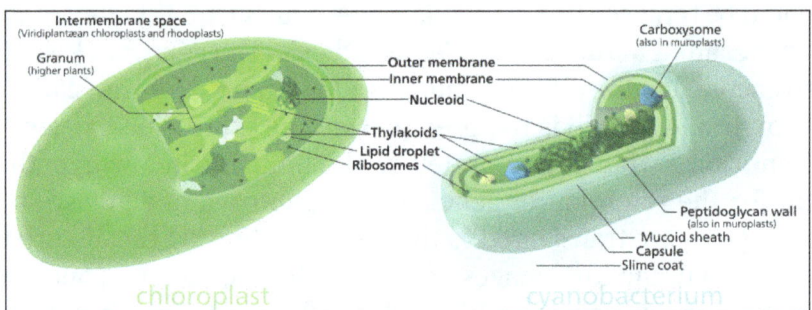

Comparison of chloroplasts and cyanobacteria showing their
similarities. Both chloroplasts and cyanobacteria have
a double membrane, DNA, ribosomes, and thylakoids.

Secondary Endosymbiosis

Primary endosymbiosis involves the engulfment of a cell by another free living organism. Secondary endosymbiosis occurs when the product of primary endosymbiosis is itself engulfed and retained by another free living eukaryote. Secondary endosymbiosis has occurred several times and has given rise to extremely diverse groups of algae and other eukaryotes. Some organisms can take opportunistic advantage of a similar process, where they engulf an alga and use the products of its photosynthesis, but once the prey item dies (or is lost) the host returns to a free living state. Obligate secondary endosymbionts become dependent on their organelles and are unable to survive in their absence.) RedToL, the Red Algal Tree of Life Initiative funded by the National Science Foundation highlights the role red algae or Rhodophyta played in the evolution of our planet through secondary endosymbiosis. One possible secondary endosymbiosis in process has been observed by

Okamoto & Inouye. The heterotrophic protist *Hatena* behaves like a predator until it ingests a green alga, which loses its flagella and cytoskeleton, while *Hatena*, now a host, switches to photosynthetic nutrition, gains the ability to move towards light and loses its feeding apparatus. The process of secondary endosymbiosis left its evolutionary signature within the unique topography of plastid membranes. Secondary plastids are surrounded by three (in euglenophytes and some dinoflagellates) or four membranes (in haptophytes, heterokonts, cryptophytes, and chlorarachniophytes). The two additional membranes are thought to correspond to the plasma membrane of the engulfed alga and the phagosomal membrane of the host cell. The endosymbiotic acquisition of a eukaryote cell is represented in the cryptophytes; where the remnant nucleus of the red algal symbiont (the nucleomorph) is present between the two inner and two outer plastid membranes. Despite the diversity of organisms containing plastids, the morphology, biochemistry, genomic organisation, and molecular phylogeny of plastid RNAs and proteins suggest a single origin of all extant plastids – although this theory is still debated. Some species including *Pediculus humanus* (lice) have multiple chromosomes in the mitochondrion. This and the phylogenetics of the genes encoded within the mitochondrion suggest that mitochondria have multiple ancestors, that these were acquired by endosymbiosis on several occasions rather than just once, and that there have been extensive mergers and rearrangements of genes on the several original mitochondrial chromosomes.

Date

The question of when the transition from prokaryotic to eukaryotic form occurred and when the first crown group eukaryotes appeared on earth is still unresolved. The oldest known body fossils that can be positively assigned to the Eukaryota are acanthomorphic acritarchs from the 1631±1 Ma Deonar Formation of India (lower Vindhyan Supergroup) of India. These fossils can still be identified as derived post-nuclear eukaryotes with a sophisticated, morphology-generating cytoskeleton sustained by mitochondria. This fossil evidence indicates that endosymbiotic acquisition of alphaproteobacteria must have occurred before 1.6 Ga. Molecular clocks have also been used to estimate the last eukaryotic common ancestor (LECA), however these methods have large inherent uncertainty and give a wide range of dates. Reasonable results for LECA include the estimate of c. 1800 Mya. A 2300 Mya estimate also seems reasonable and has the added attraction of coinciding with one of the most pronounced biogeochemical perturbations in Earth history (the Great Oxygenation Event). The marked increase in atmospheric oxygen concentrations during the early Palaeoproterozoic Great Oxidation Event has been invoked as a contributing cause of eukaryogenesis – by inducing the evolution of oxygen-detoxifying mitochondria. Alternatively, the Great Oxidation Event might be a consequence of eukaryogenesis and its impact on the export and burial of organic carbon.

LIPID BILAYER

The lipid bilayer (or phospholipid bilayer) is a thin polar membrane made of two layers of lipid molecules. These membranes are flat sheets that form a continuous barrier around all cells. The cell membranes of almost all organisms and many viruses are made of a lipid bilayer, as are the nuclear membrane surrounding the cell nucleus, and other membranes surrounding sub-cellular structures. The lipid bilayer is the barrier that keeps ions, proteins and other molecules where they are needed and prevents them from diffusing into areas where they should not be. Lipid bilayers are ideally suited to this role, even though they are only a few nanometers in width, they are impermeable to most water-soluble (hydrophilic) molecules. Bilayers are particularly impermeable to ions, which allows cells to regulate salt concentrations and pH by transporting ions across their membranes using proteins called ion pumps.

Biological bilayers are usually composed of amphiphilic phospholipids that have a hydrophilic phosphate head and a hydrophobic tail consisting of two fatty acid chains. Phospholipids with certain head groups can alter the surface chemistry of a bilayer and can, for example, serve as signals as well as "anchors" for other molecules in the membranes of cells. Just like the heads, the tails of lipids can also affect membrane properties, for instance by determining the phase of the bilayer. The bilayer can adopt a solid gel phase state at lower temperatures but undergo phase transition to a fluid state at higher temperatures, and the chemical properties of the lipids' tails influence at which temperature this happens. The packing of lipids within the bilayer also affects its mechanical properties, including its resistance to stretching and bending. Many of these properties have been studied with the use of artificial "model" bilayers produced in a lab. Vesicles made by model bilayers have also been used clinically to deliver drugs.

Biological membranes typically include several types of molecules other than phospholipids. A particularly important example in animal cells is cholesterol, which helps strengthen the bilayer and decrease its permeability. Cholesterol also helps regulate the activity of certain integral membrane proteins. Integral membrane proteins function when incorporated into a lipid bilayer, and they are held tightly to lipid bilayer with the help of an annular lipid shell. Because bilayers define the boundaries of the cell and its compartments, these membrane proteins are involved in many intra- and inter-cellular signaling processes. Certain kinds of membrane proteins are involved in the process of fusing two bilayers together. This fusion allows the joining of two distinct structures as in the fertilization of an egg by sperm or the entry of a virus into a cell. Because lipid bilayers are quite fragile and invisible in a traditional microscope, they are a challenge to study. Experiments on bilayers often require advanced techniques like electron microscopy and atomic force microscopy.

Structure and Organization

When phospholipids are exposed to water, they self-assemble into a two-layered sheet with the hydrophobic tails pointing toward the center of the sheet. This arrangement results in two "leaflets" that are each a single molecular layer. The center of this bilayer contains almost no water and excludes molecules like sugars or salts that dissolve in water. The assembly process is driven by interactions between hydrophobic molecules (also called the hydrophobic effect). An increase in interactions between hydrophobic molecules (causing clustering of hydrophobic regions) allows water molecules to bond more freely with each other, increasing the entropy of the system. This complex process includes non-covalent interactions such as van der Waals forces, electrostatic and hydrogen bonds.

Cross Section Analysis

Schematic cross sectional profile of a typical lipid bilayer. There are three distinct regions: the fully hydrated headgroups, the fully dehydrated alkane core and a short intermediate region with partial hydration. Although the head groups are neutral, they have significant dipole moments that influence the molecular arrangement.

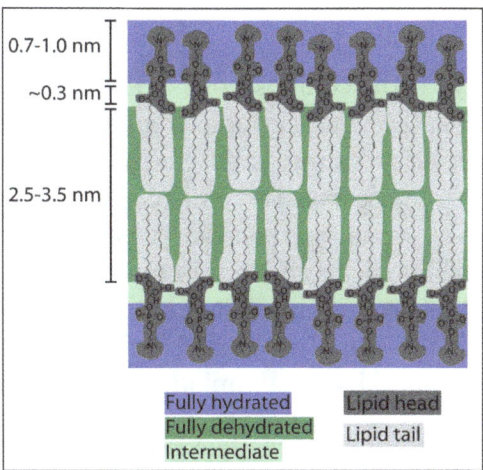

The lipid bilayer is very thin compared to its lateral dimensions. If a typical mammalian cell (diameter ~10 micrometers) were magnified to the size of a watermelon (~1 ft/30 cm), the lipid bilayer making up the plasma membrane would be about as thick as a piece of office paper. Despite being only a few nanometers thick, the bilayer is composed of several distinct chemical regions across its cross-section. These regions and their interactions with the surrounding water have been characterized over the past several decades with x-ray reflectometry, neutron scattering and nuclear magnetic resonance techniques.

The first region on either side of the bilayer is the hydrophilic headgroup. This portion of the membrane is completely hydrated and is typically around 0.8-0.9 nm thick. In phospholipid bilayers the phosphate group is located within this hydrated region, approximately 0.5 nm outside the hydrophobic core. In some cases, the hydrated region can extend much further, for instance in lipids with a large protein or long sugar chain grafted to the head. One common example of such a modification in nature is the lipopolysaccharide coat on a bacterial outer membrane, which helps retain a water layer around the bacterium to prevent dehydration.

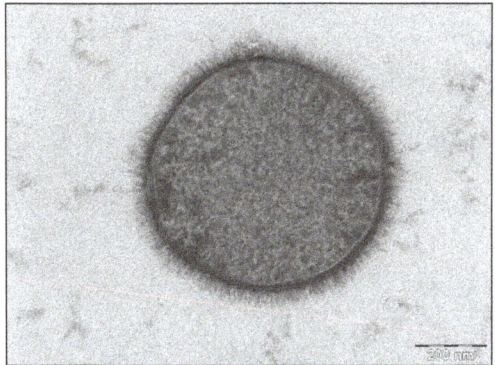

TEM image of a bacterium. The furry appearance on the outside is due to a coat of long-chain sugars attached to the cell membrane. This coating helps trap water to prevent the bacterium from becoming dehydrated.

Next to the hydrated region is an intermediate region that is only partially hydrated. This boundary layer is approximately 0.3 nm thick. Within this short distance, the water concentration drops from 2M on the headgroup side to nearly zero on the tail (core) side. The hydrophobic core of the bilayer is typically 3-4 nm thick, but this value varies with chain length and chemistry. Core thickness also varies significantly with temperature, in particular near a phase transition.

Asymmetry

In many naturally occurring bilayers, the compositions of the inner and outer membrane leaflets are different. In human red blood cells, the inner (cytoplasmic) leaflet is composed mostly of phosphatidylethanolamine, phosphatidylserine and phosphatidylinositol and its phosphorylated derivatives. By contrast, the outer (extracellular) leaflet is based on phosphatidylcholine, sphingomyelin and a variety of glycolipids, In some cases, this asymmetry is based on where the lipids are made in the cell and reflects their initial orientation. The biological functions of lipid asymmetry are imperfectly understood, although it is clear that it is used in several different situations. For example, when a cell undergoes apoptosis, the phosphatidylserine — normally localised to the cytoplasmic leaflet — is transferred to the outer surface: There, it is recognised by a macrophage that then actively scavenges the dying cell.

Lipid asymmetry arises, at least in part, from the fact that most phospholipids are synthesised and initially inserted into the inner monolayer: those that constitute the outer monolayer are then transported from the inner monolayer by a class of enzymes called flippases. Other lipids, such as sphingomyelin, appear to be synthesised at the external leaflet. Flippases are members of a larger family of lipid transport molecules that also includes floppases, which transfer lipids in the opposite direction, and scramblases, which randomize lipid distribution across lipid bilayers (as in apoptotic cells). In any case, once lipid asymmetry is established, it does not normally dissipate quickly because spontaneous flip-flop of lipids between leaflets is extremely slow.

It is possible to mimic this asymmetry in the laboratory in model bilayer systems. Certain types of very small artificial vesicle will automatically make themselves slightly asymmetric, although the mechanism by which this asymmetry is generated is very different from that in cells. By utilizing two different monolayers in Langmuir-Blodgett deposition or a combination of Langmuir-Blodgett and vesicle rupture deposition it is also possible to synthesize an asymmetric planar bilayer. This asymmetry may be lost over time as lipids in supported bilayers can be prone to flip-flop.

Phases and Phase Transitions

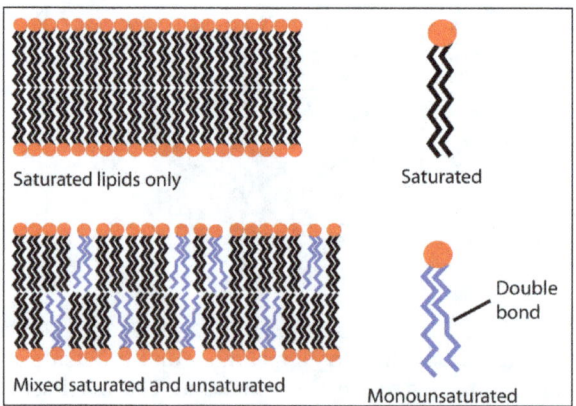

Diagram showing the effect of unsaturated lipids on a bilayer. The lipids with an unsaturated tail (blue) disrupt the packing of those with only saturated tails (black). The resulting bilayer has more free space and is, as a consequence, more permeable to water and other small molecules.

At a given temperature a lipid bilayer can exist in either a liquid or a gel (solid) phase. All lipids have a characteristic temperature at which they transition (melt) from the gel to liquid phase. In both phases the lipid molecules are prevented from flip-flopping across the bilayer, but in liquid phase bilayers a given lipid will exchange locations with its neighbor millions of times a second. This random walk exchange allows lipid to diffuse and thus wander across the surface of the membrane. Unlike liquid phase bilayers, the lipids in a gel phase bilayer have less mobility.

The phase behavior of lipid bilayers is determined largely by the strength of the attractive Van der Waals interactions between adjacent lipid molecules. Longer-tailed lipids have more area over which to interact, increasing the strength of this interaction and, as a consequence, decreasing the lipid mobility. Thus, at a given temperature, a short-tailed lipid will be more fluid than an otherwise identical long-tailed lipid. Transition temperature can also be affected by the degree of unsaturation of the lipid tails. An unsaturated double bond can produce a kink in the alkane chain, disrupting the lipid packing. This disruption creates extra free space within the bilayer that allows additional flexibility in the adjacent chains. An example of this effect can be noted in everyday life as butter, which has a large percentage saturated fats, is solid at room temperature while vegetable oil, which is mostly unsaturated, is liquid.

Most natural membranes are a complex mixture of different lipid molecules. If some of the components are liquid at a given temperature while others are in the gel phase, the two phases can coexist in spatially separated regions, rather like an iceberg floating in the ocean. This phase separation plays a critical role in biochemical phenomena because membrane components such as proteins can partition into one or the other phase and thus be locally concentrated or activated. One particularly important component of many mixed phase systems is cholesterol, which modulates bilayer permeability, mechanical strength, and biochemical interactions.

Surface Chemistry

While lipid tails primarily modulate bilayer phase behavior, it is the headgroup that determines the bilayer surface chemistry. Most natural bilayers are composed primarily of phospholipids, but sphingolipids and sterols such as cholesterol are also important components. Of the phospholipids, the most common headgroup is phosphatidylcholine (PC), accounting for about half the phospholipids in most mammalian cells. PC is a zwitterionic headgroup, as it has a negative charge on the phosphate group and a positive charge on the amine but, because these local charges balance, no net charge.

Other headgroups are also present to varying degrees and can include phosphatidylserine (PS) phosphatidylethanolamine (PE) and phosphatidylglycerol (PG). These alternate headgroups often confer specific biological functionality that is highly context-dependent. For instance, PS presence on the extracellular membrane face of erythrocytes is a marker of cell apoptosis, whereas PS in growth plate vesicles is necessary for the nucleation of hydroxyapatite crystals and subsequent bone mineralization. Unlike PC, some of the other headgroups carry a net charge, which can alter the electrostatic interactions of small molecules with the bilayer.

Biological Roles

Containment and Separation

The primary role of the lipid bilayer in biology is to separate aqueous compartments from their surroundings. Without some form of barrier delineating "self" from "non-self," it is difficult to even define the concept of an organism or of life. This barrier takes the form of a lipid bilayer in all known life forms except for a few species of archaea that utilize a specially adapted lipid monolayer. It has even been proposed that the very first form of life may have been a simple lipid vesicle with virtually its sole biosynthetic capability being the production of more phospholipids. The partitioning ability of the lipid bilayer is based on the fact that hydrophilic molecules cannot easily cross the hydrophobic bilayer core. The nucleus, mitochondria and chloroplasts have two lipid bilayers, while other sub-cellular structures are surrounded by a single lipid bilayer (such as the plasma membrane, endoplasmic reticula, Golgi apparatus and lysosomes).

Prokaryotes have only one lipid bilayer- the cell membrane (also known as the plasma membrane). Many prokaryotes also have a cell wall, but the cell wall is composed of proteins or long chain carbohydrates, not lipids. In contrast, eukaryotes have a range of organelles including the nucleus, mitochondria, lysosomes and endoplasmic reticulum. All of these sub-cellular compartments are surrounded by one or more lipid bilayers and, together, typically comprise the majority of the bilayer area present in the cell. In liver hepatocytes for example, the plasma membrane accounts for only two percent of the total bilayer area of the cell, whereas the endoplasmic reticulum contains more than fifty percent and the mitochondria a further thirty percent.

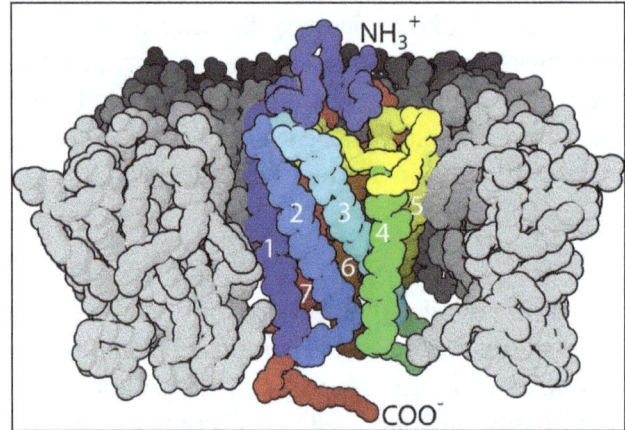

Illustration of a GPCR signaling protein. In response to a molecule such as
a hormone binding to the exterior domain (blue) the GPCR changes shape and catalyzes
a chemical reaction on the interior domain (red). The gray feature is the surrounding bilayer.

Signaling

Probably the most familiar form of cellular signaling is synaptic transmission, whereby a nerve impulse that has reached the end of one neuron is conveyed to an adjacent neuron via the release of neurotransmitters. This transmission is made possible by the action of synaptic vesicles loaded with the neurotransmitters to be released. These vesicles fuse with the cell membrane at the pre-synaptic terminal and release its contents to the exterior of the cell. The contents then diffuse across the synapse to the post-synaptic terminal.

Lipid bilayers are also involved in signal transduction through their role as the home of integral membrane proteins. This is an extremely broad and important class of biomolecule. It is estimated that up to a third of the human proteome are membrane proteins. Some of these proteins are linked to the exterior of the cell membrane. An example of this is the CD59 protein, which identifies cells as "self" and thus inhibits their destruction by the immune system. The HIV virus evades the immune system in part by grafting these proteins from the host membrane onto its own surface. Alternatively, some membrane proteins penetrate all the way through the bilayer and serve to relay individual signal events from the outside to the inside of the cell. The most common class of this type of protein is the G protein-coupled receptor (GPCR). GPCRs are responsible for much of the cell's ability to sense its surroundings and, because of this important role, approximately 40% of all modern drugs are targeted at GPCRs.

In addition to protein- and solution-mediated processes, it is also possible for lipid bilayers to participate directly in signaling. A classic example of this is phosphatidylserine-triggered phago-cytosis. Normally, phosphatidylserine is asymmetrically distributed in the cell membrane and is present only on the interior side. During programmed cell death a protein called a scramblase equilibrates this distribution, displaying phosphatidylserine on the extracellular bilayer face. The presence of phosphatidylserine then triggers phagocytosis to remove the dead or dying cell.

Characterization Methods

The lipid bilayer is a very difficult structure to study because it is so thin and fragile. In spite of these limitations dozens of techniques have been developed over the last seventy years to allow investigations of its structure and function.

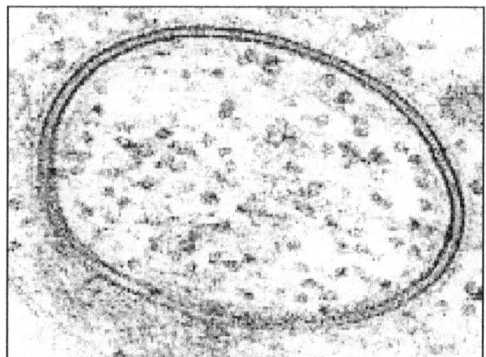

Transmission Electron Microscope (TEM) image of a lipid vesicle. The two
dark bands around the edge are the two leaflets of the bilayer.

Electrical Measurements

Electrical measurements are a straightforward way to characterize an important function of a bilay-er: its ability to segregate and prevent the flow of ions in solution. By applying a voltage across the bi-layer and measuring the resulting current, the resistance of the bilayer is determined. This resistance is typically quite high (10^8 Ohm-cm^2 or more) since the hydrophobic core is impermeable to charged species. The presence of even a few nanometer-scale holes results in a dramatic increase in current. The sensitivity of this system is such that even the activity of single ion channels can be resolved.

Fluorescence Microscopy

Electrical measurements do not provide an actual picture like imaging with a microscope can. Lipid bilayers cannot be seen in a traditional microscope because they are too thin. In order to see bilay-ers, researchers often use fluorescence microscopy. A sample is excited with one wavelength of light and observed in a different wavelength, so that only fluorescent molecules with a matching exci-tation and emission profile will be seen. Natural lipid bilayers are not fluorescent, so a dye is used that attaches to the desired molecules in the bilayer. Resolution is usually limited to a few hundred nanometers, much smaller than a typical cell but much larger than the thickness of a lipid bilayer.

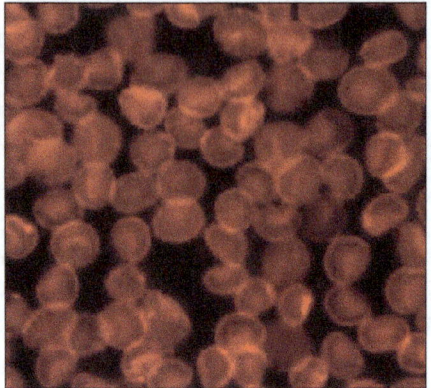

Human red blood cells viewed through a fluorescence microscope.
The cell membrane has been stained with a fluorescent dye. Scale bar is 20μm.

Electron Microscopy

Electron microscopy offers a higher resolution image. In an electron microscope, a beam of focused electrons interacts with the sample rather than a beam of light as in traditional microscopy. In conjunction with rapid freezing techniques, electron microscopy has also been used to study the mechanisms of inter- and intracellular transport, for instance in demonstrating that exocytotic vesicles are the means of chemical release at synapses.

Nuclear Magnetic Resonance Spectroscopy

^{31}P-NMR(nuclear magnetic resonance) spectroscopy is widely used for studies of phospholipid bilayers and biological membranes in native conditions. The analysis of ^{31}P-NMR spectra of lipids could provide a wide range of information about lipid bilayer packing, phase transitions (gel phase, physiological liquid crystal phase, ripple phases, non bilayer phases), lipid head group orientation/dynamics, and elastic properties of pure lipid bilayer and as a result of binding of proteins and other biomolecules.

Atomic Force Microscopy

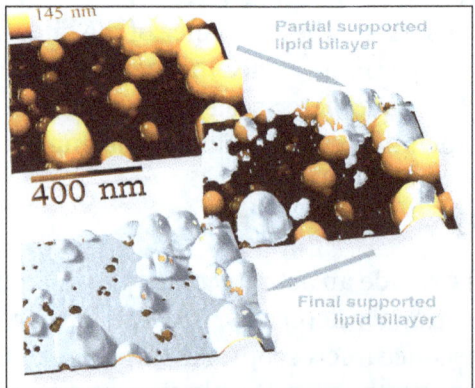

3d-Adapted AFM images showing formation of transmembrane
pores (holes) in supported lipid bilayer.

A new method to study lipid bilayers is Atomic force microscopy (AFM). Rather than using a beam of light or particles, a very small sharpened tip scans the surface by making physical contact with the bilayer and moving across it, like a record player needle. AFM is a promising technique because it has the potential to image with nanometer resolution at room temperature and even under water or physiological buffer, conditions necessary for natural bilayer behavior. Utilizing this capability, AFM has been used to examine dynamic bilayer behavior including the formation of transmembrane pores (holes) and phase transitions in supported bilayers. Another advantage is that AFM does not require fluorescent or isotopic labeling of the lipids, since the probe tip interacts mechanically with the bilayer surface. Because of this, the same scan can image both lipids and associated proteins, sometimes even with single-molecule resolution. AFM can also probe the mechanical nature of lipid bilayers.

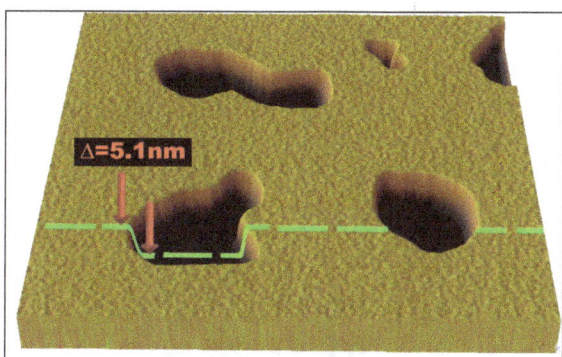

Illustration of a typical AFM scan of a supported lipid bilayer. The pits are defects in the bilayer, exposing the smooth surface of the substrate underneath.

Dual Polarisation Interferometry

Lipid bilayers exhibit high levels of birefringence where the refractive index in the plane of the bilayer differs from that perpendicular by as much as 0.1 refractive index units. This has been used to characterise the degree of order and disruption in bilayers using dual polarisation interferometry to understand mechanisms of protein interaction.

Quantum Chemical Calculations

Lipid bilayers are complicated molecular systems with many degrees of freedom. Thus atomistic simulation of membrane and in particular abinitio calculations of its properties is difficult and computationally expensive. Quantum chemical calculations has recently been successfully performed to estimate dipole and quadrupole moments of lipid membranes.

Transport across the Bilayer

Passive Diffusion

Most polar molecules have low solubility in the hydrocarbon core of a lipid bilayer and, as a consequence, have low permeability coefficients across the bilayer. This effect is particularly pronounced for charged species, which have even lower permeability coefficients than neutral polar molecules. Anions typically have a higher rate of diffusion through bilayers than cations. Compared to ions, water molecules actually have a relatively large permeability through the bilayer, as evidenced by osmotic swelling. When a cell or vesicle with a high interior salt concentration is placed in a solution with a low salt concentration it will swell and eventually burst. Such a result would not be observed unless water was able to pass through the bilayer with relative ease. The anomalously large permeability of water through bilayers is still not completely understood and continues to be the subject of active debate. Small uncharged apolar molecules diffuse through lipid bilayers many orders of magnitude faster than ions or water. This applies both to fats and organic solvents like chloroform and ether. Regardless of their polar character larger molecules diffuse more slowly across lipid bilayers than small molecules.

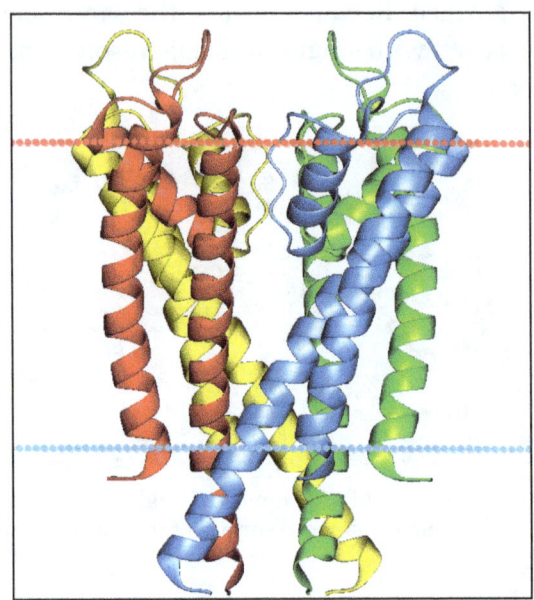

Structure of a potassium ion channel. The alpha helices
penetrate the bilayer (boundaries indicated by red and blue lines),
opening a hole through which potassium ions can flow.

Ion pumps and Channels

Two special classes of protein deal with the ionic gradients found across cellular and sub-cellular membranes in nature- ion channels and ion pumps. Both pumps and channels are integral membrane proteins that pass through the bilayer, but their roles are quite different. Ion pumps are the proteins that build and maintain the chemical gradients by utilizing an external energy source to move ions against the concentration gradient to an area of higher chemical potential. The energy source can be ATP, as is the case for the Na^+-K^+ ATPase. Alternatively, the energy source can be another chemical gradient already in place, as in the Ca^{2+}/Na^+ antiporter. It is through the action of ion pumps that cells are able to regulate pH via the pumping of protons.

In contrast to ion pumps, ion channels do not build chemical gradients but rather dissipate them in order to perform work or send a signal. Probably the most familiar and best studied example is the voltage-gated Na^+ channel, which allows conduction of an action potential along neurons. All ion pumps have some sort of trigger or "gating" mechanism. In the previous example it was electrical bias, but other channels can be activated by binding a molecular agonist or through a conformational change in another nearby protein.

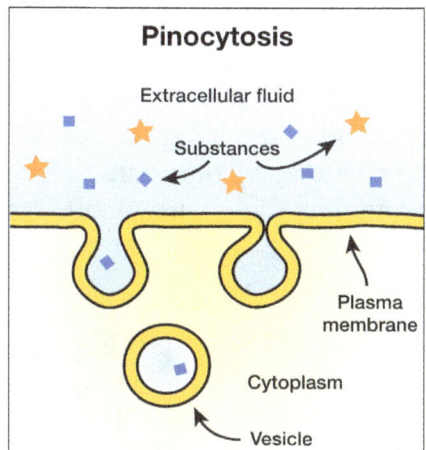

Schematic illustration of pinocytosis, a type of endocytosis.

Endocytosis and Exocytosis

Some molecules or particles are too large or too hydrophilic to pass through a lipid bilayer. Other molecules could pass through the bilayer but must be transported rapidly in such large numbers that channel-type transport is impractical. In both cases, these types of cargo can be moved across the cell membrane through fusion or budding of vesicles. When a vesicle is produced inside the cell and fuses with the plasma membrane to release its contents into the extracellular space, this process is known as exocytosis. In the reverse process, a region of the cell membrane will dimple inwards and eventually pinch off, enclosing a portion of the extracellular fluid to transport it into the cell. Endocytosis and exocytosis rely on very different molecular machinery to function, but the two processes are intimately linked and could not work without each other. The primary mechanism of this interdependence is the large amount of lipid material involved. In a typical cell, an area of bilayer equivalent to the entire plasma membrane will travel through the endocytosis/exocytosis cycle in about half an hour. If these two processes were not balancing each other, the cell would either balloon outward to an unmanageable size or completely deplete its plasma membrane within a short time.

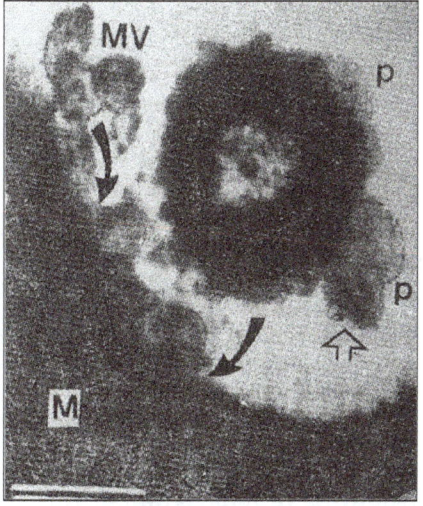

Exocytosis of outer membrane vesicles (MV) liberated from inflated periplasmic pockets (p) on surface of human *Salmonella* 3,10:r:- pathogens docking on plasma. membrane of macrophage cells (M) in chicken ileum, for host-pathogen signaling *in vivo*.

Exocytosis in prokaryotes: Membrane vesicular exocytosis, popularly known as membrane vesicle trafficking, a Nobel prize-winning process, is traditionally regarded as a prerogative of eukaryotic cells. This *myth* was however broken with the revelation that nanovesicles, popularly known as bacterial outer membrane vesicles, released by gram-negative microbes, translocate bacterial signal molecules to host or target cells to carry out multiple processes in favour of the secreting microbe e.g., in *host cell invasion* and microbe-environment interactions, in general.

Electroporation

Electroporation is the rapid increase in bilayer permeability induced by the application of a large artificial electric field across the membrane. Experimentally, electroporation is used to introduce hydrophilic molecules into cells. It is a particularly useful technique for large highly charged molecules such as DNA, which would never passively diffuse across the hydrophobic bilayer core. Because of this, electroporation is one of the key methods of transfection as well as bacterial transformation. It has even been proposed that electroporation resulting from lightning strikes could be a mechanism of natural horizontal gene transfer.

This increase in permeability primarily affects transport of ions and other hydrated species, indicating that the mechanism is the creation of nm-scale water-filled holes in the membrane. Although electroporation and dielectric breakdown both result from application of an electric field, the mechanisms involved are fundamentally different. In dielectric breakdown the barrier material is ionized, creating a conductive pathway. The material alteration is thus chemical in nature. In contrast, during electroporation the lipid molecules are not chemically altered but simply shift position, opening up a pore that acts as the conductive pathway through the bilayer as it is filled with water.

Mechanics

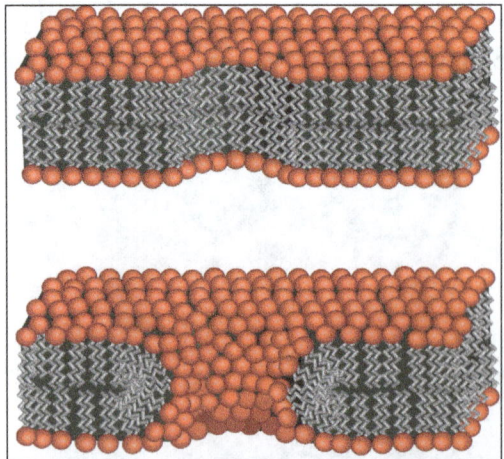

Schematic showing two possible conformations of the lipids at the edge of a pore.
In the top image the lipids have not rearranged, so the pore wall is hydrophobic.
In the bottom image some of the lipid heads have bent over, so the pore wall is hydrophilic.

Lipid bilayers are large enough structures to have some of the mechanical properties of liquids or solids. The area compression modulus K_a, bending modulus K_b, and edge energy Λ, can be used to describe them. Solid lipid bilayers also have a shear modulus, but like any liquid, the shear modulus is zero for fluid bilayers. These mechanical properties affect how the membrane functions. K_a and K_b affect the ability of proteins and small molecules to insert into the bilayer, and bilayer mechanical properties have been shown to alter the function of mechanically activated ion channels. Bilayer mechanical properties also govern what types of stress a cell can withstand without tearing. Although lipid bilayers can easily bend, most cannot stretch more than a few percent before rupturing.

The hydrophobic attraction of lipid tails in water is the primary force holding lipid bilayers together. Thus, the elastic modulus of the bilayer is primarily determined by how much extra area is exposed to water when the lipid molecules are stretched apart. It is not surprising given this understanding of the forces involved that studies have shown that K_a varies strongly with osmotic pressure but only weakly with tail length and unsaturation. Because the forces involved are so small, it is difficult to experimentally determine K_a. Most techniques require sophisticated microscopy and very sensitive measurement equipment.

In contrast to K_a, which is a measure of how much energy is needed to stretch the bilayer, K_b is a measure of how much energy is needed to bend or flex the bilayer. Formally, bending modulus is defined as the energy required to deform a membrane from its intrinsic curvature to some other curvature. Intrinsic curvature is defined by the ratio of the diameter of the head group to that of the tail group. For two-tailed PC lipids, this ratio is nearly one so the intrinsic curvature is nearly zero. If a particular lipid has too large a deviation from zero intrinsic curvature it will not form a bilayer and will instead form other phases such as micelles or inverted micelles. Addition of *small hydrophilic molecules* like *sucrose* into mixed lipid *lamellar liposomes* made from galactolipid-rich thylakoid membranes destabilises bilayers into micellar phase. Typically, K_b is not measured experimentally but rather is calculated from measurements of K_a and bilayer thickness, since the three parameters are related.

Λ (\displaystyle \Lambda) is a measure of how much energy it takes to expose a bilayer edge to water by tearing the bilayer or creating a hole in it. The origin of this energy is the fact that creating such an interface exposes some of the lipid tails to water, but the exact orientation of these border lipids is unknown. There is some evidence that both hydrophobic (tails straight) and hydrophilic (heads curved around) pores can coexist.

Fusion

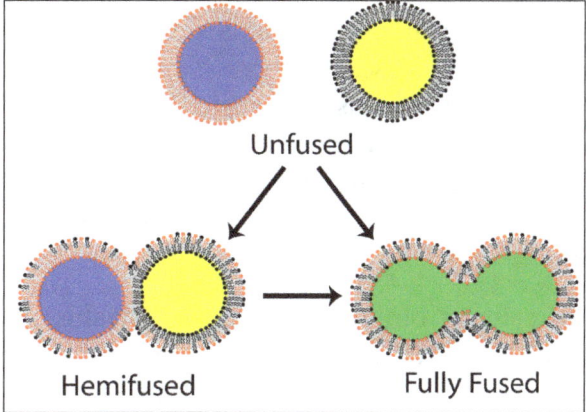

Illustration of lipid vesicles fusing showing two possible outcomes: hemifusion
and full fusion. In hemifusion, only the outer bilayer leaflets mix.
In full fusion both leaflets as well as the internal contents mix.

Fusion is the process by which two lipid bilayers merge, resulting in one connected structure. If this fusion proceeds completely through both leaflets of both bilayers, a water-filled bridge is formed and the solutions contained by the bilayers can mix. Alternatively, if only one leaflet from each bilayer is involved in the fusion process, the bilayers are said to be hemifused. Fusion is involved in many cellular processes, in particular in eukaryotes, since the eukaryotic cell is extensively sub-divided by lipid bilayer membranes. Exocytosis, fertilization of an egg by sperm and transport of waste products to the lysozome are a few of the many eukaryotic processes that rely on some form of fusion. Even the entry of pathogens can be governed by fusion, as many bilayer-coated viruses have dedicated fusion proteins to gain entry into the host cell.

There are four fundamental steps in the fusion process. First, the involved membranes must aggregate, approaching each other to within several nanometers. Second, the two bilayers must come into very close contact (within a few angstroms). To achieve this close contact, the two surfaces must become at least partially dehydrated, as the bound surface water normally present causes bilayers to strongly repel. The presence of ions, in particular divalent cations like magnesium and calcium, strongly affects this step. One of the critical roles of calcium in the body is regulating membrane fusion. Third, a destabilization must form at one point between the two bilayers, locally distorting their structures. The exact nature of this distortion is not known. One theory is that a highly curved "stalk" must form between the two bilayers. Proponents of this theory believe that it explains why phosphatidylethanolamine, a highly curved lipid, promotes fusion. Finally, in the last step of fusion, this point defect grows and the components of the two bilayers mix and diffuse away from the site of contact.

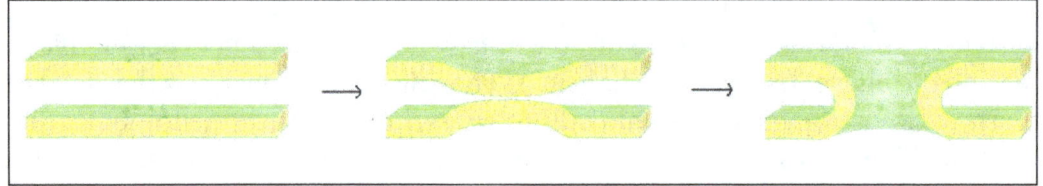

Schematic illustration of the process of fusion through stalk formation.

The situation is further complicated when considering fusion *in vivo* since biological fusion is almost always regulated by the action of membrane-associated proteins. The first of these proteins to be studied were the viral fusion proteins, which allow an enveloped virus to insert its genetic material into the host cell (enveloped viruses are those surrounded by a lipid bilayer; some others have only a protein coat). Eukaryotic cells also use fusion proteins, the best-studied of which are the SNAREs. SNARE proteins are used to direct all vesicular intracellular trafficking. Despite years of study, much is still unknown about the function of this protein class. In fact, there is still an active debate regarding whether SNAREs are linked to early docking or participate later in the fusion process by facilitating hemifusion.

In studies of molecular and cellular biology it is often desirable to artificially induce fusion. The addition of polyethylene glycol (PEG) causes fusion without significant aggregation or biochemical disruption. This procedure is now used extensively, for example by fusing B-cells with myeloma cells. The resulting "hybridoma" from this combination expresses a desired antibody as determined by the B-cell involved, but is immortalized due to the melanoma component. Fusion can also be artificially induced through electroporation in a process known as electrofusion. It is believed that this phenomenon results from the energetically active edges formed during electroporation, which can act as the local defect point to nucleate stalk growth between two bilayers.

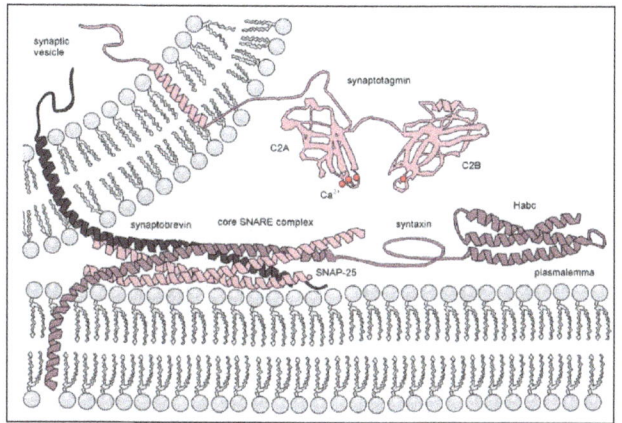

Diagram of the action of SNARE proteins docking a vesicle for exocytosis. Complementary versions of the protein on the vesicle and the target membrane bind and wrap around each other, drawing the two bilayers close together in the process.

Model Systems

Lipid bilayers can be created artificially in the lab to allow researchers to perform experiments that cannot be done with natural bilayers. They can also be used in the field of Synthetic Biology, to define the boundaries of artificial cells. These synthetic systems are called model lipid bilayers. There are many different types of model bilayers, each having experimental advantages and disadvantages. They can be made with either synthetic or natural lipids. Among the most common model systems are:

- Black lipid membranes (BLM),

- Supported lipid bilayers (SLB),

- Tethered Bilayer Lipid Membranes (t-BLM),

- Vesicles,

- Droplet Interface Bilayers (DIBs).

Commercial Applications

To date, the most successful commercial application of lipid bilayers has been the use of liposomes for drug delivery, especially for cancer treatment. (the term "liposome" is in essence synonymous with "vesicle" except that vesicle is a general term for the structure whereas liposome refers to only artificial not natural vesicles) The basic idea of liposomal drug delivery is that the drug is encapsulated in solution inside the liposome then injected into the patient. These drug-loaded liposomes travel through the system until they bind at the target site and rupture, releasing the drug. In theory, liposomes should make an ideal drug delivery system since they can isolate nearly any hydrophilic drug, can be grafted with molecules to target specific tissues and can be relatively non-toxic since the body possesses biochemical pathways for degrading lipids.

The first generation of drug delivery liposomes had a simple lipid composition and suffered from several limitations. Circulation in the bloodstream was extremely limited due to both renal clearing and phagocytosis. Refinement of the lipid composition to tune fluidity, surface charge density, and surface hydration resulted in vesicles that adsorb fewer proteins from serum and thus are less readily recognized by the immune system. The most significant advance in this area was the grafting of polyethylene glycol (PEG) onto the liposome surface to produce "stealth" vesicles, which circulate over long times without immune or renal clearing.

The first stealth liposomes were passively targeted at tumor tissues. Because tumors induce rapid and uncontrolled angiogenesis they are especially "leaky" and allow liposomes to exit the bloodstream at a much higher rate than normal tissue would. More recently work has been undertaken to graft antibodies or other molecular markers onto the liposome surface in the hope of actively binding them to a specific cell or tissue type. Some examples of this approach are already in clinical trials.

Another potential application of lipid bilayers is the field of biosensors. Since the lipid bilayer is the barrier between the interior and exterior of the cell, it is also the site of extensive signal transduction. Researchers over the years have tried to harness this potential to develop a bilayer-based device for clinical diagnosis or bioterrorism detection. Progress has been slow in this area and, although a few companies have developed automated lipid-based detection systems, they are still targeted at the research community. These include Biacore (now GE Healthcare Life Sciences), which offers a disposable chip for utilizing lipid bilayers in studies of binding kinetics and Nanion Inc., which has developed an automated patch clamping system. Other, more exotic applications are also being pursued such as the use of lipid bilayer membrane pores for DNA sequencing by Oxford Nanolabs. To date, this technology has not proven commercially viable.

A supported lipid bilayer (SLB) as described above has achieved commercial success as a screening technique to measure the permeability of drugs. This parallel artificial membrane permeability assay PAMPA technique measures the permeability across specifically formulated lipid cocktail(s) found to be highly correlated with Caco-2 cultures, the gastrointestinal tract, blood–brain barrier and skin.

References

- Corning, Peter A. (2010). Holistic Darwinism: Synergy, Cybernetics, and the Bioeconomics of Evolution. Chicago: University of Chicago Press. P. 81. ISBN 978-0-22611-633-4
- Cell-theory: biologydictionary.net, Retrieved 3 January, 2019
- Martin EA, Hine R (2008). A dictionary of biology (6th ed.). Oxford: Oxford University Press. ISBN 9780199204625. OCLC 176818780
- Keeling, P. J.; Archibald, J.M. (2008). "Organelle evolution: what's in a name?". Current Biology. 18 (8): 345–347. Doi:10.1016/j.cub.2008.02.065. PMID 18430636
- Griffiths AJ (2012). Introduction to genetic analysis (10th ed.). New York: W.H. Freeman and Co. ISBN 9781429229432. OCLC 698085201
- Nagle JF, Tristram-Nagle S (November 2000). "Structure of lipid bilayers". Biochim. Biophys. Acta. 1469 (3): 159–95. Doi:10.1016/S0304-4157(00)00016-2. PMC 2747654. PMID 11063882
- Marieb EN (2000). Essentials of human anatomy and physiology (6th ed.). San Francisco: Benjamin Cummings. ISBN 978-0805349405. OCLC 41266267

3

Different Types of Cells

There are two types of cells: eukaryotic and prokaryotic cells. Eukaryotic cells contain a nucleus and can be either single-celled or multicellular. Prokaryotic cells do not comprise of a nucleus and are single-celled organisms. The topics elaborated in this chapter will help in gaining a better perspective about these two categories of cells.

Because of the millions of diverse species of life on Earth, which grow and change gradually over time, there are countless differences between the countless extant types of cells.

However, here we will look at the two major types of cells, and two important sub-categories of each.

Prokaryotes

Prokaryotes are the simpler and older of the two major types of cells. Prokaryotes are single-celled organisms. Bacteria and archaebacteria are examples of prokaryotic cells.

Prokaryotic cells have a cell membrane, and one or more layers of additional protection from the outside environment. Many prokaryotes have a cell membrane made of phospholipids, enclosed by a cell wall made of a rigid sugar. The cell wall may be enclosed by another thick "capsule" made of sugars.

Many prokaryotic cells also have cilia, tails, or other ways in which the cell can control its movement.

These characteristics, as well as the cell wall and capsule, reflect the fact that prokaryotic cells are going it alone in the environment. They are not part of a multicellular organism, which might have whole layers of cells devoted to protecting other cells from the environment, or to creating motion.

Prokaryotic cells have a single chromosome which contains all of the cell's essential hereditary material and operating instructions. This single chromosome is usually round. There is no nucleus, or any other internal membranes or organelles. The chromosome just floats in the cell's cytoplasm.

Additional genetic traits and information might be contained in other gene units within the cytoplasm, called "plasmids," but these are usually genes that are passed back and forth by prokaryotes though the process of "horizontal gene transfer," which is when one cell gives genetic material to another. Plasmids contain non-essential DNA that the cell can live without, and which is not necessarily passed on to offspring.

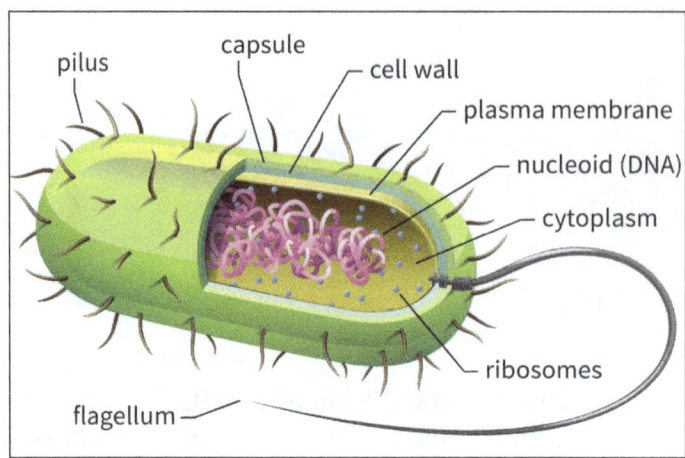

Prokaryote cell.

When a prokaryotic cell is ready to reproduce, it makes a copy of its single chromosome. Then the cell splits in half, apportioning one copy of its chromosome and a random assortment of plasmids to each daughter cell.

There are two major types of prokaryotes known to scientists to date: archaebacteria, which are a very old lineage of life with some biochemical differences from bacteria and eukaryotes, and bacteria, sometimes called "eubacteria," or "true bacteria" to differentiate them from archaebacteria.

Bacteria are thought to be more "modern" descendants of archaebacteria. Both families have "bacteria" in the name because the differences between them were not understood prior to the invention of modern biochemical and genetic analysis techniques.

When scientists began to examine the biochemistry and genetics of prokaryotes in detail, they discovered these two very different groups, who probably have different relationships to eukaryotes and different evolutionary histories! Some scientists think that eukaryotes like humans are more closely related to bacteria, since eukaryotes have similar cell membrane chemistry to bacteria. Others think that archaebacteria are more closely related to us eukaryotes, since they use similar proteins to reproduce their chromosomes.

Still others think that we might be descended from both – that eukaryotic cells might have come into existence when archaebacteria started living inside of a bacterial cell, or vice versa! This would explain how we have important genetic and chemical attributes of both, and why we have multiple internal compartments such as the nucleus, chloroplasts, and mitochondria!

Eukaryotes

Eukaryotic cells are thought to be the most modern major cell type. All multicellular organisms, including you, your cat, and your houseplants, are eukaryotes. Eukaryotic cells seem to have "learned" to work together to create multicellular organisms, while prokaryotes seem unable to do this.

Eukaryotic cells usually have more than one chromosome, which contains large amounts of genetic information. Within the body of a multicellular organism, different genes within these chromosomes may be switched "on" and "off," allowing for cells that have different traits and perform different functions within the same organism.

Eukaryotic cells also have one or more internal membranes, which has led scientists to the conclusion that eukaryotic cells likely evolved when one or more types of prokaryote began living in symbiotic relationships inside of other cells.

Organelles with interior membranes found in eukaryotic cells typically include:

- For animal cells: Mitochondria, which liberate the energy from sugar and turn it into ATP in an extremely efficient way.

- Mitochondria even have their own DNA, separate from the cells' nuclear DNA, which gives further support for the theory that they used to be independent bacteria.

- For plant cells: Chloroplasts, which perform photosynthesis, making ATP and sugar from sunlight and air.

- Chloroplasts also have their own DNA, suggesting that they may have originated as photosynthetic bacteria.

- Nucleus: In eukaryotic cells, the nucleus contains the essential DNA blueprints and operating instructions for the cell.

- The nuclear envelope is thought to provide an extra layer of protection for the DNA against toxins or invaders which might damage it.

- It is unknown whether the nucleus might also have been an endosymbiotic prokaryote at one time, or whether its membrane simply evolved as an extra layer of protection for the cell's DNA.

- Endoplasmic reticulum: This complex internal membrane is a major site of protein creation for cells. The evolutionary origin of the endoplasmic reticulum is not known.

- Golgi apparatus: This internal membrane complex can be thought of like the endoplasmic reticulum's "post office." It receives proteins from the ER, packages and "labels" them by attaching sugars as needed, and then ships them off to their final destinations.

- Others: Many eukaryotic cells can create temporary internal membrane "sacs," called "vacuoles," to store waste, or to package important materials.

Some cells, for example have special vacuoles called "lysosomes" which are full of corrosive substances and digestive enzymes. Cells simply dump their "trash" into lysosomes, where the harsh environment breaks them down into simpler components that can be re-used.

Examples of Cells

Archaebacteria

Archaebacteria are a very old form of prokaryotic cells. Biologists actually put them in their own "domain" of life, separate from other bacteria.

Key ways in which archaebacteria differ from other bacteria include:

- Their cell membranes, which are made of a type of lipid not found in either bacteria or eukaryotic cell membranes.

- Their DNA replication enzymes, which are more similar to those of eukaryotes than those of bacteria, suggesting that bacteria and archae are only distantly related, and archaebacteria may actually be more closely related to us than to modern bacteria.

- Some archaebacteria have the ability to produce methane, which is a metabolic process not found in any bacteria or any eukaryotes.

Archaebacteria's unique chemical attributes allow them to live in extreme environments, such as superheated water, extremely salty water, and some environments which are toxic to all other life forms.

Scientists became very excited in recent years at the discovery of Lokiarchaeota – a type of archaebacteria which shares many genes with eukaryotes that had never before been found in prokaryotic cells. It is now thought that Lokiarchaeota may be our closest living relative in the prokaryotic world.

Bacteria

You are most likely familiar with the type of bacteria that can make you sick. Indeed, common pathogens like Streptococcus and Staphylococcus are prokaryotic bacterial cells.

But there are also many types of helpful bacteria – including those that break down dead waste to turn useless materials into fertile soil, and bacteria that live in our own digestive tract and help us digest food. Bacterial cells can commonly be found living in symbiotic relationships with multicellular organisms like ourselves, in the soil, and anywhere else that's not too extreme for them to live!

Plant Cells

Plant cells are eukaryotic cells that are part of multicellular, photosynthetic organisms. Plants cells have chloroplast organelles, which contain pigments that absorb photons of light and harvest the energy of those photons. Chloroplasts have the remarkable ability to turn light energy into cellular fuel, and use this energy to take carbon dioxide from the air and turn it into sugars that can be used by living things as fuel or building material.

In addition to having chloroplasts, plant cells also typically have a cell wall made of a rigid sugars, to enable plant tissues to maintain their upright structures such as leaves, stems, and tree trunks. Plant cells also have the usual eukaryotic organelles including a nucleus, endoplasmic reticulum, and Golgi apparatus.

Animal Cells

Like all animal cells, it has mitochondria which perform cellular respiration, turning oxygen and sugar into large amounts of ATP to power cellular functions.

It also has the same organelles as most animal cells: a nucleus, endoplasmic reticulum, Golgi apparatus, etc. But as part of a multicellular organism, your liver cell also expresses unique genes, which give it unique traits and abilities.

Liver cells in particular contain enzymes that break down many toxins, which is what allows the liver to purify your blood and break down dangerous bodily waste.

The liver cell is an excellent example of how multicellular organisms can be more efficient by having different cell types work together.

Your body could not survive without liver cells to break down certain toxins and waste products, but the liver cell itself could not survive without nerve and muscle cells that help you find food, and a digestive tract to break down that food into easily digestible sugars.

And all of these cell types contain the information to make all the other cell types! It's simply a matter of which genes are switched "on" or "off" during development.

EUKARYOTIC CELL

Eukaryotic cells are cells that contain a membrane-bound nucleus, a structural feature that is not present in bacterial or archaeal cells. In addition to the nucleus, eukaryotic cells are characterized by numerous membrane-bound organelles such as the endoplasmic reticulum, Golgi apparatus, chloroplasts, mitochondria, and others.

Previously, we began to consider the Design Challenge of making cells larger than a small bacterium—more precisely, growing cells to sizes at which, in the eyes of natural selection, relying on diffusion of substances for transport through a highly viscous cytosol comes with inherent functional trade-offs that offset most selective benefits of getting larger. In the lectures and readings on bacterial cell structure, we discovered some morphological features of large bacteria that allow them to effectively overcome diffusion-limited size barriers (e.g., filling the cytoplasm with a large storage vacuole maintains a small volume for metabolic activity that remains compatible with diffusion-driven transport).

As we transition our focus to eukaryotic cells, we want you to approach the study by constantly returning to the Design Challenge. We will cover a large number of subcellular structures that are unique to eukaryotes, and you will certainly be expected to know the names of these structures or organelles, to associate them with one or more "functions", and to identify them on a canonical cartoon representation of a eukaryotic cell. This memorization exercise is necessary but not sufficient. We will also ask you to start thinking a bit deeper about some of the functional and evolutionary costs and benefits (trade-offs) of both evolving eukaryotic cells and various eukaryotic organelles, as well as how a eukaryotic cell might coordinate the functions of different organelles.

Our hypotheses may sometimes come in the form of statements like, "Thing A exists because of rationale B." To be completely honest, however, in many cases, we don't actually know all of the selective pressures that led to the creation or maintenance of certain cellular structures, and the likelihood that one explanation will fit all cases is slim in biology. The causal linkage/relationship implied by the use of terms like "because" should be treated as good hypotheses rather than objective, concrete, undisputed, factual knowledge. We want you to understand these hypotheses and to be able to discuss the ideas presented in class, but we also want you to indulge your own curiosity and to begin thinking critically about these ideas yourself. Try using the Design Challenge rubric to explore some of your ideas. In the following, we will try to seed questions to encourage this activity.

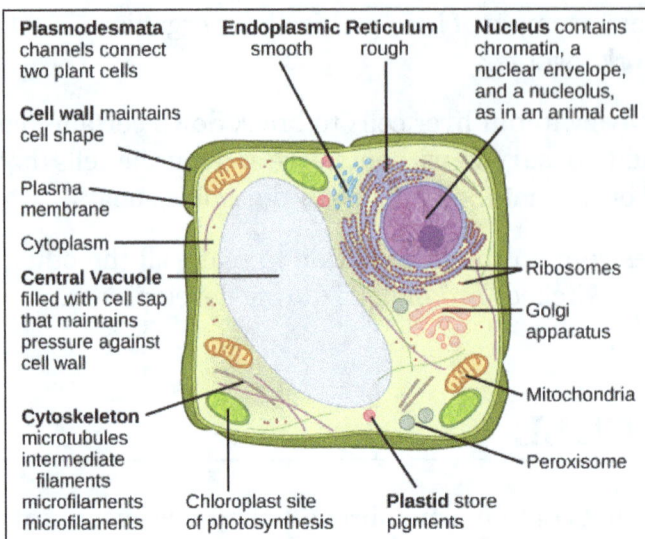

These figures show the major organelles and other cell components of (a) a typical animal cell and (b) a typical eukaryotic plant cell. The plant cell has a cell wall, chloroplasts, plastids, and a central vacuole—structures not found in animal cells. Plant cells do not have lysosomes or centrosomes.

Plasma Membrane

Like bacteria and archaea, eukaryotic cells have a plasma membrane, a phospholipid bilayer with embedded proteins that separates the internal contents of the cell from its surrounding environment. The plasma membrane controls the passage of organic molecules, ions, water, and oxygen into and out of the cell. Wastes (such as carbon dioxide and ammonia) also leave the cell by passing through the plasma membrane, usually with some help of protein transporters.

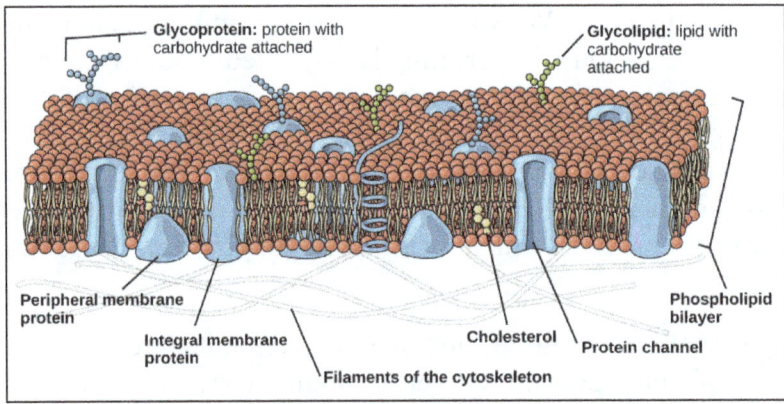

The eukaryotic plasma membrane is a phospholipid
bilayer with proteins and cholesterol embedded in it.

The plasma membranes of eukaryotic cells may also adopt unique conformations. For instance, the plasma membrane of cells that, in multicellular organisms, specialize in absorption are often folded into fingerlike projections called microvilli (singular = microvillus). The "folding" of the membrane into microvilli effectively increases the surface area for absorption while minimally impacting the cytosolic volume. Such cells can be found lining the small intestine, the organ that absorbs nutrients from digested food.

An aside: People with celiac disease have an immune response to gluten, a protein found in wheat, barley, and rye. The immune response damages microvilli. As a consequence, afflicted individuals have an impaired ability to absorb nutrients. This can lead to malnutrition, cramping, and diarrhea.

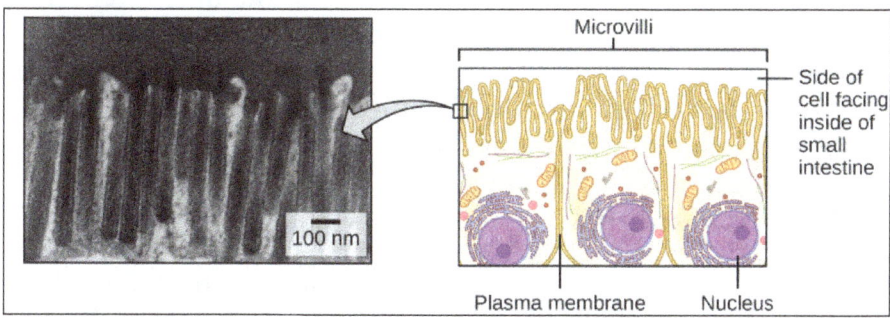

Microvilli, shown here as they appear on cells lining the small intestine, increase the surface area available for absorption. These microvilli are only found on the area of the plasma membrane that faces the cavity from which substances will be absorbed.

The Cytoplasm

The cytoplasm refers to the entire region of a cell between the plasma membrane and the nuclear envelope. It is composed of organelles suspended in the gel-like cytosol, the cytoskeleton, and various chemicals. Even though the cytoplasm consists of 70 to 80 percent water, it nevertheless has a semisolid consistency. It is crowded in there. Proteins, simple sugars, polysaccharides, amino acids, nucleic acids, fatty acids, ions and many other water-soluble molecules are all competing for space and water.

The Nucleus

Typically, the nucleus is the most prominent organelle in a cell when viewed through a microscope. The nucleus (plural = nuclei) houses the cell's DNA. Let's look at it in more detail.

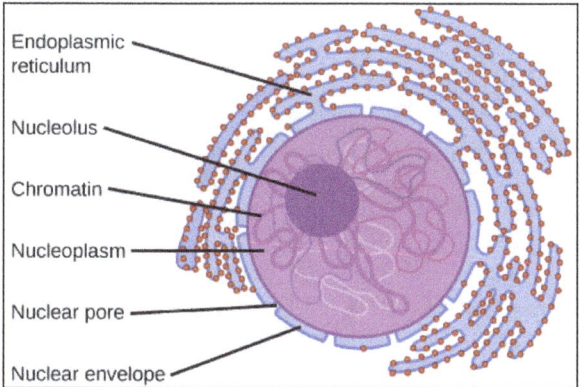

The nucleus stores chromatin (DNA plus proteins) in a gel-like substance called the nucleoplasm. The nucleolus is a condensed region of chromatin where ribosome synthesis occurs. The boundary of the nucleus is called the nuclear envelope. It consists of two phospholipid bilayers: an outer membrane and an inner membrane. The nuclear membrane is continuous with the endoplasmic reticulum. Nuclear pores allow substances to enter and exit the nucleus.

Nuclear Envelope

The nuclear envelope, a structure that constitutes the outermost boundary of the nucleus, is a double-membrane—both the inner and outer membranes of the nuclear envelope are phospholipid bilayers. The nuclear envelope is also punctuated with protein-based pores that control the passage of ions, molecules, and RNA between the nucleoplasm and cytoplasm. The nucleoplasm is the semisolid fluid inside the nucleus where we find the chromatin and the nucleolus, a condensed region of chromatin where ribosome synthesis occurs.

Chromatin and Chromosomes

To understand chromatin, it is helpful to first consider chromosomes. Chromosomes are structures within the nucleus that are made up of DNA, the hereditary material. You may remember that in bacteria and archaea, DNA is typically organized into one or more circular chromosome(s). In eukaryotes, chromosomes are linear structures. Every eukaryotic species has a specific number of chromosomes in the nuclei of its cells. In humans, for example, the chromosome number is 23, while in fruit flies, it is 4.

Chromosomes are only clearly visible and distinguishable from one another by visible optical microscopy when the cell is preparing to divide and the DNA is tightly packed by proteins into easily distinguishable shapes. When the cell is in the growth and maintenance phases of its life cycle, numerous proteins are still associated with the nucleic acids, but the DNA strands more closely resemble an unwound, jumbled bunch of threads. The term chromatin is used to describe chromosomes (the protein-DNA complexes) when they are both condensed and decondensed.

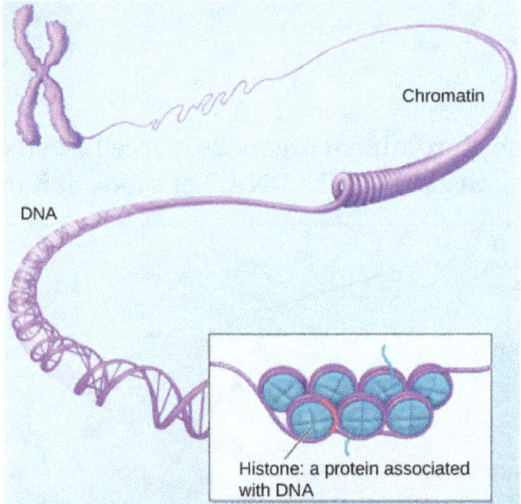

(a) This image shows various levels of the organization of chromatin
(DNA and protein). (b) This image shows paired chromosomes.

The Nucleolus

Some chromosomes have sections of DNA that encode ribosomal RNA. A darkly staining area within the nucleus called the nucleolus (plural = nucleoli) aggregates the ribosomal RNA with associated proteins to assemble the ribosomal subunits that are then transported out to the cytoplasm through the pores in the nuclear envelope.

Ribosomes

Ribosomes are the cellular structures responsible for protein synthesis. When viewed through an electron microscope, ribosomes appear either as clusters (polyribosomes) or single, tiny dots that float freely in the cytoplasm. They may be attached to the cytoplasmic side of the plasma membrane or the cytoplasmic side of the endoplasmic reticulum and the outer membrane of the nuclear envelope.

Electron microscopy has shown us that ribosomes, which are large complexes of protein and RNA, consist of two subunits, aptly called large and small. Ribosomes receive their "instructions" for protein synthesis from the nucleus, where the DNA is transcribed into messenger RNA (mRNA). The mRNA travels to the ribosomes, which translate the code provided by the sequence of the nitrogenous bases in the mRNA into a specific order of amino acids in a protein.

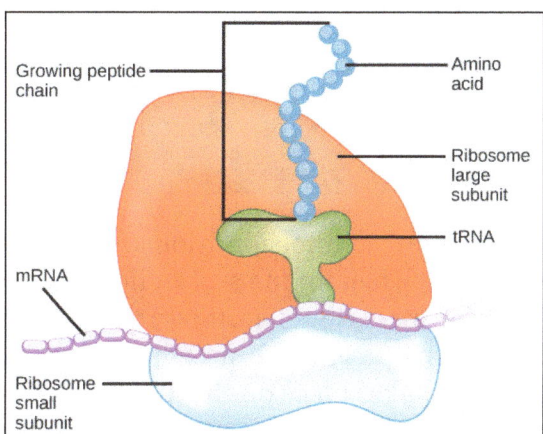

Ribosomes are made up of a large subunit (top) and a small subunit (bottom). During protein synthesis, ribosomes assemble amino acids into proteins.

Mitochondria

Mitochondria (singular = mitochondrion) are often called the "powerhouses" or "energy factories" of a cell because they are the primary site of metabolic respiration in eukaryotes. Depending on the species and the type of mitochondria found in those cells, the respiratory pathways may be anaerobic or aerobic. By definition, when respiration is aerobic, the terminal electron is oxygen; when respiration is anaerobic, a compound other than oxygen functions as the terminal electron acceptor. In either case, the result of these respiratory processes is the production of ATP via oxidative phosphorylation, hence the use of terms "powerhouse" and/or "energy factory" to describe this organelle. Nearly all mitochondria also possess a small genome that encodes genes whose functions are typically restricted to the mitochondrion.

In some cases, the number of mitochondria per cell is tunable, depending, typically, on energy demand. It is for instance possible muscle cells that are used—that by extension have a higher demand for ATP—may often be found to have a significantly higher number of mitochondria than cells that do not have a high energy load.

The structure of the mitochondria can vary significantly depending on the organism and the state of the cell cycle which one is observing. The typical textbook image, however, depicts mitochondria as oval-shaped organelles with a double inner and outer membrane; learn to recognize this generic representation. Both the inner and outer membranes are phospholipid bilayers embedded with proteins that mediate transport across them and catalyze various other biochemical reactions. The inner membrane layer has folds called cristae that increase the surface area into which respiratory chain proteins can be embedded. The region within the cristae is called the mitochondrial matrix and contains—among other things—enzymes of the TCA cycle. During respiration, protons are pumped by respiratory chain complexes from the matrix into a region known as the intermembrane space (between the inner and outer membranes).

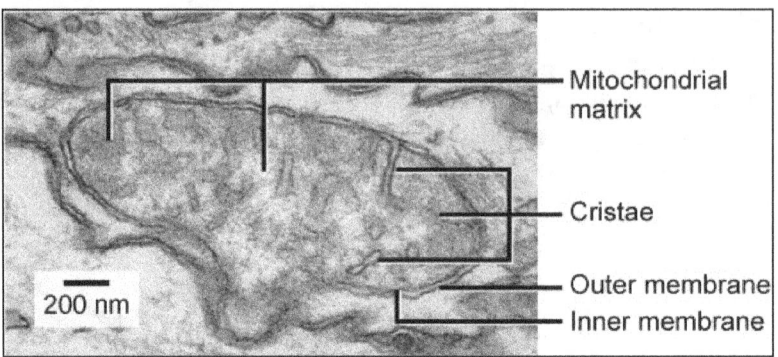

In this figure electron micrograph shows a mitochondrion as viewed with a transmission electron microscope. This organelle has an outer membrane and an inner membrane. The inner membrane contains folds, called cristae, which increase its surface area. The space between the two membranes is called the intermembrane space, and the space inside the inner membrane is called the mitochondrial matrix. ATP synthesis takes place on the inner membrane.

Peroxisomes

Peroxisomes are small, round organelles enclosed by single membranes. These organelles carry out redox reactions that oxidize and break down fatty acids and amino acids. They also help to detoxify many toxins that may enter the body. Many of these redox reactions release hydrogen peroxide, H_2O_2, which would be damaging to cells; however, when these reactions are confined to peroxisomes, enzymes safely break down the H_2O_2 into oxygen and water. For example, alcohol is detoxified by peroxisomes in liver cells. Glyoxysomes, which are specialized peroxisomes in plants, are responsible for converting stored fats into sugars.

Vesicles and Vacuoles

Vesicles and vacuoles are membrane-bound sacs that function in storage and transport. Other than the fact that vacuoles are somewhat larger than vesicles, there is a very subtle distinction between them: the membranes of vesicles can fuse with either the plasma membrane or other membrane systems within the cell. Additionally, some agents such as enzymes within plant vacuoles break down macromolecules. The membrane of a vacuole does not fuse with the membranes of other cellular components.

Animal Cells versus Plant Cells

At this point, you know that each eukaryotic cell has a plasma membrane, cytoplasm, a nucleus, ribosomes, mitochondria, peroxisomes, and in some, vacuoles. There are some striking differences between animal and plant cells worth noting. Here is a brief list of differences that we want you to be familiar with and a slightly expanded description below:

While all eukaryotic cells use microtubule and motor protein the based mechanisms to segregate chromosomes during cell division, the structures used to organize these microtubules differ in plants versus animal and yeast cells. Animal and yeast cells organize and anchor their microtubules into structures called microtubule organizing centers (MTOCs). These structures are composed of structures called centrioles that are composed largely of α-tubulin, β-tubulin, and other proteins. Two centrioles organize into a structure called a centrosome. By contrast, in plants, while microtubules also organize into discrete bundles, there are no conspicuous structures similar to the MTOCs seen in animal and yeast cells. Rather, depending on the organism, it appears that there can be several places where these bundles of microtubules can nucleate from places called acentriolar (without centriole) microtubule organizing centers. A third type of tubulin, γ-tubulin, appears to be implicated, but our knowledge of the precise mechanisms used by plants to organize microtubule spindles is still spotty.

Animal cells typically have organelles called lysosomes responsible for degradation of biomolecules. Some plant cells contain functionally similar degradative organelles, but there is a debate as to how they should be named. Some plant biologists call these organelles lysosomes while others lump them into the general category of plastids and do not give them a specific name.

Plant cells have a cell wall, chloroplasts and other specialized plastids, and a large central vacuole, whereas animal cells do not.

The Centrosome

The centrosome is a microtubule-organizing center found near the nuclei of animal cells. It contains a pair of centrioles, two structures that lie perpendicular to each other. Each centriole is a cylinder of nine triplets of microtubules.

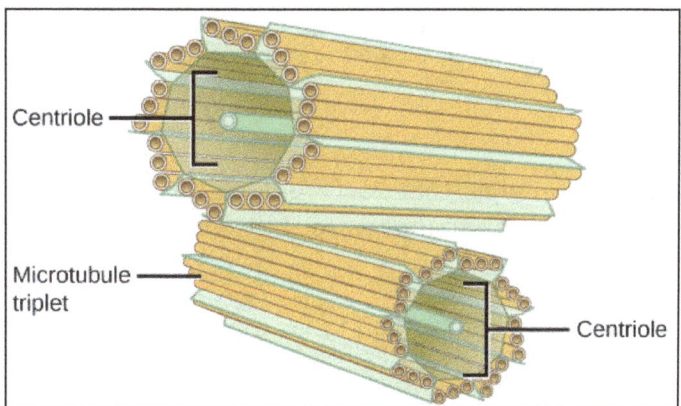

The centrosome consists of two centrioles that lie at right angles
to each other. Each centriole is a cylinder made up of nine triplets of microtubules.
Nontubulin proteins (indicated by the green lines) hold the microtubule triplets together.

The centrosome (the organelle where all microtubules originate in animal and yeast) replicates itself before a cell divides, and the centrioles appear to have some role in pulling the duplicated chromosomes to opposite ends of the dividing cell. However, the exact function of the centrioles in cell division remains unclear, as cells that have had their centrosome removed can still divide, and plant cells, which lack centrosomes, are capable of cell division.

Lysosomes

Animal cells have another set of organelles not found in plant cells: lysosomes. Colloquially, the lysosomes are sometimes called the cell's "garbage disposal". Enzymes within the lysosomes aid the breakdown of proteins, polysaccharides, lipids, nucleic acids, and even "worn-out" organelles. These enzymes are active at a much lower pH than that of the cytoplasm. Therefore, the pH within lysosomes is more acidic than the pH of the cytoplasm. In plant cells, many of the same digestive processes take place in vacuoles.

Cell Wall

If you examine the diagram above depicting plant and animal cells, you will see in the diagram of a plant cell a structure external to the plasma membrane called the cell wall. The cell wall is a rigid covering that protects the cell, provides structural support, and gives shape to the cell. Fungal and protistan cells also have cell walls. While the chief component of bacterial cell walls is peptidoglycan, the major organic molecule in the plant cell wall is cellulose, a polysaccharide made up of glucose subunits.

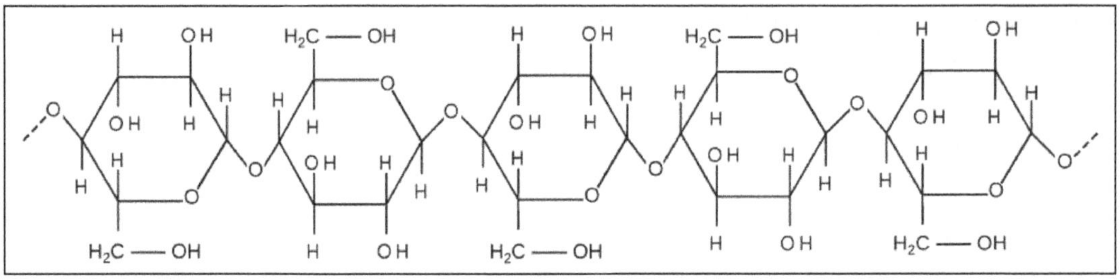

Cellulose is a long chain of β-glucose molecules connected by a 1-4 linkage.
The dashed lines at each end of the figure indicate a series of many more glucose units.
The size of the page makes it impossible to portray an entire cellulose molecule.

Chloroplasts

Chloroplasts are plant cell organelles that carry out photosynthesis. Like the mitochondria, chloroplasts have their own DNA and ribosomes, but chloroplasts have an entirely different function.

Like mitochondria, chloroplasts have outer and inner membranes, but within the space enclosed by a chloroplast's inner membrane is a set of interconnected and stacked fluid-filled membrane sacs called thylakoids. Each stack of thylakoids is called a granum (plural = grana). The fluid enclosed by the inner membrane that surrounds the grana is called the stroma.

The chloroplasts contain a green pigment called chlorophyll, which captures the light energy that drives the reactions of photosynthesis. Like plant cells, photosynthetic protists also have chloroplasts. Some bacteria perform photosynthesis, but their chlorophyll is not relegated to an organelle.

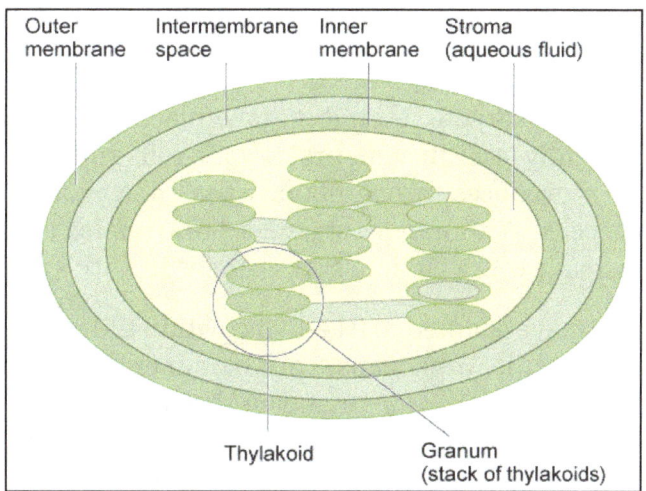

The chloroplast has an outer membrane, an inner membrane, and membrane structures called thylakoids that are stacked into grana. The space inside the thylakoid membranes is called the thylakoid space. The light harvesting reactions take place in the thylakoid membranes, and the synthesis of sugar takes place in the fluid inside the inner membrane, which is called the stroma. Chloroplasts also have their own genome, which is contained on a single circular chromosome.

Central Vacuole

Previously, we mentioned vacuoles as essential components of plant cells. The central vacuole plays a key role in regulating the cell's concentration of water in changing environmental conditions.

Silly vacuole factoid: Have you ever noticed that if you forget to water a plant for a few days, it wilts? That's because as the water concentration in the soil becomes lower than the water concentration in the plant, water moves out of the central vacuoles and cytoplasm. As the central vacuole shrinks, it leaves the cell wall unsupported. This loss of support to the cell walls of plant cells results in the wilted appearance of the plant.

The central vacuole also supports the expansion of the cell. When the central vacuole holds more water, the cell gets larger without having to invest a lot of energy in synthesizing new cytoplasm.

PROKARYOTIC CELL

Prokaryotic cells are cells that do not have a true nucleus or membrane-bound organelles. Organisms within the domains Bacteria and Archaea have prokaryotic cells, while other forms of life are eukaryotic. However, organisms with prokaryotic cells are abundant and make up much of Earth's biomass.

Organisms that have prokaryotic cells are unicellular and are called prokaryotes. Prokaryotic cells can be contrasted with eukaryotic cells, which are more complex. Eukaryotic cells have a nucleus surrounded by a nuclear membrane and also have other organelles that perform specific functions in the cell. A prokaryotic cell contains only a single membrane, which surrounds the cell as an outer membrane.

All of the reactions within a prokaryote, therefore, take place within the cytoplasm of the cell. While this makes the cells slightly less efficient, prokaryotic cells still have a remarkable reproductive capacity. A prokaryote reproduces through binary fission, a process which simply splits duplicated DNA into separate cells. Without any organelles or complex chromosomes to reproduce, most prokaryotic cells can divide every 24 hours, or even faster with an adequate supply of food.

While many prokaryotic cells have adapted to free-living within the environment, many more have adapted to live within the gut of other organisms. These commensal organisms survive by breaking down molecules inside the gut and allow the organism they are living within the ability to digest a wider variety of foods. For example, the human gut contains 2-3 pounds of bacteria, which have evolved to help us digest complex carbohydrates, proteins, and fats.

Examples of Prokaryotic Cells

Bacterial Cells

Bacteria are single-celled microorganisms that are found nearly everywhere on Earth, and they are very diverse in their shapes and structures. There are about 5×10^{30} bacteria living on Earth, including in our own bodies; in the human gut, bacteria outnumber human cells 10:1.

The cell walls of bacteria contain peptidoglycan, a molecule made of sugars and amino acids that gives the cell wall its structure and is thicker in some bacteria than others. Bacteria contain certain structures unique to them as previously mentioned, such as the capsule, flagella, and pili. Most bacteria have just one chromosome that is circular, which can range from about 160,000 base pairs (bp) to 12,200,000 bp. They also contain plasmids, which are small circular pieces of DNA that replicate independently of the chromosome.

Some bacteria can form endospores. These are tough, dormant structures that the bacteria can reduce themselves to under starvation conditions when not enough nutrients are available. They do not need nutrients and are resistant to extreme temperatures, UV rays, and chemicals. When environmental conditions become favorable again, the endospore can reactivate.

Archaeal Cells

Archaea are similar in size and shape to bacteria, and they are also unicellular. Since bacteria and archaea are the two types of prokaryotes, this means that all prokaryotes are unicellular. Some archaea are found in extreme environments, such as hot springs, but they can be found in a variety of locations, such as soils, oceans, marshlands, and inside other organisms, including humans.

Like bacteria, archaea can have a cell wall and flagella. However, the structure of these organelles is different. For example, archaeal cell walls do not contain peptidoglycan. In addition, the flagella of archaea work the same way as those of bacteria, but they evolved from different structures. Membranes of archaea are very different than those of all other lifeforms; they contain different lipids, which have a different stereochemistry. Archaea usually have one circular chromosome, as bacteria do. The archaeal chromosome can range from less than 491,000 bp to about 5,700,000 bp. They can also contain plasmids. Less is known about archaea than bacteria; they were not classified as a separate group of prokaryotes until 1977.

Prokaryotic Cell Structure

Prokaryotic cells do not have a true nucleus that contains their genetic material as eukaryotic cells do. Instead, prokaryotic cells have a nucleoid region, which is an irregularly-shaped region that contains the cell's DNA and is not surrounded by a nuclear envelope. Some other parts of prokaryotic cells are similar to those in eukaryotic cells, such as a cell wall surrounding the cell (which is also found in plant cells, although it has a different composition).

Like eukaryotic cells, prokaryotic cells have cytoplasm, a gel-like substance that makes up the "filling" of the cell, and a cytoskeleton that holds components of the cell in place. Both prokaryotic cells and eukaryotic cells have ribosomes, which are organelles that produce proteins, and vacuoles, small spaces in cells that store nutrients and help eliminate waste.

Some prokaryotic cells have flagella, which are tail-like structures that enable the organism to move around. They may also have pili, small hair-like structures that help bacteria adhere to surfaces and can allow DNA to be transferred between two prokaryotic cells in a process known as conjugation. Another part that is found in some bacteria is the capsule. The capsule is a sticky layer of carbohydrates that helps the bacterium adhere to surfaces in its surroundings.

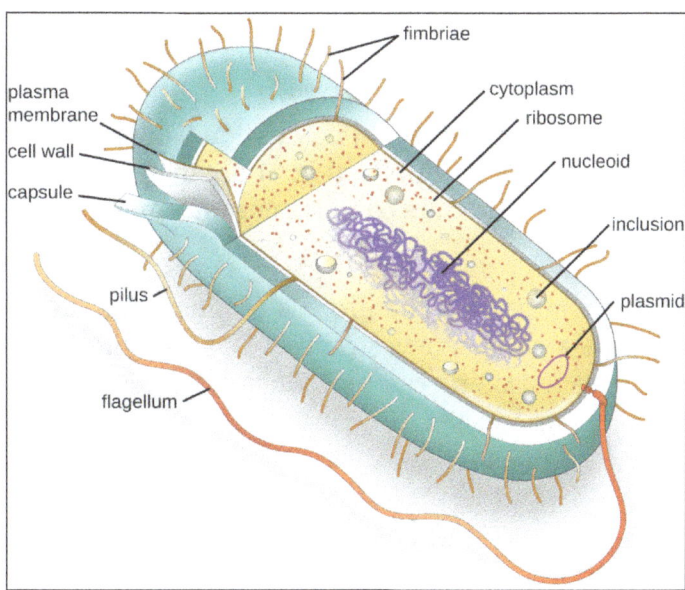

Prokaryote cell.

Characteristics of Prokaryotic Cells

All prokaryotic cells have a nucleoid region, DNA and RNA as their genetic material, ribosomes that make proteins, and cytoplasm that contains a cytoskeleton, which organizes and supports the parts of the cell. Prokaryotic cells are simpler than eukaryotic cells, and an organism that is a prokaryote is unicellular; it is made up of only one prokaryotic cell.

Prokaryotic cells are usually between 0.1 to 5 micrometers in length (.00001 to .0005 cm). Eukaryotic cells are generally much larger, between 10 and 100 micrometers. Prokaryotic cells have a higher surface-area-to-volume ratio because they are smaller, which makes them able to obtain a larger amount of nutrients via their plasma membrane.

Prokaryotic Cell Parts

Unlike eukaryotic cells, prokaryotic cells have no distinct organelles bound by membranes. Instead, the many reactions the cell conducts happen within the cytoplasm of the cell. In fact, there are 2 main components that are present within all prokaryotic cells.

The first is a cell membrane. This is a layer of phospholipid molecules which separate the inside of the cell from the outside. While not present in all prokaryotes, many secrete a cell wall, used to protect and house the cell in an extra layer of proteins and structural molecules.

The second part found in all prokaryotic cells is DNA. DNA is the basic blueprint for all life and is found within all cells. In prokaryotes, the DNA often takes the form of a large circular genome. This can be compared to the organized chromosomes which are typically found within eukaryotes. This large circle of DNA directs which proteins the cell creates, and regulates the actions of the cell.

Other prokaryotic cells can have a large number of different parts, such as cilia and flagella to help them move around. While these structures are similar in function to those found within eukaryotes, they often have a different structure. This suggests that the two types of cell have undergone very different selection processes and have independently involved the structures.

Prokaryotic Cells Division

Prokaryotic cells divide through the process of binary fission. Unlike mitosis, this process does not involve the condensation of DNA or the duplication of organelles. Prokaryotic cells have only a small amount of DNA, which is not stored in complex chromosomes. Further, there are no organelles so there is nothing to divide.

When a prokaryote grows to a large size, the process of binary fission takes place. This process duplicates the DNA, then separates each new strand of DNA into individual cells. This process is simpler than mitosis, and as such bacteria can reproduce much faster.

4

Cell Structure

There are numerous structures which make up a cell. A few of the major structures are cell membrane, cell wall, cytoskeleton, cytoplasm, centriole, plasmid, cytosol and thylakoid. This chapter has been carefully written to provide an easy understanding of the varied aspects of these cell structures.

CELL MEMBRANE

The cell membrane, also known as the plasma membrane, is a double layer of lipids and proteins that surrounds a cell and separates the cytoplasm (the contents of the cell) from its surrounding environment. It is selectively permeable, which means that it only lets certain molecules enter and exit. It can also control the amount of some substances that go into or out of the cell. All cells have a cell membrane.

Function of the Cell Membrane

The cell membrane gives the cell its structure and regulates the materials that enter and leave the cell. Like a drawbridge intended to protect a castle and keep out enemies, the cell membrane only allows certain molecules to enter or exit. Oxygen, which cells need in order to carry out metabolic functions such as cellular respiration, and carbon dioxide, a byproduct of these functions, can easily enter and exit through the membrane. Water can also freely cross the membrane, although it does so at a slower rate. However, highly charged molecules, like ions, cannot directly pass through, nor can large macromolecules like carbohydrates or amino acids. Instead, these molecules must pass through proteins that are embedded in the membrane. In this way, the cell can control the rate of diffusion of these substances.

Another way the cell membrane can bring molecules inside it is through endocytosis. This includes phagocytosis ("cell eating") and pinocytosis ("cell drinking"). During these processes, the cell membrane forms a depression and surrounds the particle that it is engulfing. It then "pinches off" to form a small sphere of membrane called a vesicle that contains the molecule and transports it to wherever it will be used in the cell. Vesicles are also created from the cell membrane when endocytosis is not occurring, and are used to transport molecules to different areas within the cell. Cells can also get rid of molecules through exocytosis, which is the opposite of endocytosis. During exocytosis, vesicles come to the surface of the cell membrane, merge with it, and release their contents to the outside of the cell. Exocytosis removes the cell's waste products– parts of molecules that are not used by the cell.

The cell membrane also plays a role in cell signaling and communication. Receptor proteins on the cell membrane can bind to molecules of substances produced by other areas of the body, such as hormones. When a molecule binds to its target receptor on the membrane, it initiates a signal transduction pathway inside the cell that transmits the signal to the appropriate molecules. Then, the cell can perform the action specified by the signal molecule, such as making or stopping production of a certain protein.

Structure of the Cell Membrane

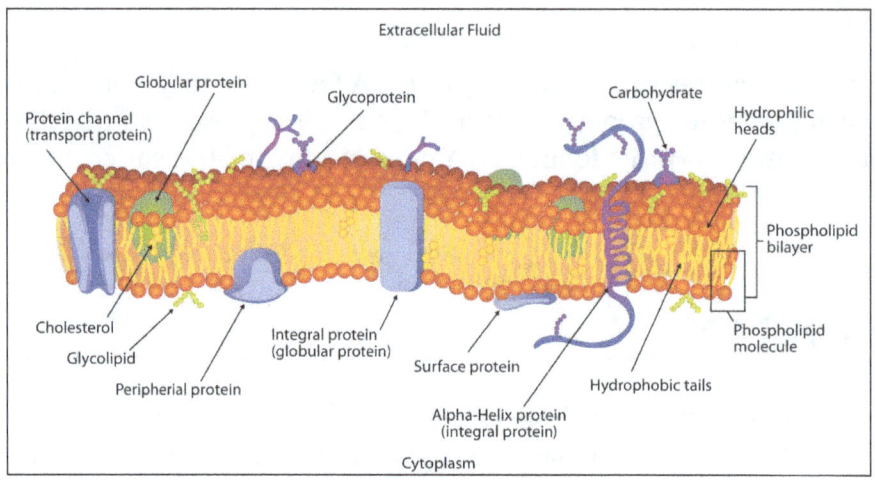

Cell membrane diagram.

Phospholipids are a main component of the cell membrane. These are lipid molecules made up of a phosphate group head and two fatty acid tails. The properties of phospholipid molecules allow them to spontaneously form a double-layered membrane. When in water or an aqueous solution, which includes the inside of the body, the hydrophilic heads of phospholipids will orient themselves to be on the outside, while the hydrophobic tails will be on the inside. The technical term for this double layer of phospholipids that forms the cell membrane is a phospholipid bilayer. Eukaryotic cells, which make up the bodies of all organisms except for bacteria and archaea, also have a nucleus that is surrounded by a phospholipid bilayer membrane.

In addition, the cell membrane contains glycolipids and sterols. One important sterol is cholesterol, which regulates the fluidity of the cell membrane in animal cells. When there is less cholesterol, membranes become more fluid, but also more permeable to molecules. The amount of cholesterol in the membrane helps maintain its permeability so that the right amount of molecules can enter the cell at a time, not too many or too few.

The cell membrane also contains many different proteins. Proteins make up about half of the cell membrane. Many of these proteins are transmembrane proteins, which are embedded in the membrane but stick out on both sides. Some of these proteins are receptors which bind to signal molecules, while others are ion channels which are the only means of allowing ions into or out of the cell. Scientists use the fluid mosaic model to describe the structure of the cell membrane. The cell membrane has a fluid consistency due to being made up in large part of phospholipids, and because of this, proteins move freely across its surface. The multitude of different proteins and lipids in the cell membrane give it the look of a mosaic.

CELL WALL

A cell wall is an outer layer surrounding certain cells that is outside of the cell membrane. All cells have cell membranes, but generally only plants, fungi, algae, most bacteria, and archaea have cells with cell walls. The cell wall provides strength and structural support to the cell, and can control to some extent what types and concentrations of molecules enter and leave the cell. The materials that make up the cell wall differ depending on the type of organism. The cell wall has evolved many different times among different groups of organisms.

Cell Wall Functions

The cell wall has a few different functions. It is flexible, but provides strength to the cell, which helps protect the cell against physical damage. It also gives the cell its shape and allows the organism to maintain a certain shape overall. The cell wall can also provide protection from pathogens such as bacteria that are trying to invade the cell. The structure of the cell wall allows many small molecules to pass through it, but not larger molecules that could harm the cell.

Cell Wall Structure

Plant Cell Walls

The main component of the plant cell wall is cellulose, a carbohydrate that forms long fibers and gives the cell wall its rigidity. Cellulose fibers group together to form bundles called microfibrils. Other important carbohydrates include hemicellulose, pectin, and liginin. These carbohydrates form a network along with structural proteins to form the cell wall. Plant cells that are in the process of growing have primary cell walls, which are thin. Once the cells are fully grown, they develop secondary cell walls. The secondary cell wall is a thick layer that is formed on the inside of the primary cell wall. This layer is what is usually meant when referring to a plant's cell wall. There is also another layer in between plant cells called the middle lamella; it is pectin-rich and helps plant cells stick together.

This diagram of a plant cell depicts the cell wall in green, surrounding the contents of the cell.

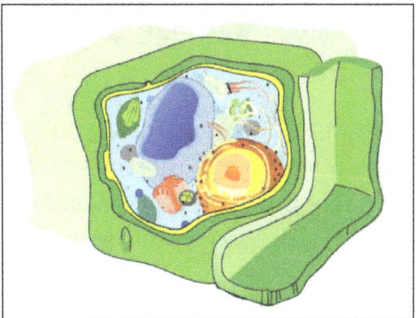

Eukaryota cell strucutre.

The cell walls of plant cells help them maintain turgor pressure, which is the pressure of the cell membrane pressing against the cell wall. Ideally, plants cells should have lots of water within them, leading to high turgidity. Whereas a cell without a cell wall, such as an animal cell, can swell and burst of too much water diffuses into it, plants need to be in hypotonic solutions (more water inside than outside, leading to lots of water entering the cell) to maintain turgor pressure and their structural shape. The cell wall efficiently holds water in so that the cell does not burst. When turgor pressure is lost, a plant will begin to wilt. Turgor pressure is what gives plant cells their characteristic square shape; the cells are full of water, so they fill up the space available and press against each other.

Algae Cell Walls

Algae are a diverse group, and the diversity in their cell walls reflects this. Some algae, such as green algae, have cell walls that are similar in structure to those of plants. Other algae, such as brown algae and red algae, have cellulose along with other polysaccharides or fibrils. Diatoms have cell walls that are made from silicic acid. Other important molecules in algal cell walls include mannans, xylans, and alginic acid.

Fungi Cell Walls

The cell walls of fungi contain chitin, which is a glucose derivative that is similar in structure to cellulose. Layers of chitin are very tough; chitin is the same molecule found in the rigid exoskeletons of animals such as insects and crustaceans. Glucans, which are other glucose polymers, are also found in the fungal cell wall along with lipids and proteins. Fungi have proteins called hydrophobins in their cell walls. Found only in fungi, hydrophobins give the cells strength, help them adhere to surfaces, and help control the movement of water into the cells. In fungi, the cell wall is the most external layer, and surrounds the cell membrane.

Bacteria and Archaea Cell Walls

The cell walls of bacteria usually contain the polysaccharide peptidoglycan, which is porous and lets small molecules through. Together, the cell membrane and cell wall are referred to as the cell envelope. The cell wall is an essential part of survival for many bacteria. It provides mechanical structure to bacteria, which are single-celled, and it also protects them from internal turgor pressure. Bacteria have higher concentration of molecules such as proteins within themselves as compared to their environment, so the cell wall stops water from rushing into the cell. Differences in cell wall thickness also make Gram staining possible. Gram staining is used for the general identification of bacteria; bacteria with thick cell walls are gram-positive, while bacteria with thinner cell walls are gram-negative.

While archaea are similar in many ways to bacteria, hardly any archaeal walls contain peptidoglycan. There are several different types of cell walls in archaea. Some are composed of pseudopeptidoglycan, some have polysaccharides, some have glycoproteins, and others have surface-layer proteins (called an S-layer, which can also be found in bacteria).

CYTOPLASM

In cell biology, the cytoplasm is all of the material within a cell, enclosed by the cell membrane, except for the cell nucleus. The material inside the nucleus and contained within the nuclear membrane is termed the nucleoplasm. The main components of the cytoplasm are cytosol – a gel-like substance, the organelles – the cell's internal sub-structures, and various cytoplasmic inclusions. The cytoplasm is about 80% water and usually colorless.

The submicroscopic ground cell substance, or cytoplasmatic matrix which remains after exclusion the cell organelles and particles is groundplasm. It is the hyaloplasm of light microscopy, and high complex, polyphasic system in which all of resolvable cytoplasmic elements of are suspended, including the larger organelles such as the ribosomes, mitochondria, the plant plastids, lipid droplets, and vacuoles.

Most cellular activities take place within the cytoplasm, such as many metabolic pathways including glycolysis, and processes such as cell division. The concentrated inner area is called the endoplasm and the outer layer is called the cell cortex or the ectoplasm. Movement of calcium ions in and out of the cytoplasm is a signaling activity for metabolic processes. In plants, movement of the cytoplasm around vacuoles is known as cytoplasmic streaming.

Physical Nature

The physical properties of the cytoplasm have been contested in recent years. It remains uncertain how the varied components of the cytoplasm interact to allow movement of particles and organelles while maintaining the cell's structure. The flow of cytoplasmic components plays an important role in many cellular functions which are dependent on the permeability of the cytoplasm. An example of such function is cell signalling, a process which is dependent on the manner in which signaling molecules are allowed to diffuse across the cell. While small signaling molecules like calcium ions are able to diffuse with ease, larger molecules and subcellular structures often require aid in moving through the cytoplasm. The irregular dynamics of such particles have given rise to various theories on the nature of the cytoplasm.

Sol-gel

There has long been evidence that the cytoplasm behaves like a sol-gel. It is thought that the component molecules and structures of the cytoplasm behave at times like a disordered colloidal solution (sol) and at other times like an integrated network, forming a solid mass (gel). This theory thus proposes that the cytoplasm exists in distinct fluid and solid phases depending on the level of interaction between cytoplasmic components, which may explain the differential dynamics of different particles observed moving through the cytoplasm.

As a Glass

Recently it has been proposed that the cytoplasm behaves like a glass-forming liquid approaching the glass transition. In this theory, the greater the concentration of cytoplasmic components, the less the cytoplasm behaves like a liquid and the more it behaves as a solid glass, freezing larger cytoplasmic components in place (it is thought that the cell's metabolic activity is able to fluidize the cytoplasm to allow the movement of such larger cytoplasmic components). A cell's ability to vitrify in the absence of metabolic activity, as in dormant periods, may be beneficial as a defence strategy. A solid glass cytoplasm would freeze subcellular structures in place, preventing damage, while allowing the transmission of very small proteins and metabolites, helping to kickstart growth upon the cell's revival from dormancy.

Other Perspectives

There has been research examining the motion of cytoplasmic particles independent of the nature of the cytoplasm. In such an alternative approach, the aggregate random forces within the cell caused by motor proteins explain the non-Brownian motion of cytoplasmic constituents.

Constituents

The three major elements of the cytoplasm are the cytosol, organelles and inclusions.

Cytosol

The cytosol is the portion of the cytoplasm not contained within membrane-bound organelles. Cytosol makes up about 70% of the cell volume and is a complex mixture of cytoskeleton filaments, dissolved molecules, and water. The cytosol's filaments include the protein filaments such as actin filaments and microtubules that make up the cytoskeleton, as well as soluble proteins and small structures such as ribosomes, proteasomes, and the mysterious vault complexes. The inner, granular and more fluid portion of the cytoplasm is referred to as endoplasm.

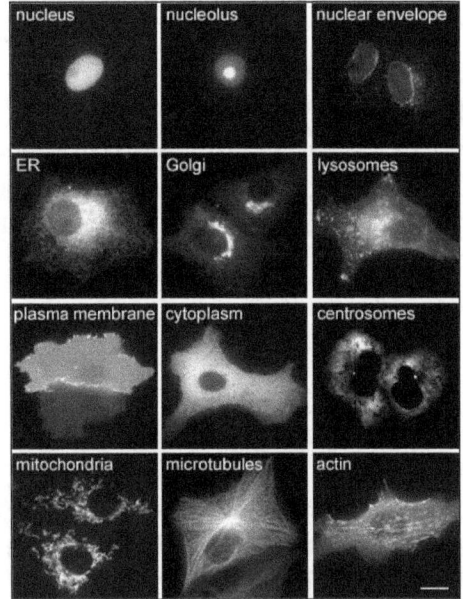

Proteins in different cellular compartments and
structures tagged with green fluorescent protein.

Due to this network of fibres and high concentrations of dissolved macromolecules, such as proteins, an effect called macromolecular crowding occurs and the cytosol does not act as an ideal solution. This crowding effect alters how the components of the cytosol interact with each other.

Organelles

Organelles (literally "little organs"), are usually membrane-bound structures inside the cell that have specific functions. Some major organelles that are suspended in the cytosol are the mitochondria, the endoplasmic reticulum, the Golgi apparatus, vacuoles, lysosomes, and in plant cells, chloroplasts.

Cytoplasmic Inclusions

The inclusions are small particles of insoluble substances suspended in the cytosol. A huge range of inclusions exist in different cell types, and range from crystals of calcium oxalate or silicon dioxide in plants, to granules of energy-storage materials such as starch, glycogen, or polyhydroxybutyrate. A particularly widespread example are lipid droplets, which are spherical droplets composed of lipids and proteins that are used in both prokaryotes and eukaryotes as a way of storing lipids such as fatty acids and sterols. Lipid droplets make up much of the volume of adipocytes, which are specialized lipid-storage cells, but they are also found in a range of other cell types.

Controversy and Research

The cytoplasm, mitochondria and most organelles are contributions to the cell from the maternal gamete. Contrary to the older information that disregards any notion of the cytoplasm being active, new research has shown it to be in control of movement and flow of nutrients in and out of the cell by viscoplastic behavior and a measure of the reciprocal rate of bond breakage within the cytoplasmic network.

The material properties of the cytoplasm remain an ongoing investigation. Recent measurements using force spectrum microscopy reveal that the cytoplasm can be likened to an elastic solid, rather than a viscoelastic fluid.

CYTOSKELETON

The cytoskeleton is a network of filaments and tubules that extends throughout a cell, through the cytoplasm, which is all of the material within a cell except for the nucleus. It is found in all cells, though the proteins that it is made of vary between organisms. The cytoskeleton supports the cell, gives it shape, organizes and tethers the organelles, and has roles in molecule transport, cell division and cell signaling.

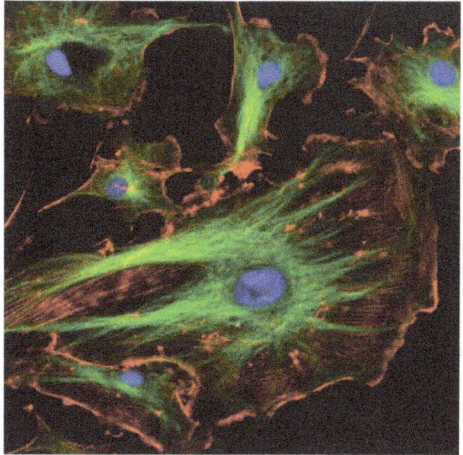

Fluorescent Cells

The microfilaments of this cell are shown in red, while microtubules are shown in green. The blue dots are nuclei.

Structure of the Cytoskeleton

All cells have a cytoskeleton, but usually the cytoskeleton of eukaryotic cells is what is meant when discussing the cytoskeleton. Eukaryotic cells are complex cells that have a nucleus and organelles. Plants, animals, fungi, and protists have eukaryotic cells. Prokaryotic cells are less complex, with no true nucleus or organelles except ribosomes, and they are found in the single-celled organisms bacteria and archaea. The cytoskeleton of prokaryotic cells was originally thought not to exist; it was not discovered until the early 1990s.

The eukaryotic cytoskeleton consists of three types of filaments, which are elongated chains of proteins: microfilaments, intermediate filaments, and microtubules.

Microfilaments

Microfilaments are also called actin filaments because they are mostly composed of the protein actin; their structure is two strands of actin wound in a spiral. They are about 7 nanometers thick, making them the thinnest filaments in the cytoskeleton. Microfilaments have many functions. They aid in cytokinesis, which is the division of a cytoplasm of a cell when it is dividing into two daughter cells. They aid in cell motility and allow single-celled organisms like amoebas to move. They are also involved in cytoplasmic streaming, which is the flowing of cytosol (the liquid part of the cytoplasm) throughout the cell. Cytoplasmic streaming transports nutrients and cell organelles. Microfilaments are also part of muscle cells and allow these cells to contract, along with myosin. Actin and myosin are the two main components of muscle contractile elements.

Intermediate Filaments

Intermediate filaments are about 8-12 nm wide; they are called intermediate because they are in-between the size of microfilaments and microtubules. Intermediate filaments are made of different proteins such as keratin (found in hair and nails, and also in animals with scales, horns, or hooves), vimentin, desmin, and lamin. All intermediate filaments are found in the cytoplasm except for lamins, which are found in the nucleus and help support the nuclear envelope that surrounds the nucleus. The intermediate filaments in the cytoplasm maintain the cell's shape, bear tension, and provide structural support to the cell.

Microtubules

Microtubules are the largest of the cytoskeleton's fibers at about 23 nm. They are hollow tubes made of alpha and beta tubulin. Microtubules form structures like flagella, which are "tails" that propel a cell forward. They are also found in structures like cilia, which are appendages that increase a cell's surface area and in some cases allow the cell to move. Most of the microtubules in an animal cell come from a cell organelle called the centrosome, which is a microtubule organizing center (MTOC). The centrosome is found near the middle of the cell, and microtubules radiate outward from it. Microtubules are important in forming the spindle apparatus (or mitotic spindle), which separates sister chromatids so that one copy can go to each daughter cell during cell division. They are also involved in transporting molecules within the cell and in the formation of the cell wall in plant cells.

Function of the Cytoskeleton

The cytoskeleton has several functions. First, it gives the cell shape. This is especially important in cells without cell walls, such as animal cells, that do not get their shape from a thick outer layer. It can also give the cell movement. The microfilaments and microtubules can disassemble, reassemble, and contract, allowing cells to crawl and migrate, and microtubules help form structures like cilia and flagella that allow for cell movement.

The cytoskeleton organizes the cell and keeps the cell's organelles in place, but it also aids in the movement of organelles throughout the cell. For example, during endocytosis when a cell engulfs a molecule, microfilaments pull the vesicle containing the engulfed particles into the cell. Similarly, the cytoskeleton helps move chromosomes during cell division.

One analogy for the cytoskeleton is the frame of a building. Like a building's frame, the cytoskeleton is the "frame" of the cell, keeping structures in place, providing support, and giving the cell a definite shape.

CENTRIOLE

Centrioles are cylindrical cell structures that are composed of groupings of microtubules, which are tube-shaped molecules or strands of protein. Without centrioles, chromosomes would not be able to move during the formation of new cells.

Centrioles help to organize the assembly of microtubules during cell division. Simplified, chromosomes use the centriole's microtubules as a highway during the cell division process.

Centriole Location

Centrioles are found in all animal cells and only a few species of lower plant cells. Two centrioles—a mother centriole and a daughter centriole—are found within the cell in a structure called a centrosome.

Composition

Most centrioles are made up of nine sets of microtubule triplets, with the exception of some species. For example, crabs have nine sets of microtubule doublets. There are a few other species that deviate from the standard centriole structure. Microtubules are composed of a single type of globular protein called tubulin.

Two Main Functions

During mitosis or cell division, the centrosome and centrioles replicate and migrate to opposite ends of the cell. Centrioles help to arrange the microtubules that move chromosomes during cell division to ensure each daughter cell receives the appropriate number of chromosomes.

Centrioles are also important for the formation of cell structures known as cilia and flagella. Cilia and flagella, found on the outside surface of cells, aid in cellular movement. A centriole combined with several additional protein structures is modified to become a basal body. Basal bodies are the anchoring sites for moving cilia and flagella.

Important Role in Cell Division

Centrioles are located outside of, but near the cell nucleus. In cell division, there are several phases, in order: interphase, prophase, metaphase, anaphase, and telophase. Centrioles have a very important role to play in all phases of cell division. The end goal is in moving replicated chromosomes into a newly created cell.

Interphase and Replication

In the first phase of mitosis, called interphase, centrioles replicate. This is the phase immediately prior to cell division, which marks the start of mitosis and meiosis in the cell cycle.

Prophase and Asters and The Mitotic Spindle

In prophase, each centrosome with centrioles migrates toward opposite ends of the cell. A single pair of centrioles is positioned at each cell pole. The mitotic spindle initially appears as structures called asters which surround each centriole pair. Microtubules form spindle fibers that extend from each centrosome, thereby separating centriole pairs and elongating the cell.

The fibers as a newly paved highway for the replicated chromosomes to move into the newly formed cell. In this analogy, the replicated chromosomes are a car along the highway.

Metaphase and Positioning of Polar Fibers

In metaphase, centrioles help to position polar fibers as they extend from the centrosome and position chromosomes along the metaphase plate. In keeping with the highway analogy, this keeps the lane straight.

Anaphase and the Sister Chromatids

In anaphase, polar fibers connected to chromosomes shorten and separate the sister chromatids (replicated chromosomes). The separated chromosomes are pulled toward opposite ends of the cell by polar fibers extending from the centrosome.

At this point in the highway analogy, its as if one car on the highway has replicated a second copy and the two cars begin moving away from each other, in opposite directions, on the same highway.

Telophase and Two Genetically Identical Daughter Cells

In telophase, the spindle fibers disperse as the chromosomes are cordoned off into distinct new nuclei. After cytokinesis, which is the division of the cell's cytoplasm, two genetically identical daughter cells are produced each containing one centrosome with one centriole pair.

CYTOSOL

The cytosol, also known as intracellular fluid (ICF) or cytoplasmic matrix, or groundplasm, is the liquid found inside cells. It is separated into compartments by membranes. For example, the mitochondrial matrix separates the mitochondrion into many compartments.

In the eukaryotic cell, the cytosol is surrounded by the cell membrane and is part of the cytoplasm, which also comprises the mitochondria, plastids, and other organelles (but not their internal fluids and structures); the cell nucleus is separate. The cytosol is thus a liquid matrix around the organelles. In prokaryotes, most of the chemical reactions of metabolism take place in the cytosol, while a few take place in membranes or in the periplasmic space. In eukaryotes, while many metabolic pathways still occur in the cytosol, others take place within organelles.

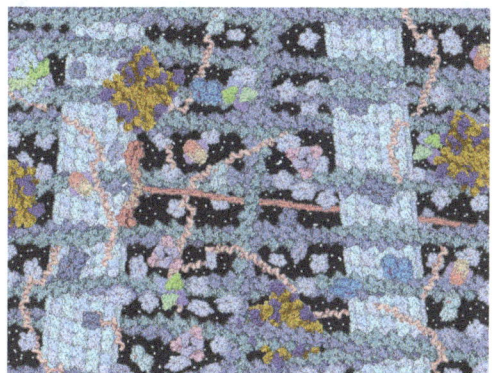

The cytosol is a crowded solution of many different types of
molecules that fills much of the volume of cells.

The cytosol is a complex mixture of substances dissolved in water. Although water forms the large majority of the cytosol, its structure and properties within cells is not well understood. The concentrations of ions such as sodium and potassium are different in the cytosol than in the extracellular fluid; these differences in ion levels are important in processes such as osmoregulation, cell signaling, and the generation of action potentials in excitable cells such as endocrine, nerve and muscle cells. The cytosol also contains large amounts of macromolecules, which can alter how molecules behave, through macromolecular crowding.

Although it was once thought to be a simple solution of molecules, the cytosol has multiple levels of organization. These include concentration gradients of small molecules such as calcium, large complexes of enzymes that act together and take part in metabolic pathways, and protein complexes such as proteasomes and carboxysomes that enclose and separate parts of the cytosol.

Properties and Composition

The proportion of cell volume that is cytosol varies: for example while this compartment forms the bulk of cell structure in bacteria, in plant cells the main compartment is the large central vacuole. The cytosol consists mostly of water, dissolved ions, small molecules, and large water-soluble molecules (such as proteins). The majority of these non-protein molecules have a molecular mass of less than 300 Da. This mixture of small molecules is extraordinarily complex, as the variety of molecules that are involved in metabolism (the metabolites) is immense. For example, up to

200,000 different small molecules might be made in plants, although not all these will be present in the same species, or in a single cell. Estimates of the number of metabolites in single cells such as *E. coli* and baker's yeast predict that under 1,000 are made.

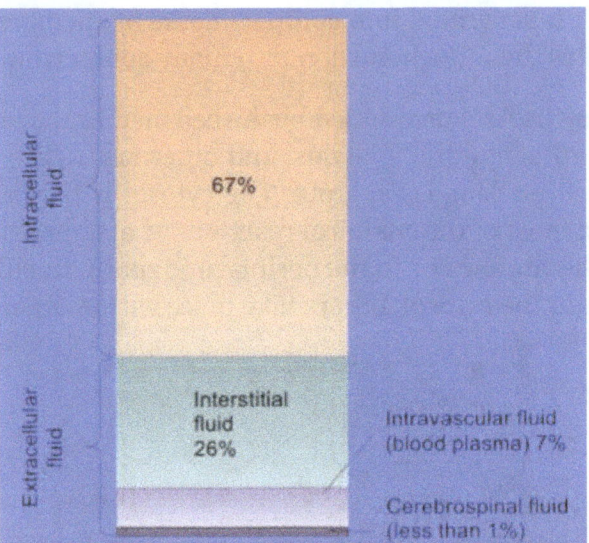

Intracellular fluid content in humans.

Water

Most of the cytosol is water, which makes up about 70% of the total volume of a typical cell. The pH of the intracellular fluid is 7.4. while human cytosolic pH ranges between 7.0 - 7.4, and is usually higher if a cell is growing. The viscosity of cytoplasm is roughly the same as pure water, although diffusion of small molecules through this liquid is about fourfold slower than in pure water, due mostly to collisions with the large numbers of macromolecules in the cytosol. Studies in the brine shrimp have examined how water affects cell functions; these saw that a 20% reduction in the amount of water in a cell inhibits metabolism, with metabolism decreasing progressively as the cell dries out and all metabolic activity halting when the water level reaches 70% below normal.

Although water is vital for life, the structure of this water in the cytosol is not well understood, mostly because methods such as nuclear magnetic resonance spectroscopy only give information on the average structure of water, and cannot measure local variations at the microscopic scale. Even the structure of pure water is poorly understood, due to the ability of water to form structures such as water clusters through hydrogen bonds.

The classic view of water in cells is that about 5% of this water is strongly bound in by solutes or macromolecules as water of solvation, while the majority has the same structure as pure water. This water of solvation is not active in osmosis and may have different solvent properties, so that some dissolved molecules are excluded, while others become concentrated. However, others argue that the effects of the high concentrations of macromolecules in cells extend throughout the cytosol and that water in cells behaves very differently from the water in dilute solutions. These ideas include the proposal that cells contain zones of low and high-density water, which could have widespread effects on the structures and functions of the other parts of the cell. However, the use of advanced nuclear magnetic resonance methods to directly measure the mobility of water in living cells contradicts this idea, as it suggests that 85% of cell water acts like that pure water, while the remainder is less mobile and probably bound to macromolecules.

Ions

The concentrations of the other ions in cytosol are quite different from those in extracellular fluid and the cytosol also contains much higher amounts of charged macromolecules such as proteins and nucleic acids than the outside of the cell structure.

Typical ion concentrations in mammalian cytosol and blood.		
Ion	Concentration in cytosol (millimolar)	Concentration in blood (millimolar)
Potassium	139	4
Sodium	12	145
Chloride	4	116
Bicarbonate	12	29
Amino acids in proteins	138	9
Magnesium	0.8	1.5
Calcium	<0.0002	1.8

In contrast to extracellular fluid, cytosol has a high concentration of potassium ions and a low concentration of sodium ions. This difference in ion concentrations is critical for osmoregulation, since if the ion levels were the same inside a cell as outside, water would enter constantly by osmosis - since the levels of macromolecules inside cells are higher than their levels outside. Instead, sodium ions are expelled and potassium ions taken up by the Na^+/K^+-ATPase, potassium ions then flow down their concentration gradient through potassium-selection ion channels, this loss of positive charge creates a negative membrane potential. To balance this potential difference, negative chloride ions also exit the cell, through selective chloride channels. The loss of sodium and chloride ions compensates for the osmotic effect of the higher concentration of organic molecules inside the cell.

Cells can deal with even larger osmotic changes by accumulating osmoprotectants such as betaines or trehalose in their cytosol. Some of these molecules can allow cells to survive being completely dried out and allow an organism to enter a state of suspended animation called cryptobiosis. In this state the cytosol and osmoprotectants become a glass-like solid that helps stabilize proteins and cell membranes from the damaging effects of desiccation.

The low concentration of calcium in the cytosol allows calcium ions to function as a second messenger in calcium signaling. Here, a signal such as a hormone or an action potential opens calcium channels so that calcium floods into the cytosol. This sudden increase in cytosolic calcium activates other signalling molecules, such as calmodulin and protein kinase C. Other ions such as chloride and potassium may also have signaling functions in the cytosol, but these are not well understood.

Macromolecules

Protein molecules that do not bind to cell membranes or the cytoskeleton are dissolved in the cytosol. The amount of protein in cells is extremely high, and approaches 200 mg/ml, occupying about 20-30% of the volume of the cytosol. However, measuring precisely how much protein is dissolved in cytosol in intact cells is difficult, since some proteins appear to be weakly associated with membranes or organelles in whole cells and are released into solution upon cell lysis. Indeed, in experiments where the plasma membrane of cells were carefully disrupted using saponin, without damaging the other cell membranes, only about one quarter of cell protein was released. These cells were also able to synthesize proteins if given ATP and amino acids, implying that many of the enzymes in cytosol are bound to the cytoskeleton. However, the idea that the majority of the proteins in cells are tightly bound in a network called the microtrabecular lattice is now seen as unlikely.

In prokaryotes the cytosol contains the cell's genome, within a structure known as a nucleoid. This is an irregular mass of DNA and associated proteins that control the transcription and replication of the bacterial chromosome and plasmids. In eukaryotes the genome is held within the cell nucleus, which is separated from the cytosol by nuclear pores that block the free diffusion of any molecule larger than about 10 nanometres in diameter.

This high concentration of macromolecules in cytosol causes an effect called macromolecular crowding, which is when the effective concentration of other macromolecules is increased, since they have less volume to move in. This crowding effect can produce large changes in both the rates and the position of chemical equilibrium of reactions in the cytosol. It is particularly important in its ability to alter dissociation constants by favoring the association of macromolecules, such as when multiple proteins come together to form protein complexes, or when DNA-binding proteins bind to their targets in the genome.

Organization

Although the components of the cytosol are not separated into regions by cell membranes, these components do not always mix randomly and several levels of organization can localize specific molecules to defined sites within the cytosol.

Concentration Gradients

Although small molecules diffuse rapidly in the cytosol, concentration gradients can still be produced within this compartment. A well-studied example of these are the "calcium sparks" that are produced for a short period in the region around an open calcium channel. These are about 2 micrometres in diameter and last for only a few milliseconds, although several sparks can merge to form larger gradients, called "calcium waves". Concentration gradients of other small molecules, such as oxygen and adenosine triphosphate may be produced in cells around clusters of mitochondria, although these are less well understood.

Protein Complexes

Proteins can associate to form protein complexes, these often contain a set of proteins with similar functions, such as enzymes that carry out several steps in the same metabolic pathway.

This organization can allow substrate channeling, which is when the product of one enzyme is passed directly to the next enzyme in a pathway without being released into solution. Channeling can make a pathway more rapid and efficient than it would be if the enzymes were randomly distributed in the cytosol, and can also prevent the release of unstable reaction intermediates. Although a wide variety of metabolic pathways involve enzymes that are tightly bound to each other, others may involve more loosely associated complexes that are very difficult to study outside the cell. Consequently, the importance of these complexes for metabolism in general remains unclear.

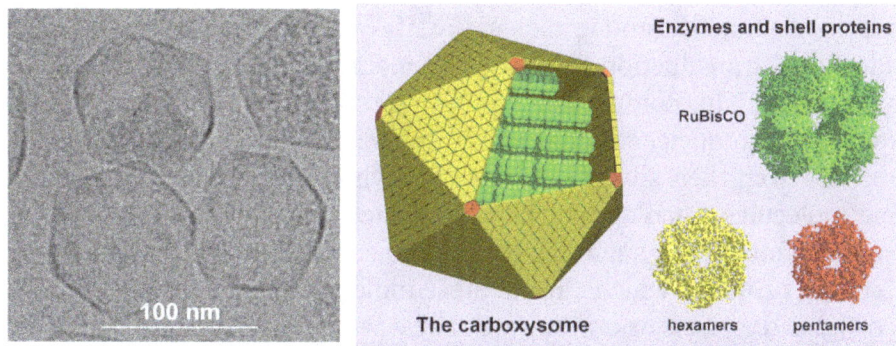

Carboxysomes are protein-enclosed bacterial microcompartments within the cytosol. On the left is an electron microscope image of carboxysomes, and on the right a model of their structure.

Protein Compartments

Some protein complexes contain a large central cavity that is isolated from the remainder of the cytosol. One example of such an enclosed compartment is the proteasome. Here, a set of subunits form a hollow barrel containing proteases that degrade cytosolic proteins. Since these would be damaging if they mixed freely with the remainder of the cytosol, the barrel is capped by a set of regulatory proteins that recognize proteins with a signal directing them for degradation (a ubiquitin tag) and feed them into the proteolytic cavity.

Another large class of protein compartments are bacterial microcompartments, which are made of a protein shell that encapsulates various enzymes. These compartments are typically about 100-200 nanometres across and made of interlocking proteins. A well-understood example is the carboxysome, which contains enzymes involved in carbon fixation such as RuBisCO.

Biomolecular Condensates

Non-membrane bound organelles can form as biomolecular condensates, which arise by clustering, oligomerisation, or polymerisation of macromolecules to drive colloidal phase separation of the cytoplasm or nucleus.

Cytoskeletal Sieving

Although the cytoskeleton is not part of the cytosol, the presence of this network of filaments restricts the diffusion of large particles in the cell. For example, in several studies tracer particles larger than about 25 nanometres (about the size of a ribosome) were excluded from parts of the cytosol around the edges of the cell and next to the nucleus. These "excluding compartments" may contain a much denser meshwork of actin fibres than the remainder of the cytosol. These microdomains could influence the distribution of large structures such as ribosomes and organelles within the cytosol by excluding them from some areas and concentrating them in others.

Function

The cytosol has no single function and is instead the site of multiple cell processes. Examples of these processes include signal transduction from the cell membrane to sites within the cell, such as the cell nucleus, or organelles. This compartment is also the site of many of the processes of cytokinesis, after the breakdown of the nuclear membrane in mitosis. Another major function of cytosol is to transport metabolites from their site of production to where they are used. This is relatively simple for water-soluble molecules, such as amino acids, which can diffuse rapidly through the cytosol. However, hydrophobic molecules, such as fatty acids or sterols, can be transported through the cytosol by specific binding proteins, which shuttle these molecules between cell membranes. Molecules taken into the cell by endocytosis or on their way to be secreted can also be transported through the cytosol inside vesicles, which are small spheres of lipids that are moved along the cytoskeleton by motor proteins.

The cytosol is the site of most metabolism in prokaryotes, and a large proportion of the metabolism of eukaryotes. For instance, in mammals about half of the proteins in the cell are localized to the cytosol. The most complete data are available in yeast, where metabolic reconstructions indicate that the majority of both metabolic processes and metabolites occur in the cytosol. Major metabolic pathways that occur in the cytosol in animals are protein biosynthesis, the pentose phosphate pathway, glycolysis and gluconeogenesis. The localization of pathways can be different in other organisms, for instance fatty acid synthesis occurs in chloroplasts in plants and in apicoplasts in apicomplexa.

PLASMID

Plasmids are small, circular molecules of DNA that are capable of replicating independently. As such, they do not rely on chromosomal DNA of the organism for replication. Because of this characteristic, they are also referred to as extra-chromosomal DNA. Although the molecule was first discovered in a member of the Enterobacteriacae, studies have shown that plasmids are naturally occurring in many types of microorganisms around the world. Although there have been debates as to whether plasmids can be regarded as microorganisms, at least using the proposed virus definition, it is worth noting that the term "plasmid" is largely used to refer to genetic elements that exist outside the chromosome (in DNA of the organism) and are capable of replicating independently.

Plasmids can be found in:

- Bacteria,

- Archaea,
- Various eukaryotes (yeast and plants).

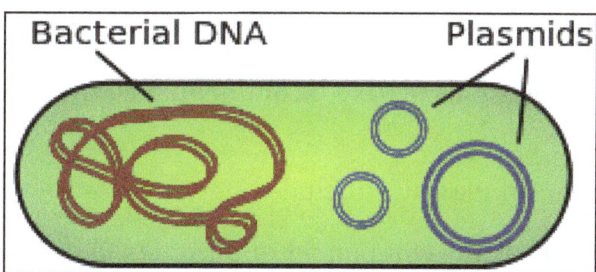

Bacterium with its chromosomal DNA and several plasmids within it.

Structure

With regards to structure, plasmids are made up of circular double chains of DNA. The circular structure of plasmids is made possible by the two ends of the double strands being joined by covalent bonds. The molecules are also small in size, especially when compared to the organisms' DNA, and measure between a few kilobases and several hundred kilobases.

Although a good number of plasmids have a covalently closed circular structure, some plasmids have a linear structure and do not form a circular shape.

Generally, plasmids are composed of three major components that include:

- Origin of replication (replicon): The origin of replication (ori) refers to a specific location in the strand at which replication begins. For plasmids, this location is largely composed of A-T base pairs that are easier to separate during replication.

   Compared to the organisms' DNA that consists of many origins of replication, plasmids have one of a few origins of replication because they are smaller in size. At the origin of replication, plasmids also contain a number of regulatory elements that contribute to the process (e.g. Rep proteins)

- Polylinker (multiple cloning sites): In plasmid, the polylinker (MCS) is one of the most important parts of the molecule. This is because it allows students to learn more about cloning. Basically, a polylinker is a short sequence of DNA consisting of a few sites for cleavage by restriction enzymes.

   As such, MCS allows for easy insertion of DNA through ligation or restriction enzyme digestion. At the site of cleavage, different polylinkers can cut the strand. Therefore, one of the restriction enzymes can cut the plasmid at given points of the sire to allow for DNA insertion.

- Antibiotic resistance gene: The antibiotic resistance gene is one of the main components of plasmids. These genes play an important role in drug resistance (to one or more antibiotics) thus making treatment of some diseases more challenging.

   Plasmids are today known for their ability to transfer from one species of bacteria to another through a process known as conjugation (contact between cells that is followed by transfer of

DNA content). In the process, they are capable of conferring antibiotic resistance properties to other species of bacteria.

While plasmid replication provides an added advantage to bacteria (resistance to certain antibiotics), it also affects cell division of bacteria due to additional replication burden. As a result, bacteria with plasmids tend to be out populated by those without plasmids due to reduced cell division.

Some of the other Components of Plasmids Include:

- A promoter region: This is the component of plasmids that is involved in recruiting transcriptional machinery

- Primer binding site: This is a short sequence of DNA on a single strand that is typically used for the purposes of PCR amplification or DNA sequencing

Although plasmids share various general characteristics, there are different types in existence.

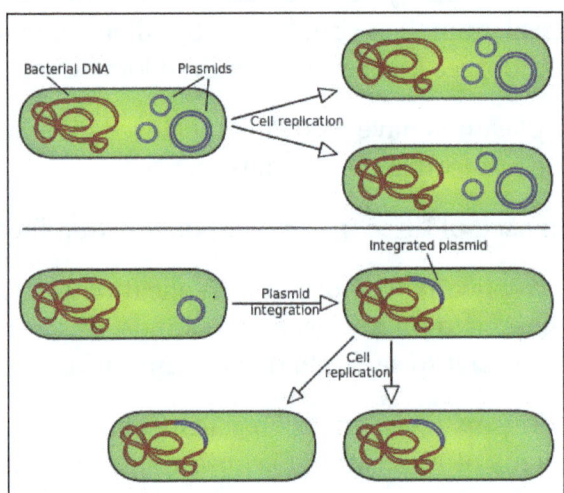

Comparing the activity of non-integrating plasmids, on the top,
with episomes, on the bottom, during cell division.

Types and Functions of Plasmids

Resistance Plasmids

Also referred to as antimicrobial resistance plasmids, resistance plasmids are a type of plasmids that carry genes that play an important role in antibiotic resistance. They are also highly involved in bacterial conjugation by producing conjugation pili which transfer the R plasmid from one bacterium to another.

Resistance plasmids are divided into two main groups that include:

- Narrow-host-range group: Often replicated within a single species.

- Broad-host-range group: Easily transferred between bacteria species. This group of resistance plasmids has been shown to carry a range of antibiotic resistance genes. Following the transfer of antibiotic resistance genes to drug-sensitive bacteria, this can cause the bacteria to develop resistance towards a variety of drugs.

Degradative Plasmids

Compared to other types of plasmids, degradative plasmids enable the host organism to degrade/break down xenobiotic compounds. Also referred to as recalcitrant substances, xenobiotic compounds include a range of compounds released into the environment as a result of human actions and are therefore not naturally occurring or common in nature.

Hosts of degradative plasmids are found in groups IncP-1, IncP-7, and IncP-9 and include such species as Ochrobactrum anthropi, Rhizobium sp, Burkholderia hospita, Escherichia coli, and Pseudomonas fluorescens among many others.

Because of the ability of the host to degrade xenobiotic compounds, researchers have attempted to use the plasmids to degrade various contaminating substances in the environment. However, given that this has not proved effective, research studies continue to be conducted to determine how to use various indigenous bacteria (as hosts of degradative plasmids) for degradation of such compounds.

While degradative plasmids contribute to the degradation of xenobiotic compounds, their behavior varies depending on a number of factors such as the capacity for replication and stability. For instance, plasmids found in IncP-1 group have not only been shown to have a broad host range, but also high transfer frequency. The differences in the behavior of different degradative plasmids have therefore been shown to result in different behaviors between them and their respective hosts. The use of biodegradative microorganisms for the purposes of removing xenobiotic compounds from contaminated environments is known as Bioaugmentation. Whereas IncP-1 plasmids have a broad range of hosts, IncP-7 has been shown to have a narrow host range. On the other hand, IncP-9 has an intermediate host range.

Fertility Plasmids

Like many other plasmids, fertility plasmids (F plasmid) have a circular structure and measures about 100 kb.

Some of the main parts of the F plasmid include:

- Transposable element (IS2, 1S3, and Tn1000),

- Replication sites (RepFIA, RepFIB, and RepFIC),

- Origin of conjugative transfer (oirT),

- Replication origin regions.

F plasmid plays an important role in reproduction given that they contain genes that code for the production of sex pilus as well as enzymes required for conjugation. F plasmid also contains genes that are involved in their own transfer. Therefore, during conjugation, they enhance their own transfer from one cell to another.

Whereas the cells that process the F plasmids are referred to as donors, those that lack this factor are the recipients. On the other hand, the plasmids that enhance the ability of the host cell to behave like a donor are known as the transfer factor.

During conjugation, the donor cell (bacteria) with sex pili (1-3 sex pili) binds to a specific protein on the outer membrane of the recipient thus initiating the mating process.

Following the initial binding, the pili retract thus allowing the two cells to bind together. This is then followed by the transfer of DNA from the donor to the recipient and consequently the transfer of the F plasmid. As a result, the recipient acquires the F factor and gains the ability to produce sex pilus involved in conjugation.

- During conjugation, only DNA is passed from the donor to the recipient. Therefore, cytoplasm and other cell material are not transferred.

- Sexual pili (sex pili) are tiny rod-like structures that allow the F-positive (cells that have the F factor) bacterial cells to attach to the F-negative (cells lacking the pili) female to promote conjugative transfer.

Col Plasmids

Col plasmids confer to bacteria the ability to produce toxic proteins known as colicines. Such bacteria as E. coli, Shigella and Salmonella use these toxins to kill other bacteria and thus thrive in their respective environments.

There are different types of Col plasmids in existence that produce different types of colicines/colicins. A few examples of Col plasmids include Col B, Col E2 and E3. Their differences are also characterized by differences in their mode of action.

For instance, whereas Col B causes damage to cell membrane of other bacteria (lacking the plasmid) Col E3 has been shown to induce degradation of the nucleic acids of the target cells.

Like fertility plasmids, some of the Col plasmids have been shown to carry elements that enhance their transmission from one cell to another. Therefore, through conjugation or the mating process, particularly for cells with the F factor (fertility plasmids) the Col plasmids can be transferred from one cell (donor) to another (recipient). As a result, the recipient acquires the ability to produce toxins that kill or inhibit the growth of the target bacteria lacking the plasmid.

- Colicins/colicines belong to a group of toxins known as bacteriocins.

- These toxins affect the target bacteria by affecting such processes as replication of DNA, translation and energy metabolism among others.

Virulence Plasmids

Compared to other harmless bacteria, bacteria that tend to be pathogenic in nature carry genes for virulence factors that allow them to invade and infect their respective hosts.

For some of these bacteria, the virulence factors are the result of the organisms' own genetic material. However, for others, this is as a result of genetic elements from extra-chromosomal DNA. Although there are other sources of such elements, e.g. transposons, plasmids are some of the most common mobile genetic elements.

With regards to pathogenicity, virulence plasmids play an important role given that they can help bacteria effectively adapt to their respective environments. This is because the virulence plasmid can enable the organism to express an array of virulence-associated functions thus providing the organism with more advantageous characteristics to thrive in their environment.

Like other types of plasmids, virulence plasmids can also be transmitted from one bacterium to another. Apart from the virulence gene, plasmids have also been shown to carry other important elements that enhance transmission and maintenance.

For this reason, they are larger in size but low in numbers. This ensures that they do not cause additional burden to the organism during cell division.

Typically, cell division and cell maintenance require the use of energy. By having low numbers of virulence plasmids, the cells are spared significant metabolic burden that would be required for maintenance and genome duplication of numerous plasmids.

Some of the other types of plasmids include:

- Recombinant plasmids: Plasmids that have been altered in the laboratory and introduced into the bacteria for the purposes of studies.

- Crptic plasmids: No known functions.

- Metabolic plasmids: Enhance metabolism of the host.

- Conjugative plasmids: Promote self-transfer.

- Suicide plasmids: Fail to replicate when transferred from one cell to another

Plasmid Vector

A vector refers to any piece of molecule that contains genetic material that can be replicated and expressed when transferred into another cell. Based on this definition, it is possible to see why the words "vector" and "plasmids" are sometimes interchanged. However, this is not to say that all plasmids are vectors.

One of the primary characteristics of plasmid vectors is that they are small in size. Apart from their size, they are characterized by an origin of replication, a selective marker as well as multiple cloning sites.

The ideal plasmid vectors have high copy numbers inside the cell. As such, it ensures high numbers of the target gene for cloning purposes. This also ensures that the gene of interest is increased during genomic division. In addition, the plasmid can have a marker gene as the visual marker to help determine whether cloning was successful.

Because of their multiple cloning sites, plasmids have been shown to be some of the best vectors for cloning. Because of this characteristic, it is possible for restriction enzymes to cleave various regions of the plasmid for cloning.

Over the years, using these vectors has allowed for recombinant DNA to be introduced into host cells for study purposes. For instance, through this type of cloning, it has become possible for researchers to sequence the genome of a range of species, study the expression of genes and even observe various cellular mechanisms.

While smaller plasmids are capable of carrying long DNA segments, reduced size can also help remove non-essential genes that are not required for cloning.

Plasmid Isolation

In order to obtain purified plasmid DNA for such procedures as cloning, PCR and transfection, plasmid isolation has to be performed. The process involves using a number of techniques to obtain the plasmid DNA from host cells in order to use it in molecular biology.

Plasmid isolation involves the following steps:

- Cell growth (growth of bacterial cells): This involves growing the bacteria that contain plasmid in a specific shaken culture. Here, given antibiotics may be used to prevent the growth of other undesired bacteria.

- Centrifugation: Bacterial growth is followed by centrifugation in order to pellet the cells. Once the supernatant has been removed, then isolation of plasmids can begin.

One of the most common techniques for isolation is the classical method that is sometimes referred to as alkaline lysis.

This involves the following steps:

- Suspension of the pellet in an isotonic solution: The bacterial pellet obtained from centrifugation is re-suspended in an isotonic solution (ethylene diamine tetraacetate) which prevents nuclease activity.

- Alkaline lysis of the cells: This involves cell lysis by using sodium dodecyl sulfate to disintegrate the lipid structure on the cell membrane.

- Precipitation of dissolved proteins using a solution of acidic potassium acetate.

- Sedimentation: Centrifugation is used for sedimentation.

- Purification: A mixture of phenol and chloroform is used for purification of the plasmid DNA. This step removes protein content.

- Add ethanol for precipitation (to be sedimented through centrifugation).

- Wash the solution with 70 percent ethanol (to remove salt content).

- Centrifuge to sediment plasmid DNA.

- Dissolve in TE solution and store.

THYLAKOID

A thylakoid is a membrane-bound compartment inside chloroplasts and cyanobacteria. They are the site of the light-dependent reactions of photosynthesis. Thylakoids consist of a thylakoid membrane surrounding a thylakoid lumen. Chloroplast thylakoids frequently form stacks of disks referred to as grana (singular: granum). Grana are connected by intergranal or stroma thylakoids, which join granum stacks together as a single functional compartment.

Structure

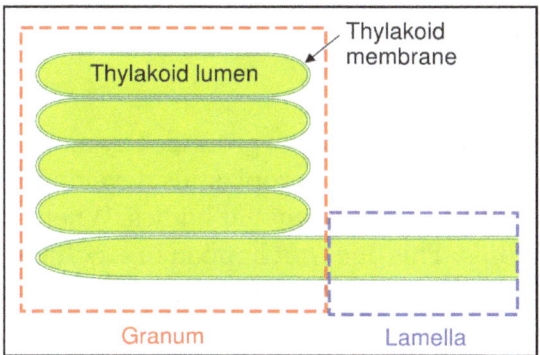

Thylakoid structures.

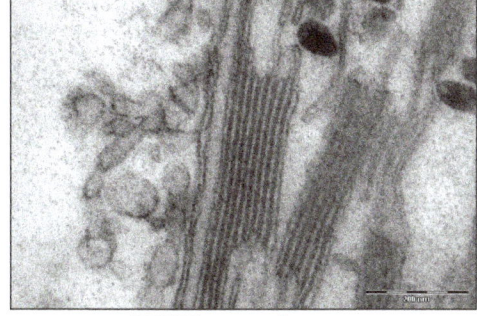

TEM image of grana.

Thylakoids are membrane-bound structures embedded in the chloroplast stroma. A stack of thylakoids is called a granum and resembles a stack of coins.

Membrane

The thylakoid membrane is the site of the light-dependent reactions of photosynthesis with the photosynthetic pigments embedded directly in the membrane. It is an alternating pattern of dark and light bands measuring each 1 nanometre. The thylakoid lipid bilayer shares characteristic features with prokaryotic membranes and the inner chloroplast membrane. For example, acidic lipids can be found in thylakoid membranes, cyanobacteria and other photosynthetic bacteria and are involved in the functional integrity of the photosystems. The thylakoid membranes of higher plants are composed primarily of phospholipids and galactolipids that are asymmetrically arranged along and across the membranes. Thylakoid membranes are richer in galactolipids rather than phospholipids; also they predominantly consist of hexagonal phase II forming monogalacotosyl diglyceride lipid. Despite this unique composition, plant thylakoid membranes have been shown to assume largely lipid-bilayer dynamic organization. Lipids forming the thylakoid membranes, richest in high-fluidity linolenic acid are synthesized in a complex pathway involving exchange of lipid precursors between the endoplasmic reticulum and inner membrane of the plastid envelope and transported from the inner membrane to the thylakoids via vesicles.

Lumen

The thylakoid lumen is a continuous aqueous phase enclosed by the thylakoid membrane. It plays an important role for photophosphorylation during photosynthesis. During the light-dependent reaction, protons are pumped across the thylakoid membrane into the lumen making it acidic down to pH 4.

Granum and Stroma Lamellae

In higher plants thylakoids are organized into a granum-stroma membrane assembly. A granum (plural grana) is a stack of thylakoid discs. Chloroplasts can have from 10 to 100 grana. Grana are connected by stroma thylakoids, also called intergranal thylakoids or lamellae. Grana thylakoids and stroma thylakoids can be distinguished by their different protein composition. Grana contribute to chloroplasts' large surface area to volume ratio. Different interpretations of electron tomography imaging of thylakoid membranes has resulted in two models for grana structure. Both posit that lamellae intersect grana stacks in parallel sheets, though whether these sheets intersect in planes perpendicular to the grana stack axis, or are arranged in a right-handed helix is debated.

Formation

Chloroplasts develop from proplastids when seedlings emerge from the ground. Thylakoid formation requires light. In the plant embryo and in the absence of light, proplastids develop into etioplasts that contain semicrystalline membrane structures called prolamellar bodies. When exposed to light, these prolamellar bodies develop into thylakoids. This does not happen in seedlings grown in the dark, which undergo etiolation. An underexposure to light can cause the thylakoids to fail. This causes the chloroplasts to fail resulting in the death of the plant.

Thylakoid formation requires the action of vesicle-inducing protein in plastids 1 (VIPP1). Plants cannot survive without this protein, and reduced VIPP1 levels lead to slower growth and paler plants with reduced ability to photosynthesize. VIPP1 appears to be required for basic thylakoid membrane formation, but not for the assembly of protein complexes of the thylakoid membrane. It is conserved in all organisms containing thylakoids, including cyanobacteria, green algae, such as Chlamydomonas, and higher plants, such as Arabidopsis thaliana.

Isolation and Fractionation

Thylakoids can be purified from plant cells using a combination of differential and gradient centrifugation. Disruption of isolated thylakoids, for example by mechanical shearing, releases the lumenal fraction. Peripheral and integral membrane fractions can be extracted from the remaining membrane fraction. Treatment with sodium carbonate (Na_2CO_3) detaches peripheral membrane proteins, whereas treatment with detergents and organic solvents solubilizes integral membrane proteins.

Proteins

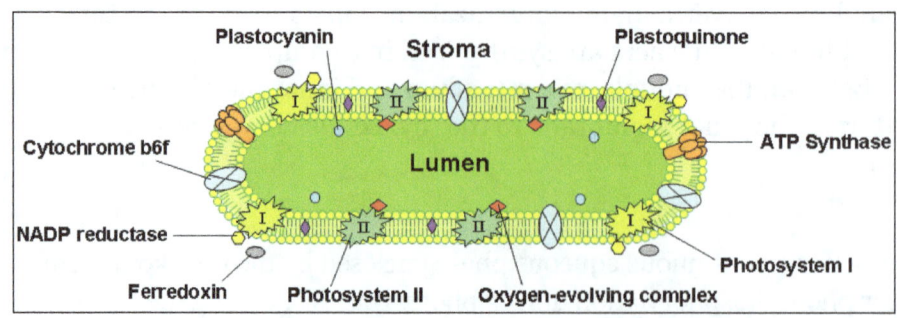

Thylakoid disc with embedded and associated proteins.

Thylakoids contain many integral and peripheral membrane proteins, as well as lumenal proteins. Recent proteomics studies of thylakoid fractions have provided further details on the protein composition of the thylakoids. These data have been summarized in several plastid protein databases that are available online.

According to these studies, the thylakoid proteome consists of at least 335 different proteins. Out of these, 89 are in the lumen, 116 are integral membrane proteins, 62 are peripheral proteins on the stroma side, and 68 peripheral proteins on the lumenal side. Additional low-abundance lumenal proteins can be predicted through computational methods. Of the thylakoid proteins with known functions, 42% are involved in photosynthesis. The next largest functional groups include proteins involved in protein targeting, processing and folding with 11%, oxidative stress response (9%) and translation (8%).

Integral Membrane Proteins

Thylakoid membranes contain integral membrane proteins which play an important role in light harvesting and the light-dependent reactions of photosynthesis. There are four major protein complexes in the thylakoid membrane:

- Photosystems I and II

- Cytochrome b6f complex

- ATP synthase

Photosystem II is located mostly in the grana thylakoids, whereas photosystem I and ATP synthase are mostly located in the stroma thylakoids and the outer layers of grana. The cytochrome b6f complex is distributed evenly throughout thylakoid membranes. Due to the separate location of the two photosystems in the thylakoid membrane system, mobile electron carriers are required to shuttle electrons between them. These carriers are plastoquinone and plastocyanin. Plastoquinone shuttles electrons from photosystem II to the cytochrome b6f complex, whereas plastocyanin carries electrons from the cytochrome b6f complex to photosystem I.

Together, these proteins make use of light energy to drive electron transport chains that generate a chemiosmotic potential across the thylakoid membrane and NADPH, a product of the terminal redox reaction. The ATP synthase uses the chemiosmotic potential to make ATP during photophosphorylation.

Photosystems

These photosystems are light-driven redox centers, each consisting of an antenna complex that uses chlorophylls and accessory photosynthetic pigments such as carotenoids and phycobiliproteins to harvest light at a variety of wavelengths. Each antenna complex has between 250 and 400 pigment molecules and the energy they absorb is shuttled by resonance energy transfer to a specialized chlorophyll a at the reaction center of each photosystem. When either of the two chlorophyll a molecules at the reaction center absorbs energy, an electron is excited and transferred to an electron-acceptor molecule. Photosystem I contains a pair of chlorophyll a molecules, designated P700, at its reaction center that maximally absorbs 700 nm light. Photosystem II contains P680 chlorophyll that absorbs 680 nm light best (note that these wavelengths correspond to deep red). The P is short for pigment and the number is the specific absorption peak in nanometers for the chlorophyll molecules in each reaction center.

Cytochrome b6f Complex

The cytochrome b6f complex is part of the thylakoid electron transport chain and couples electron transfer to the pumping of protons into the thylakoid lumen. Energetically, it is situated between the two photosystems and transfers electrons from photosystem II-plastoquinone to plastocyanin-photosystem I.

ATP Synthase

The thylakoid ATP synthase is a CF1FO-ATP synthase similar to the mitochondrial ATPase. It is integrated into the thylakoid membrane with the CF1-part sticking into stroma. Thus, ATP synthesis occurs on the stromal side of the thylakoids where the ATP is needed for the light-independent reactions of photosynthesis.

Lumen Proteins

The electron transport protein plastocyanin is present in the lumen and shuttles electrons from the cytochrome b6f protein complex to photosystem I. While plastoquinones are lipid-soluble and therefore move within the thylakoid membrane, plastocyanin moves through the thylakoid lumen.

The lumen of the thylakoids is also the site of water oxidation by the oxygen evolving complex associated with the lumenal side of photosystem II. Lumenal proteins can be predicted computationally based on their targeting signals. In Arabidopsis, out of the predicted lumenal proteins possessing the Tat signal, the largest groups with known functions are 19% involved in protein processing (proteolysis and folding), 18% in photosynthesis, 11% in metabolism, and 7% redox carriers and defense.

Protein Expression

Chloroplasts have their own genome, which encodes a number of thylakoid proteins. However, during the course of plastid evolution from their cyanobacterial endosymbiotic ancestors, extensive gene transfer from the chloroplast genome to the cell nucleus took place. This results in the four major thylakoid protein complexes being encoded in part by the chloroplast genome and in part by the nuclear genome. Plants have developed several mechanisms to co-regulate the expression of the different subunits encoded in the two different organelles to assure the proper stoichiometry and assembly of these protein complexes. For example, transcription of nuclear genes encoding parts of the photosynthetic apparatus is regulated by light. Biogenesis, stability and turnover of thylakoid protein complexes are regulated by phosphorylation via redox-sensitive kinases in the thylakoid membranes. The translation rate of chloroplast-encoded proteins is controlled by the presence or absence of assembly partners (control by epistasy of synthesis). This mechanism involves negative feedback through binding of excess protein to the 5' untranslated region of the chloroplast mRNA. Chloroplasts also need to balance the ratios of photosystem I and II for the electron transfer chain. The redox state of the electron carrier plastoquinone in the thylakoid membrane directly affects the transcription of chloroplast genes encoding proteins of the reaction centers of the photosystems, thus counteracting imbalances in the electron transfer chain.

Protein Targeting to the Thylakoids

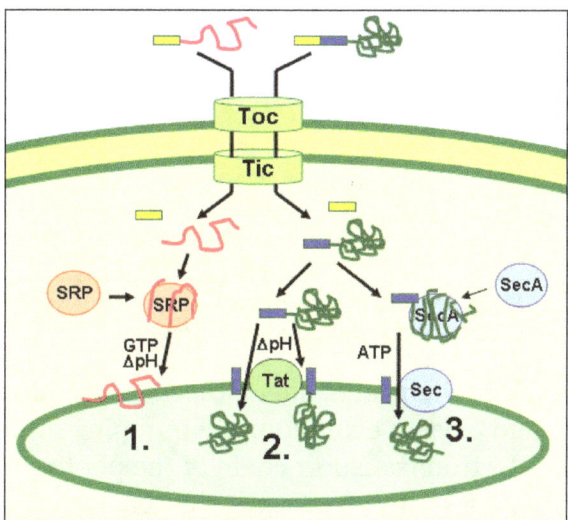

Schematic representation of thylakoid protein targeting pathways.

Thylakoid proteins are targeted to their destination via signal peptides and prokaryotic-type secretory pathways inside the chloroplast. Most thylakoid proteins encoded by a plant's nuclear genome need two targeting signals for proper localization: An N-terminal chloroplast targeting peptide, followed by a thylakoid targeting peptide (shown in blue). Proteins are imported through the translocon of outer and inner membrane (Toc and Tic) complexes. After entering the chloroplast, the first targeting peptide is cleaved off by a protease processing imported proteins. This unmasks the second targeting signal and the protein is exported from the stroma into the thylakoid in a second targeting step. This second step requires the action of protein translocation components of the thylakoids and is energy-dependent. Proteins are inserted into the membrane via the SRP-dependent pathway (1), the Tat-dependent pathway (2), or spontaneously via their transmembrane domains. Lumenal proteins are exported across the thylakoid membrane into the lumen by either the Tat-dependent pathway (2) or the Sec-dependent pathway (3) and released by cleavage from the thylakoid targeting signal. The different pathways utilize different signals and energy sources. The Sec (secretory) pathway requires ATP as energy source and consists of SecA, which binds to the imported protein and a Sec membrane complex to shuttle the protein across. Proteins with a twin arginine motif in their thylakoid signal peptide are shuttled through the Tat (twin arginine translocation) pathway, which requires a membrane-bound Tat complex and the pH gradient as an energy source. Some other proteins are inserted into the membrane via the SRP (signal recognition particle) pathway.

The chloroplast SRP can interact with its target proteins either post-translationally or co-translationally, thus transporting imported proteins as well as those that are translated inside the chloroplast. The SRP pathway requires GTP and the pH gradient as energy sources. Some transmembrane proteins may also spontaneously insert into the membrane from the stromal side without energy requirement.

Function

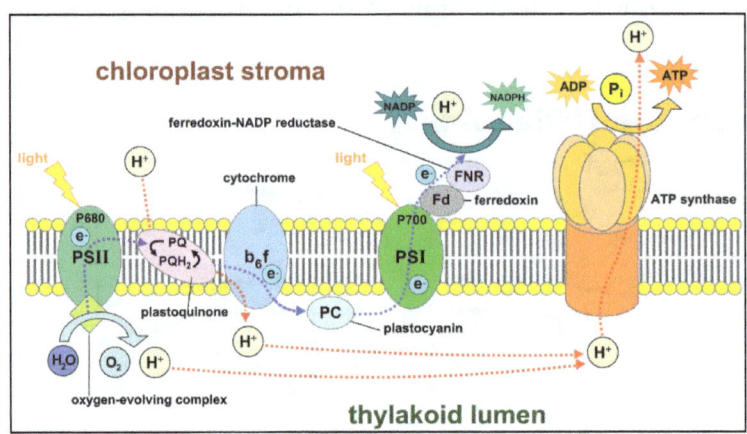

Light-dependent reactions of photosynthesis
at the thylakoid membrane.

The thylakoids are the site of the light-dependent reactions of photosynthesis. These include light-driven water oxidation and oxygen evolution, the pumping of protons across the thylakoid membranes coupled with the electron transport chain of the photosystems and cytochrome complex, and ATP synthesis by the ATP synthase utilizing the generated proton gradient.

Water Photolysis

The first step in photosynthesis is the light-driven reduction (splitting) of water to provide the electrons for the photosynthetic electron transport chains as well as protons for the establishment of a proton gradient. The water-splitting reaction occurs on the lumenal side of the thylakoid membrane and is driven by the light energy captured by the photosystems. This oxidation of water conveniently produces the waste product O_2 that is vital for cellular respiration. The molecular oxygen formed by the reaction is released into the atmosphere.

Electron Transport Chains

Two different variations of electron transport are used during photosynthesis:

- Noncyclic electron transport or Non-cyclic photophosphorylation produces NADPH + H^+ and ATP.

- Cyclic electron transport or Cyclic photophosphorylation produces only ATP.

The noncyclic variety involves the participation of both photosystems, while the cyclic electron flow is dependent on only photosystem I.

- Photosystem I uses light energy to reduce $NADP^+$ to NADPH + H^+, and is active in both noncyclic and cyclic electron transport. In cyclic mode, the energized electron is passed down a chain that ultimately returns it (in its base state) to the chlorophyll that energized it.

- Photosystem II uses light energy to oxidize water molecules, producing electrons (e^-), protons (H^+), and molecular oxygen (O_2), and is only active in noncyclic transport.

Electrons in this system are not conserved, but are rather continually entering from oxidized $2H_2O$ ($O_2 + 4 H^+ + 4 e^-$) and exiting with $NADP^+$ when it is finally reduced to NADPH.

Chemiosmosis

A major function of the thylakoid membrane and its integral photosystems is the establishment of chemiosmotic potential. The carriers in the electron transport chain use some of the electron's energy to actively transport protons from the stroma to the lumen. During photosynthesis, the lumen becomes acidic, as low as pH 4, compared to pH 8 in the stroma. This represents a 10,000 fold concentration gradient for protons across the thylakoid membrane.

Source of Proton Gradient

The protons in the lumen come from three primary sources:

- Photolysis by photosystem II oxidises water to oxygen, protons and electrons in the lumen.

- The transfer of electrons from photosystem II to plastoquinone during non-cyclic electron transport consumes two protons from the stroma. These are released in the lumen when the reduced plastoquinol is oxidized by the cytochrome b6f protein complex on the lumen side of the thylakoid membrane. From the plastoquinone pool, electrons pass through the cytochrome b6f complex. This integral membrane assembly resembles cytochrome bc1.

- The reduction of plastoquinone by ferredoxin during cyclic electron transport also transfers two protons from the stroma to the lumen.

The proton gradient is also caused by the consumptions of protons in the stroma to make NADPH from NADP+ at the NADP reductase.

ATP Generation

The molecular mechanism of ATP (Adenosine triphosphate) generation in chloroplasts is similar to that in mitochondria and takes the required energy from the proton motive force (PMF). However, chloroplasts rely more on the chemical potential of the PMF to generate the potential energy required for ATP synthesis. The PMF is the sum of a proton chemical potential (given by the proton concentration gradient) and a transmembrane electrical potential (given by charge separation across the membrane). Compared to the inner membranes of mitochondria, which have a significantly higher membrane potential due to charge separation, thylakoid membranes lack a charge gradient. To compensate for this, the 10,000 fold proton concentration gradient across the thylakoid membrane is much higher compared to a 10 fold gradient across the inner membrane of mitochondria. The resulting chemiosmotic potential between the lumen and stroma is high enough to drive ATP synthesis using the ATP synthase. As the protons travel back down the gradient through channels in ATP synthase, ADP + P_i are combined into ATP. In this manner, the light-dependent reactions are coupled to the synthesis of ATP via the proton gradient.

Thylakoid Membranes in Cyanobacteria

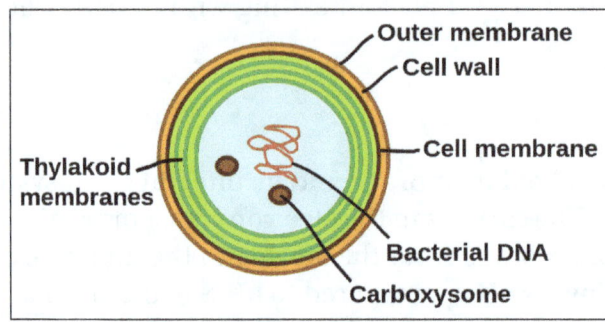

Thylakoids (green) inside a cyanobacterium (*Synechocystis*).

Cyanobacteria are photosynthetic prokaryotes with highly differentiated membrane systems. Cyanobacteria have an internal system of thylakoid membranes where the fully functional electron transfer chains of photosynthesis and respiration reside. The presence of different membrane systems lends these cells a unique complexity among bacteria. Cyanobacteria must be able to reorganize the membranes, synthesize new membrane lipids, and properly target proteins to the correct membrane system. The outer membrane, plasma membrane, and thylakoid membranes each have specialized roles in the cyanobacterial cell. Understanding the organization, functionality, protein composition and dynamics of the membrane systems remains a great challenge in cyanobacterial cell biology.

The thylakoid membranes of the cyanobacteria are not differentiated into granum and stroma regions as observed in plants. They form stacks of parallel sheets close to the cytoplasmic membrane with a low packing density. The relatively large distance between the thylakoids provides space for the external light harvesting antennae, the phycobilisomes. This macrostructure, as in the case of higher plants, shows some flexibility during changes in the physicochemical environment.

References

- Van wijk k (2004). "plastid proteomics". Plant physiol biochem. 42 (12): 963–77. Doi:10.1016/j.plaphy.2004.10.015. Pmid 15707834

- Cell-membrane: biologydictionary.net, Retrieved 30 January, 2019

- Van zon a, mossink mh, scheper rj, sonneveld p, wiemer ea (september 2003). "the vault complex". Cellular and molecular life sciences. 60 (9): 1828–37. Doi:10.1007/s00018-003-3030-y. Pmid 14523546

- Cell-wall: biologydictionary.net, Retrieved 29 June, 2019

- Shearer j, graham te (april 2002). "new perspectives on the storage and organization of muscle glycogen". Canadian journal of applied physiology. 27(2): 179–203. Doi:10.1139/h02-012. Pmid 12179957

- Cytoskeleton: biologydictionary.net, Retrieved 13 March, 2019

- Brown, thomas a. (2011). Rapid review physiology. Elsevier health sciences. P. 2. Isbn 978-0323072601

- Centrioles-373538: thoughtco.com, Retrieved 23 February, 2019

- Bowsher cg, tobin ak (april 2001). "compartmentation of metabolism within mitochondria and plastids". J. Exp. Bot. 52 (356): 513–27. Doi:10.1093/jexbot/52.356.513. Pmid 11373301

- Plasmids: microscopemaster.com, Retrieved 3 January, 2019

- Benning c, xu c, awai k (2006). "non-vesicular and vesicular lipid trafficking involving plastids". Curr opin plant biol. 9 (3): 241–7. Doi:10.1016/j.pbi.2006.03.012. Pmid 16603410

5

Cell Organelles

Cell organelles are the specialized sub-units within a cell that have well-defined functions. They can be either membrane-bound or non-membrane bound. Various cell organelles include ribosome, lysosome, endosome, vacuole, cell nucleus, chloroplast, mitochondrion, etc. This chapter has been carefully written to provide an easy understanding of these cell organelles.

ORGANELLE

An organelle is a tiny cellular structure that performs specific functions within a cell. Organelles are embedded within the cytoplasm of eukaryotic and prokaryotic cells. In the more complex eukaryotic cells, organelles are often enclosed by their own membrane. Analogous to the body's internal organs, organelles are specialized and perform valuable functions necessary for normal cellular operation. Organelles have a wide range of responsibilities that include everything from generating energy for a cell to controlling the cell's growth and reproduction.

Eukaryotic Organelles

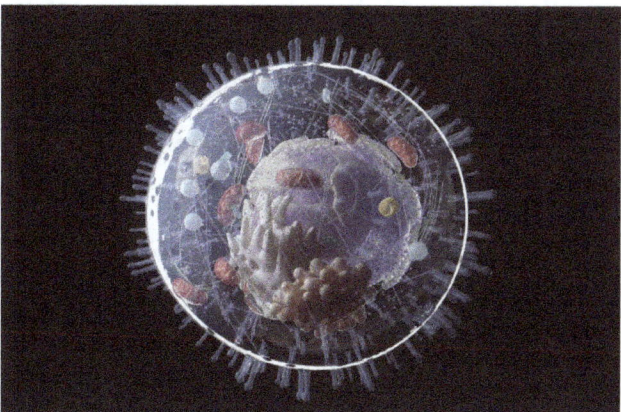

Cellular Organelles in a Human Cell.

Eukaryotic cells are cells with a nucleus. The nucleus is an organelle that is surrounded by a double membrane called the nuclear envelope. The nuclear envelope separates the contents of the nucleus from the rest of the cell. Eukaryotic cells also have a cell membrane (plasma membrane), cytoplasm, cytoskeleton, and various cellular organelles. Animals, plants, fungi, and protists are examples of eukaryotic organisms. Animal and plant cells contain many of the same kinds or organelles.

There are also certain organelles found in plant cells that are not found in animal cells and vice versa. Examples of organelles found in plant cells and animal cells include:

- Nucleus: A membrane bound structure that contains the cell's hereditary (DNA) information and controls the cell's growth and reproduction. It is commonly the most prominent organelle in the cell.

- Mitochondria: As the cell's power producers, mitochondria convert energy into forms that are usable by the cell. They are the sites of cellular respiration which ultimately generates fuel for the cell's activities. Mitochondria are also involved in other cell processes such as cell division and growth, as well as cell death.

- Endoplasmic Reticulum: Extensive network of membranes composed of both regions with ribosomes (rough ER) and regions without ribosomes (smooth ER). This organelle manufactures membranes, secretory proteins, carbohydrates, lipids, and hormones.

- Golgi complex: Also called the Golgi apparatus, this structure is responsible for manufacturing, warehousing, and shipping certain cellular products, particularly those from the endoplasmic reticulum (ER).

- Ribosomes: These organelles consist of RNA and proteins and are responsible for protein production. Ribosomes are found suspended in the cytosol or bound to the endoplasmic reticulum.

- Lysosomes: These membranous sacs of enzymes recycle the cell's organic material by digesting cellular macromolecules, such as nucleic acids, polysaccharides, fats, and proteins.

- Peroxisomes: Like lysosomes, peroxisomes are bound by a membrane and contain enzymes. Peroxisomes help to detoxify alcohol, form bile acid, and break down fats.

- Vacuole: These fluid-filled, enclosed structures are found most commonly in plant cells and fungi. Vacuoles are responsible for a wide variety of important functions in a cell including nutrient storage, detoxification, and waste exportation.

- Chloroplast: This chlorophyll containing plastid is found in plant cells, but not animal cells. Chloroplasts absorb the sun's light energy for photosynthesis.

- Cell Wall: This rigid outer wall is positioned next to the cell membrane in most plant cells. Not found in animal cells, the cell wall helps to provide support and protection for the cell.

- Centrioles: These cylindrical structures are found in animal cells, but not plant cells. Centrioles help to organize the assembly of microtubules during cell division.

- Cilia and Flagella: Cilia and flagella are protrusions from some cells that aid in cellular locomotion. They are formed from specialized groupings of microtubules called basal bodies.

Prokaryotic Cells

Prokaryotic cells have a structure that is less complex than eukaryotic cells since they are the most primitive and earliest forms of life on the planet. They do not have a nucleus or region where the

DNA is bound by a membrane. Prokaryotic DNA is coiled up in a region of the cytoplasm called the nucleoid. Like eukaryotic cells, prokaryotic cells contain a plasma membrane, cell wall, and cytoplasm. Unlike eukaryotic cells, prokaryotic cells do not contain membrane-bound organelles. However, they do contain some non-membranous organelles such as ribosomes, flagella, and plasmids (circular DNA structures that are not involved in reproduction). Examples of prokaryotic cells include bacteria and archaeans.

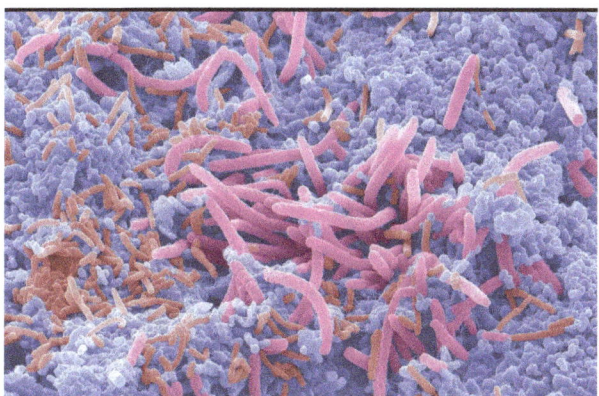

Prokaryotic cells like these bacteria on the tongue,
do not have membrane-based organelles.

ENDOMEMBRANE SYSTEM

The endomembrane system (endo- = "within") is a group of membranes and organelles in eukaryotic cells that works together to modify, package, and transport lipids and proteins. It includes a variety of organelles, such as the nuclear envelope and lysosomes, which you may already know, and the endoplasmic reticulum and Golgi apparatus, which we will cover shortly.

Although it's not technically inside the cell, the plasma membrane is also part of the endomembrane system. The plasma membrane interacts with the other endomembrane organelles, and it's the site where secreted proteins (like the pancreatic enzymes in the intro) are exported. Important note: the endomembrane system does not include mitochondria, chloroplasts, or peroxisomes.

Let's take a closer look at the different parts of the endomembrane system and how they function in the shipping of proteins and lipids.

Lysosomes

The lysosome is an organelle that contains digestive enzymes and acts as the organelle-recycling facility of an animal cell. It breaks down old and unnecessary structures so their molecules can be reused. Lysosomes are part of the endomembrane system, and some vesicles that leave the Golgi are bound for the lysosome.

Lysosomes can also digest foreign particles that are brought into the cell from outside. As an example, let's consider a class of white blood cells called macrophages, which are part of the human immune system. In a process known as phagocytosis, a section of the macrophage's plasma membrane invaginates—folds inward—to engulf a pathogen, as shown below.

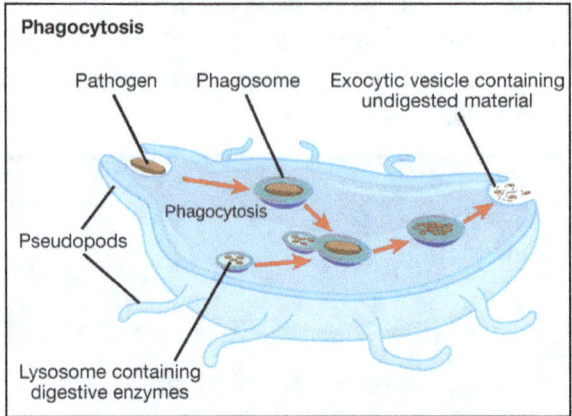

The endomembrane system and proteins.

The invaginated section, with the pathogen inside, pinches off from the plasma membrane to form a structure called a phagosome. The phagosome then fuses with a lysosome, forming a combined compartment where digestive enzymes destroy the pathogen.

Vacuoles

Plants cells are unique because they have a lysosome-like organelle called the vacuole. The large central vacuole stores water and wastes, isolates hazardous materials, and has enzymes that can break down macromolecules and cellular components, like those of a lysosome. Plant vacuoles also function in water balance and may be used to store compounds such as toxins and pigments (colored particles).

Lysosomes vs. Peroxisomes

One point that can be confusing is the difference between lysosomes and peroxisomes. Both types of organelles are involved in breaking down molecules and neutralizing hazards to the cell. Also, both usually show up as small, round blobs in diagrams.

However, the peroxisome is a different organelle with its own unique properties and role in the cell. It houses enzymes involved in oxidation reactions, which produce hydrogen peroxide (H_2O_2) as a by-product. The enzymes break down fatty acids and amino acids, and they also detoxify some substances that enter the body. For example, alcohol is detoxified by peroxisomes found in liver cells. Importantly, peroxisomes—unlike lysosomes—are not part of the endomembrane system. That means they don't receive vesicles from the Golgi apparatus.

ENDOPLASMIC RETICULUM

Endoplasmic reticulum is an organelle found in eukaryotic cells. About 50% of the total membrane surface in an animal cell is provided by endoplasmic reticulum (ER). The organelle called 'endoplasmic reticulum' occurs in both plants and animals and is a very important manufacturing site for lipids (fats) and many proteins. Many of these products are made for and exported to other organelles.

There are two types of endoplasmic reticulum: rough endoplasmic reticulum (rough ER) and smooth endoplasmic reticulum (smooth ER). Both types are present in plant and animal cells. The two types of ER often appear as if separate, but they are sub-compartments of the same organelle. Cells specialising in the production of proteins will tend to have a larger amount of rough ER whilst cells producing lipids (fats) and steroid hormones will have a greater amount of smooth ER.

Part of the ER is contiguous with the nuclear envelope. The Golgi apparatus is also closely associated with the ER and recent observations suggest that parts of the two organelles, i.e. the ER and the Golgi complex, are so close that some chemical products probably pass directly between them instead of being packaged into vesicles (droplets enclosed within a membrane) and transported to them through the cytoplasm.

Rough Endoplasmic Reticulum

This is an extensive organelle composed of greatly convoluted but flattish sealed sacs, which are contiguous with the nuclear membrane. It is called 'rough' endoplasmic reticulum because it is studded on its outer surface (the surface in contact with the cytosol) with ribosomes. These are called membrane bound ribosomes and are firmly attached to the outer cytosolic side of the ER About 13 million ribosomes are present on the RER in the average liver cell. Rough ER is found throughout the cell but the density is higher near the nucleus and the Golgi apparatus.

Ribosomes on the rough endoplasmic reticulum are called 'membrane bound' and are responsible for the assembly of many proteins. This process is called translation. Certain cells of the pancreas and digestive tract produce a high volume of protein as enzymes. Many of the proteins are produced in quantity in the cells of the pancreas and the digestive tract and function as digestive enzymes.

The rough ER working with membrane bound ribosomes takes polypeptides and amino acids from the cytosol and continues protein assembly including, at an early stage, recognising a 'destination label' attached to each of them. Proteins are produced for the plasma membrane, Golgi apparatus, secretory vesicles, plant vacuoles, lysosomes, endosomes and the endoplasmic reticulum itself. Some of the proteins are delivered into the lumen or space inside the ER whilst others are processed within the ER membrane itself. In the lumen some proteins have sugar groups added to them to form glycoproteins. Some have metal groups added to them. It is in the rough ER for example that four polypeptide chains are brought together to form haemoglobin.

Protein Folding Unit

It is in the lumen of the rough ER that proteins are folded to produce the highly important biochemical architecture which will provide 'lock and key' and other recognition and linking sites.

Protein Quality Control Section

It is also in the lumen that an amazing process of quality control checking is carried out. Proteins are subjected to a quality control check and any that are found to be incorrectly formed or incorrectly folded are rejected. These rejects are stored in the lumen or sent for recycling for eventual breakdown to amino acids. A type of emphysema (a lung problem) is caused by the ER quality control section continually rejecting an incorrectly folded protein. The protein is wrongly folded as a result of receiving an altered genetic message. The required protein is never exported from the lumen of rough ER. Research into protein structure failures relating to HIV are also focusing on reactions in the ER.

Rigorous Quality Control Plays a Part in Cystic Fibrosis

A form of cystic fibrosis is caused by a missing single amino acid, phenylanaline, in a particular position in the protein construction. The protein might work well without the amino acid but the very exacting service provided by the quality control section spots the error and rejects the protein retaining it in the lumen of the rough ER. In this case the customer (the person with cystic fibrosis) loses out completely due to high standards when a slightly poorer product would have been better than no product at all.

Rough ER to Golgi

In most cases proteins are transferred to the Golgi apparatus for 'finishing'. They are conveyed in vesicles or possibly directly between the ER and Golgi surfaces. After 'finishing' they are delivered to specific locations.

Smooth Endoplasmic Reticulum

Smooth ER is more tubular than rough ER and forms an interconnecting network sub-compartment of ER. It is found fairly evenly distributed throughout the cytoplasm. It is not studded with ribosomes hence 'smooth' ER.

Smooth ER is devoted almost exclusively to the manufacture of lipids and in some cases to the metabolism of them and associated products. In liver cells for example smooth ER enables glycogen that is stored as granules on the external surface of smooth ER to be broken down to glucose. Smooth ER is also involved in the production of steroid hormones in the adrenal cortex and endocrine glands.

Smooth ER – The Detox Stop

Smooth ER also plays a large part in detoxifying a number of organic chemicals converting them to safer water-soluble products. Large amounts of smooth ER are found in liver cells where one of its main functions is to detoxify products of natural metabolism and to endeavour to detoxify overloads of ethanol derived from excess alcoholic drinking and also barbiturates from drug overdose. To assist with this, smooth ER can double its surface area within a few days, returning to its normal size when the assault has subsided.

The contraction of muscle cells is triggered by the orderly release of calcium ions. These ions are released from the smooth endoplasmic reticulum.

NUCLEAR ENVELOPE

The nuclear membrane, also called the nuclear envelope, is a double membrane layer that separates the contents of the nucleus from the rest of the cell. It is found in both animal and plant cells. A cell has many jobs, such as building proteins, converting molecules into energy, and removing waste products. The nuclear envelope protects the cell's genetic material from the chemical reactions that take place outside the nucleus. It also contains many proteins that are used in organizing DNA and regulating genes.

Function of the Nuclear Membrane

The nuclear membrane is a barrier that physically protects the cell's DNA from the chemical reactions that are occurring elsewhere in the cell. If molecules that stay in the cytoplasm were to enter the nucleus, they could destroy part of the cell's DNA, which would stop it from functioning properly and could even lead to cell death. The envelope also contains a network of proteins that keep the genetic material in place inside the nucleus.

It also manages what materials can enter and exit the nucleus. It does so by being selectively permeable. Only certain proteins can physically pass through the double layer. This protects genetic information from mixing with other parts of the cell, and allows different cellular activities to occur inside the nucleus and outside the nucleus in the cytoplasm, where all other cellular structures are located.

Parts of the Nuclear Membrane

The nuclear membrane surrounds the nucleus of the cell.

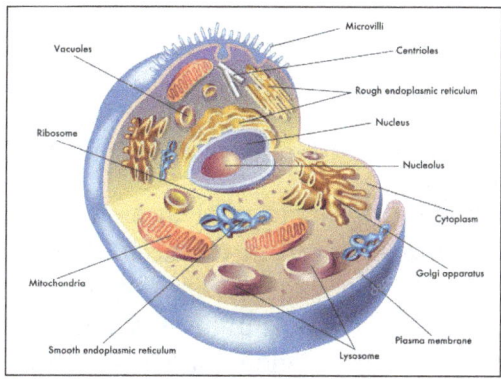
Diagram human cell nucleus.

Outer Membrane

Like the cell membrane, the nuclear membrane is a lipid bilayer, meaning that it consists of two layers of lipid molecules. The outer layer of lipids has ribosomes, structures that make proteins, on its surface. It is connected to the endoplasmic reticulum, a cell structure that packages and transports proteins.

Inner Membrane

The inner membrane contains proteins that help organize the nucleus and tether genetic material in place. This network of fibers and proteins attached to the inner membrane is called the nuclear lamina. It structurally supports the nucleus, plays a role in repairing DNA, and regulates events in the cell cycle such as cell division and the replication of DNA. The nuclear lamina is only found in animal cells, although plant cells may have some similar proteins on the inner membrane.

Nuclear Pores

Nuclear pores pass through both the outer and inner membranes of the nuclear membrane. They are made up of large complexes of proteins and allow certain molecules to pass through the nuclear membrane. Each nuclear pore is made up of about 30 different proteins that work together to transport materials. They also connect the outer and inner membranes.

During cell division, more nuclear pores are formed in the nuclear membrane in preparation for cell division. The nuclear membrane eventually breaks down and is reformed around the nuclei of each of the two daughter cells.

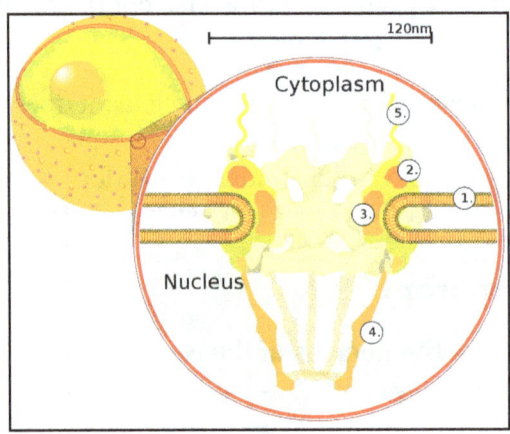

Nuclear Pores.

Differences between Nuclear Membranes in Plant and Animal Cells

Much more is known about animal and yeast cell nuclear membranes than those of plant cells, but the knowledge gap is decreasing thanks to recent research. Plant nuclear membranes lack many of the proteins that are found on the nuclear membranes of animal cells, but they have other pore membrane proteins that are unique to plants. Animal cells have centrosomes, structures that help organize DNA when the cell is preparing to divide; plants lack these structures and appear to rely entirely on the nuclear membrane for organization during cell division. With further research, scientists may better understand the uniqueness of plant cell nuclear membranes.

LYSOSOME

A lysosome is a membrane-bound organelle found in many animal cells. They are spherical vesicles that contain hydrolytic enzymes that can break down many kinds of biomolecules. A lysosome has a specific composition, of both its membrane proteins, and its lumenal proteins. The lumen's pH (~4.5–5.0) is optimal for the enzymes involved in hydrolysis, analogous to the activity of the stomach. Besides degradation of polymers, the lysosome is involved in various cell processes, including secretion, plasma membrane repair, cell signaling, and energy metabolism.

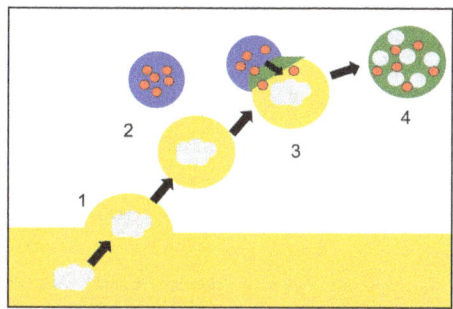

Lysosomes digest materials taken into the cell and recycle intracellular materials. Step one shows material entering a food vacuole through the plasma membrane, a process known as endocytosis. In step two a lysosome with an active hydrolytic enzyme comes into the picture as the food vacuole moves away from the plasma membrane. Step three consists of the lysosome fusing with the food vacuole and hydrolytic enzymes entering the food vacuole. In the final step, step four, hydrolytic enzymes digest the food particles.

Lysosomes act as the waste disposal system of the cell by digesting obsolete or un-used materials in the cytoplasm, from both inside and outside the cell. Material from outside the cell is taken-up through endocytosis, while material from the inside of the cell is digested through autophagy. The sizes of the organelles vary greatly—the larger ones can be more than 10 times the size of the smaller ones. They were discovered and named by Belgian biologist Christian de Duve, who eventually received the Nobel Prize in Physiology or Medicine in 1974.

Lysosomes are known to contain more than 60 different enzymes, and have more than 50 membrane proteins. Enzymes of the lysosomes are synthesised in the rough endoplasmic reticulum. The enzymes are imported from the Golgi apparatus in small vesicles, which fuse with larger acidic vesicles. Enzymes destined for a lysosome are specifically tagged with the molecule mannose 6-phosphate, so that they are properly sorted into acidified vesicles.

Synthesis of lysosomal enzymes is controlled by nuclear genes. Mutations in the genes for these enzymes are responsible for more than 30 different human genetic disorders, which are collectively known as lysosomal storage diseases. These diseases result from an accumulation of specific substrates, due to the inability to break them down. These genetic defects are related to several neurodegenerative disorders, cancers, cardiovascular diseases, and ageing-related diseases. Lysosomes should not be confused with liposomes, or with micelles.

Function and Structure

Lysosomes contain a variety of enzymes, enabling the cell to break down various biomolecules it engulfs, including peptides, nucleic acids, carbohydrates, and lipids (lysosomal lipase). The enzymes responsible for this hydrolysis require an acidic environment for optimal activity.

In addition to being able to break down polymers, lysosomes are capable of fusing with other organelles & digesting large structures or cellular debris; through cooperation with phagosomes, they are able to conduct autophagy, clearing out damaged structures. Similarly, they are able to break-down virus particles or bacteria in phagocytosis of macrophages.

The size of lysosomes varies from 0.1 μm to 1.2 μm. With a pH ranging from ~4.5–5.0, the interior of the lysosomes is acidic compared to the slightly basic cytosol (pH 7.2). The lysosomal membrane protects the cytosol, and therefore the rest of the cell, from the degradative enzymes within the lysosome. The cell is additionally protected from any lysosomal acid hydrolases that drain into the cytosol, as these enzymes are pH-sensitive and do not function well or at all in the alkaline environment of the cytosol. This ensures that cytosolic molecules and organelles are not destroyed in case there is leakage of the hydrolytic enzymes from the lysosome.

The lysosome maintains its pH differential by pumping in protons (H^+ ions) from the cytosol across the membrane via proton pumps and chloride ion channels. Vacuolar-ATPases are responsible for transport of protons, while the counter transport of chloride ions is performed by ClC-7 Cl^-/H^+ antiporter. In this way a steady acidic environment is maintained.

It sources its versatile capacity for degradation by import of enzymes with specificity for different substrates; cathepsins are the major class of hydrolytic enzymes, while lysosomal alpha-glucosidase is responsible for carbohydrates, and lysosomal acid phosphatase is necessary to release phosphate groups of phospholipids.

Formation

The lysosome is shown in purple, as an endpoint in Endocytotic sorting. AP2 is necessary for vesicle formation, whereas the Mannose-6-receptor is necessary for sorting Hydrolase into the Lysosome's lumen.

Many components of animal cells are recycled by transferring them inside or embedded in sections of membrane. For instance, in endocytosis (more specifically, macropinocytosis), a portion of the cell's plasma membrane pinches off to form vesicles that will eventually fuse with an organelle within the cell. Without active replenishment, the plasma membrane would continuously decrease in size. It is thought that lysosomes participate in this dynamic membrane exchange system and are formed by a gradual maturation process from endosomes.

The production of lysosomal proteins suggests one method of lysosome sustainment. Lysosomal protein genes are transcribed in the nucleus. mRNA transcripts exit the nucleus into the cytosol, where they are translated by ribosomes. The nascent peptide chains are translocated into the rough endoplasmic reticulum, where they are modified. Lysosomal soluble proteins exit the endoplasmic reticulum via CLN8-mediated recruitment in COPII-coated vesicles and enter the Golgi apparatus, where a specific lysosomal tag, mannose 6-phosphate, is added to the peptides. The presence of these tags allow for binding to mannose 6-phosphate receptors in the Golgi apparatus, a phenomenon that is crucial for proper packaging into vesicles destined for the lysosomal system.

Upon leaving the Golgi apparatus, the lysosomal enzyme-filled vesicle fuses with a late endosome, a relatively acidic organelle with an approximate pH of 5.5. This acidic environment causes dissociation of the lysosomal enzymes from the mannose 6-phosphate receptors. The enzymes are packed into vesicles for further transport to established lysosomes. The late endosome itself can eventually grow into a mature lysosome, as evidenced by the transport of endosomal membrane components from the lysosomes back to the endosomes.

Pathogen Entry

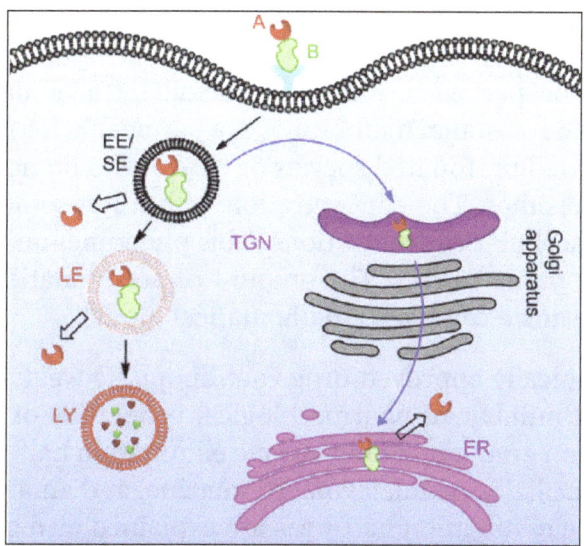

Cholera gaining entry into a cell via endocytosis.

As the endpoint of endocytosis, the lysosome also acts as a safeguard in preventing pathogens from being able to reach the cytoplasm before being degraded. Pathogens often hijack endocytotic pathways such as pinocytosis in order to gain entry into the cell. The lysosome prevents easy entry into the cell by hydrolyzing the biomolecules of pathogens necessary for their replication strategies; reduced Lysosomal activity results in an increase in viral infectivity, including HIV. In addition, AB_5 toxins such as cholera hijack the endosomal pathway while evading lysosomal degradation.

Clinical Significance

Lysosomes are involved in a group of genetically inherited deficiencies, or mutations called lysosomal storage diseases (LSD), inborn errors of metabolism caused by a dysfunction of one of the enzymes. The rate of incidence is estimated to be 1 in 5,000 births, and the true figure expected to be higher as many cases are likely to be undiagnosed or misdiagnosed. The primary cause is deficiency of an acid hydrolase. Other conditions are due to defects in lysosomal membrane proteins that fail to transport the enzyme, non-enzymatic soluble lysosomal proteins. The initial effect of such disorders is accumulation of specific macromolecules or monomeric compounds inside the endosomal–autophagic–lysosomal system. This results in abnormal signaling pathways, calcium homeostasis, lipid biosynthesis and degradation and intracellular trafficking, ultimately leading to pathogenetic disorders. The organs most affected are brain, viscera, bone and cartilage.

There is no direct medical treatment to cure LSDs. The most common LSD is Gaucher's disease, which is due to deficiency of the enzyme glucocerebrosidase. Consequently, the enzyme substrate, the fatty acid glucosylceramide accumulates, particularly in white blood cells, which in turn affects spleen, liver, kidneys, lungs, brain and bone marrow. The disease is characterized by bruises, fatigue, anaemia, low blood platelets, osteoporosis, and enlargement of the liver and spleen. As of 2017, enzyme replacement therapy is available for treating 8 of the 50-60 known LDs.

The most severe and rarely found, lysosomal storage disease is inclusion cell disease.

Metachromatic leukodystrophy is another lysosomal storage disease that also affects sphingolipid metabolism.

Lysosomotropism

Weak bases with lipophilic properties accumulate in acidic intracellular compartments like lysosomes. While the plasma and lysosomal membranes are permeable for neutral and uncharged species of weak bases, the charged protonated species of weak bases do not permeate biomembranes and accumulate within lysosomes. The concentration within lysosomes may reach levels 100 to 1000 fold higher than extracellular concentrations. This phenomenon is called lysosomotropism, "acid trapping" or "proton pump" effect. The amount of accumulation of lysosomotropic compounds may be estimated using a cell-based mathematical model.

A significant part of the clinically approved drugs are lipophilic weak bases with lysosomotropic properties. This explains a number of pharmacological properties of these drugs, such as high tissue-to-blood concentration gradients or long tissue elimination half-lifes; these properties have been found for drugs such as haloperidol, levomepromazine, and amantadine. However, high tissue concentrations and long elimination half-lives are explained also by lipophilicity and absorption of drugs to fatty tissue structures. Important lysosomal enzymes, such as acid sphingomyelinase, may be inhibited by lysosomally accumulated drugs. Such compounds are termed FIASMAs (functional inhibitor of acid sphingomyelinase) and include for example fluoxetine, sertraline, or amitriptyline.

Ambroxol is a lysosomotropic drug of clinical use to treat conditions of productive cough for its mucolytic action. Ambroxol triggers the exocytosis of lysosomes via neutralization of lysosomal pH and calcium release from acidic calcium stores. Presumably for this reason, Ambroxol was also found to improve cellular function in some disease of lysosomal origin such as Parkinson's or lysosomal storage disease.

Systemic Lupus Erythematosus

Impaired lysosome function is prominent in systemic lupus erythematosus preventing macrophages and monocytes from degrading neutrophil extracellular traps and immune complexes. The failure to degrade internalized immune complexes stems from chronic mTORC2 activity, which impairs lysosome acidification. As a result, immune complexes in the lysosome recycle to the surface of macrophages causing an accumulation of nuclear antigens upstream of multiple lupus-associated pathologies.

ENDOSOME

In cell biology, an endosome is a membrane-bound compartment inside eukaryotic cells. It is a compartment of the endocytic membrane transport pathway originating from the *trans* Golgi membrane. Molecules or ligands internalized from the plasma membrane can follow this pathway all the way to lysosomes for degradation, or they can be recycled back to the plasma membrane. Molecules are also transported to endosomes from the *trans*-Golgi network and either continue to lysosomes or recycle back to the Golgi. Endosomes can be classified as early, sorting, or late depending on their stage post internalization. Endosomes represent a major sorting compartment of the endomembrane system in cells. In HeLa cells, endosomes are approximately 500 nm in diameter when fully mature.

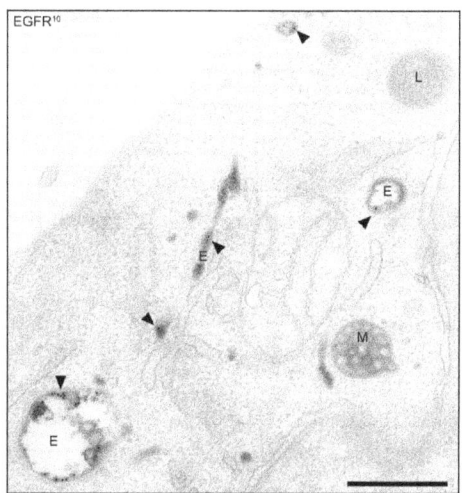

Electron micrograph of endosomes in human HeLa cells. Early endosomes (E - labeled for EGFR, 5 minutes after internalisation, and transferrin), late endosomes/MVBs (M) and lysosomes (L) are visible. Bar, 500 nm.

Function

Endosomes provide an environment for material to be sorted before it reaches the degradative lysosome. For example, LDL is taken into the cell by binding to the LDL receptor at the cell surface. Upon reaching early endosomes, the LDL dissociates from the receptor, and the receptor can be recycled to the cell surface. The LDL remains in the endosome and is delivered to lysosomes for processing. LDL dissociates because of the slightly acidified environment of the early endosome, generated by a vacuolar membrane proton pump V-ATPase. On the other hand, EGF and the EGF receptor have a pH-resistant bond that persists until it is delivered to lysosomes for their degradation. The mannose 6-phosphate receptor carries ligands from the Golgi destined for the lysosome by a similar mechanism.

Types

There are three different types of endosomes: *early endosomes*, *late endosomes*, and *recycling endosomes*. They are distinguished by the time it takes for endocytosed material to reach them, and by markers such as rabs. They also have different morphology. Once endocytic vesicles have uncoated, they fuse with early endosomes. Early endosomes then *mature* into late endosomes before fusing with lysosomes.

Early endosomes mature in several ways to form late endosomes. They become increasingly acidic mainly through the activity of the V-ATPase. Many molecules that are recycled are removed by concentration in the tubular regions of early endosomes. Loss of these tubules to recycling pathways means that late endosomes mostly lack tubules. They also increase in size due to the homotypic fusion of early endosomes into larger vesicles. Molecules are also sorted into smaller vesicles that bud from the perimeter membrane into the endosome lumen, forming *lumenal vesicles*; this leads to the multivesicular appearance of late endosomes and so they are also known as *multivesicular bodies* (MVBs). Removal of recycling molecules such as transferrin receptors and mannose 6-phosphate receptors continues during this period, probably via budding of vesicles out of endosomes. Finally, the endosomes lose RAB5A and acquire RAB7A, making them competent for fusion with lysosomes.

Fusion of late endosomes with lysosomes has been shown to result in the formation of a 'hybrid' compartment, with characteristics intermediate of the two source compartments. For example, lysosomes are more dense than late endosomes, and the hybrids have an intermediate density. Lysosomes reform by recondensation to their normal, higher density. However, before this happens, more late endosomes may fuse with the hybrid.

Some material recycles to the plasma membrane directly from early endosomes, but most traffics via recycling endosomes.

- Early endosomes consist of a dynamic tubular-vesicular network (vesicles up to 1 μm in diameter with connected tubules of approx. 50 nm diameter). Markers include RAB5A and RAB4, Transferrin and its receptor and EEA1.

- Late endosomes, also known as MVBs, are mainly spherical, lack tubules, and contain many close-packed lumenal vesicles. Markers include RAB7, RAB9, and mannose 6-phosphate receptors.

- Recycling endosomes are concentrated at the microtubule organizing center and consist of a mainly tubular network. Marker; RAB11.

More subtypes exist in specialized cells such as polarized cells and macrophages. Phagosomes, macropinosomes and autophagosomes mature in a manner similar to endosomes, and may require fusion with normal endosomes for their maturation. Some intracellular pathogens subvert this process, for example, by preventing RAB7 acquisition.

Late endosomes/MVBs are sometimes called endocytic carrier vesicles, but this term was used to describe vesicles that bud from early endosomes and fuse with late endosomes. However, several observations have now demonstrated that it is more likely that transport between these two compartments occurs by a maturation process, rather than vesicle transport.

Another unique identifying feature that differs between the various classes of endosomes is the lipid composition in their membranes. Phosphatidyl inositol phosphates (PIPs), one of the most important lipid signaling molecules, is found to differ as the endosomes mature from early to late. $PI(4,5)P_2$ is present on plasma membranes, $PI(3)P$ on early endosomes, $PI(3,5)P_2$ on late endosomes and $PI(4)P$ on the trans Golgi network. These lipids on the surface of the endosomes help in the specific recruitment of proteins from the cytosol, thus providing them an identity. The inter-conversion of these lipids is a result of the concerted action of phosphoinositide kinases and phosphatases that are strategically localized.

Pathways

Diagram of the pathways that intersect endosomes in the endocytic pathway of animal cells. Examples of molecules that follow some of the pathways are shown, including receptors for EGF, transferrin, and lysosomal hydrolases. Recycling endosomes, and compartments and pathways found in more specialized cells, are not shown.

There are three main compartments that have pathways that connect with endosomes. More pathways exist in specialized cells, such as melanocytes and polarized cells. For example, in epithelial cells, a special process called transcytosis allows some materials to enter one side of a cell and exit from the opposite side. Also, in some circumstances, late endosomes/MVBs fuse with the plasma membrane instead of with lysosomes, releasing the lumenal vesicles, now called exosomes, into the extracellular medium.

There is no consensus as to the exact nature of these pathways, and the sequential route taken by any given cargo in any given situation will tend to be a matter of debate.

Golgi to/from Endosomes

Vesicles pass between the Golgi and endosomes in both directions. The GGAs and AP-1 clathrin-coated vesicle adaptors make vesicles at the Golgi that carry molecules to endosomes. In the opposite direction, retromer generates vesicles at early endosomes that carry molecules back to the Golgi. Some studies describe a retrograde traffic pathway from late endosomes to the Golgi that is mediated by Rab9 and TIP47, but other studies dispute these findings. Molecules that follow these pathways include the mannose-6-phosphate receptors that carry lysosomal hydrolases to the endocytic pathway. The hydrolases are released in the acidic environment of endosomes, and the receptor is retrieved to the Golgi by retromer and Rab9.

Plasma Membrane to/from Early Endosomes (Via Recycling Endosomes)

Molecules are delivered from the plasma membrane to early endosomes in endocytic vesicles. Molecules can be internalized via receptor-mediated endocytosis in clathrin-coated vesicles. Other types of vesicles also form at the plasma membrane for this pathway, including ones utilising caveolin. Vesicles also transport molecules directly back to the plasma membrane, but many molecules are transported in vesicles that first fuse with recycling endosomes. Molecules following this recycling pathway are concentrated in the tubules of early endosomes. Molecules that follow these pathways include the receptors for LDL, the growth factor EGF, and the iron transport protein transferrin. Internalization of these receptors from the plasma membrane occurs by receptor-mediated endocytosis. LDL is released in endosomes because of the lower pH, and the receptor is recycled to the cell surface. Cholesterol is carried in the blood primarily by (LDL), and transport by the LDL receptor is the main mechanism by which cholesterol is taken up by cells. EGFRs are activated when EGF binds. The activated receptors stimulate their own internalization and degradation in lysosomes. EGF remains bound to the EGFR once it is endocytosed to endosomes. The activated EGFRs stimulate their own ubiquitination, and this directs them to lumenal vesicles and so they are not recycled to the plasma membrane. This removes the signaling portion of the protein from the cytosol and thus prevents continued stimulation of growth - in cells not stimulated with EGF, EGFRs have no EGF bound to them and therefore recycle if they reach endosomes.

Transferrin also remains associated with its receptor, but, in the acidic endosome, iron is released from the transferrin, and then the iron-free transferrin (still bound to the transferrin receptor) returns from the early endosome to the cell surface, both directly and via recycling endosomes.

Late Endosomes to Lysosomes

Transport from late endosomes to lysosomes is, in essence, unidirectional, since a late endosome is "consumed" in the process of fusing with a lysosome. Hence, soluble molecules in the lumen of endosomes will tend to end up in lysosomes, unless they are retrieved in some way. Transmembrane proteins can be delivered to the perimeter membrane or the lumen of lysosomes. Transmembrane proteins destined for the lysosome lumen are sorted into the vesicles that bud from the perimeter membrane into endosomes, a process that begins in early endosomes. When the endosome has matured into a late endosome/MVB and fuses with a lysosome, the vesicles in the lumen are delivered to the lysosome lumen. Proteins are marked for this pathway by the addition of ubiquitin. The endosomal sorting complexes required for transport (ESCRTs) recognise this ubiquitin and sort the protein into the forming lumenal vesicles. Molecules that follow these pathways include LDL and the lysosomal hydrolases delivered by mannose-6-phosphate receptors. These soluble molecules remain in endosomes and are therefore delivered to lysosomes. Also, the transmembrane EGFRs, bound to EGF, are tagged with ubiquitin and are therefore sorted into lumenal vesicles by the ESCRTs.

CELL NUCLEUS

In cell biology, the nucleus is a membrane-bound organelle found in eukaryotic cells. Eukaryotes usually have a single nucleus, but a few cell types, such as mammalian red blood cells, have no nuclei, and a few others including osteoclasts have many.

The cell nucleus contains all of the cell's genome, except for a small fraction of mitochondrial DNA, organized as multiple long linear DNA molecules in a complex with a large variety of proteins, such as histones, to form chromosomes. The genes within these chromosomes are structured in such a way to promote cell function. The nucleus maintains the integrity of genes and controls the activities of the cell by regulating gene expression—the nucleus is, therefore, the control center of the cell. The main structures making up the nucleus are the nuclear envelope, a double membrane that encloses the entire organelle and isolates its contents from the cellular cytoplasm, and the nuclear matrix (which includes the nuclear lamina), a network within the nucleus that adds mechanical support, much like the cytoskeleton, which supports the cell as a whole.

Because the nuclear envelope is impermeable to large molecules, nuclear pores are required to regulate nuclear transport of molecules across the envelope. The pores cross both nuclear membranes, providing a channel through which larger molecules must be actively transported by carrier proteins while allowing free movement of small molecules and ions. Movement of large molecules such as proteins and RNA through the pores is required for both gene expression and the maintenance of chromosomes. Although the interior of the nucleus does not contain any membrane-bound subcompartments, its contents are not uniform, and a number of nuclear bodies exist, made up of unique proteins, RNA molecules, and particular parts of the chromosomes. The best-known of these is the nucleolus, which is mainly involved in the assembly of ribosomes. After being produced in the nucleolus, ribosomes are exported to the cytoplasm where they translate mRNA.

Structures

The nucleus is the largest organelle in animal cells. In mammalian cells, the average diameter of the nucleus is approximately 6 micrometres (μm), which occupies about 10% of the total cell volume. The contents of the nucleus are held in the nucleoplasm similar to the cytoplasm in the rest of the cell. The fluid component of this is termed the nucleosol, similar to the cytosol in the cytoplasm.

In most types of granulocyte, a white blood cell, the nucleus is lobated and can be bi-lobed, tri-lobed or multi-lobed.

Nuclear Envelope and Pores

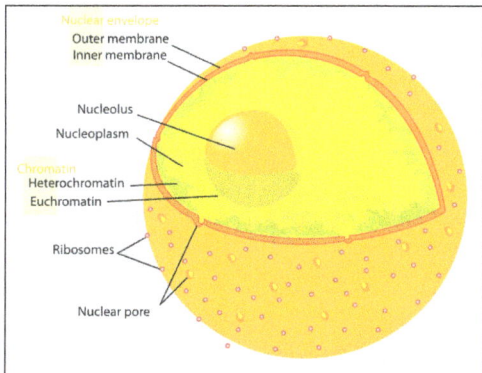

The eukaryotic cell nucleus. Visible in this diagram are the ribosome-studded double membranes of the nuclear envelope, the DNA (complexed as chromatin), and the nucleolus. Within the cell nucleus is a viscous liquid called nucleoplasm, similar to the cytoplasm found outside the nucleus.

The nuclear envelope, otherwise known as nuclear membrane, consists of two cellular membranes, an inner and an outer membrane, arranged parallel to one another and separated by 10 to 50 nanometres (nm). The nuclear envelope completely encloses the nucleus and separates the cell's genetic material from the surrounding cytoplasm, serving as a barrier to prevent macromolecules from diffusing freely between the nucleoplasm and the cytoplasm. The outer nuclear membrane is continuous with the membrane of the rough endoplasmic reticulum (RER), and is similarly studded with ribosomes. The space between the membranes is called the perinuclear space and is continuous with the RER lumen.

Nuclear pores, which provide aqueous channels through the envelope, are composed of multiple proteins, collectively referred to as nucleoporins. The pores are about 125 million daltons in molecular weight and consist of around 50 (in yeast) to several hundred proteins (in vertebrates). The pores are 100 nm in total diameter; however, the gap through which molecules freely diffuse is only about 9 nm wide, due to the presence of regulatory systems within the center of the pore. This size selectively allows the passage of small water-soluble molecules while preventing larger molecules, such as nucleic acids and larger proteins, from inappropriately entering or exiting the nucleus. These large molecules must be actively transported into the nucleus instead. The nucleus of a typical mammalian cell will have about 3000 to 4000 pores throughout its envelope, each of which contains an eightfold-symmetric ring-shaped structure at a position where the inner and outer membranes fuse. Attached to the ring is a structure called the *nuclear basket* that extends into the nucleoplasm, and a series of filamentous extensions that reach into the cytoplasm. Both structures serve to mediate binding to nuclear transport proteins.

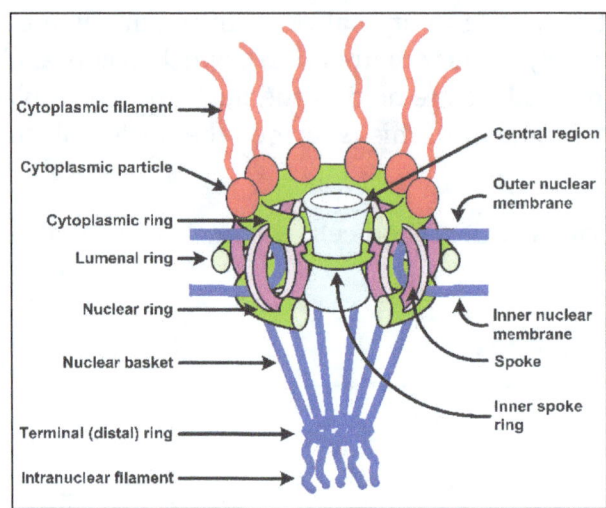

A cross section of a nuclear pore on the surface of the nuclear envelope.

Most proteins, ribosomal subunits, and some DNAs are transported through the pore complexes in a process mediated by a family of transport factors known as karyopherins. Those karyopherins that mediate movement into the nucleus are also called importins, whereas those that mediate movement out of the nucleus are called exportins. Most karyopherins interact directly with their cargo, although some use adaptor proteins. Steroid hormones such as cortisol and aldosterone, as well as other small lipid-soluble molecules involved in intercellular signaling, can diffuse through the cell membrane and into the cytoplasm, where they bind nuclear receptor proteins that are trafficked into the nucleus. There they serve as transcription factors when bound to their ligand; in the absence of a ligand, many such receptors function as histone deacetylases that repress gene expression.

Nuclear Lamina

In animal cells, two networks of intermediate filaments provide the nucleus with mechanical support: The nuclear lamina forms an organized meshwork on the internal face of the envelope, while less organized support is provided on the cytosolic face of the envelope. Both systems provide structural support for the nuclear envelope and anchoring sites for chromosomes and nuclear pores.

The nuclear lamina is composed mostly of lamin proteins. Like all proteins, lamins are synthesized in the cytoplasm and later transported to the nucleus interior, where they are assembled before being incorporated into the existing network of nuclear lamina. Lamins found on the cytosolic face of the membrane, such as emerin and nesprin, bind to the cytoskeleton to provide structural support. Lamins are also found inside the nucleoplasm where they form another regular structure, known as the nucleoplasmic veil, that is visible using fluorescence microscopy. The actual function of the veil is not clear, although it is excluded from the nucleolus and is present during interphase.

Lamin structures that make up the veil, such as LEM3, bind chromatin and disrupting their structure inhibits transcription of protein-coding genes.

Like the components of other intermediate filaments, the lamin monomer contains an alpha-helical domain used by two monomers to coil around each other, forming a dimer structure called a coiled coil. Two of these dimer structures then join side by side, in an antiparallel arrangement, to form a tetramer called a *protofilament*. Eight of these protofilaments form a lateral arrangement that is twisted to form a ropelike *filament*. These filaments can be assembled or disassembled in a dynamic manner, meaning that changes in the length of the filament depend on the competing rates of filament addition and removal.

Mutations in lamin genes leading to defects in filament assembly cause a group of rare genetic disorders known as *laminopathies*. The most notable laminopathy is the family of diseases known as progeria, which causes the appearance of premature aging in its sufferers. The exact mechanism by which the associated biochemical changes give rise to the aged phenotype is not well understood.

Chromosomes

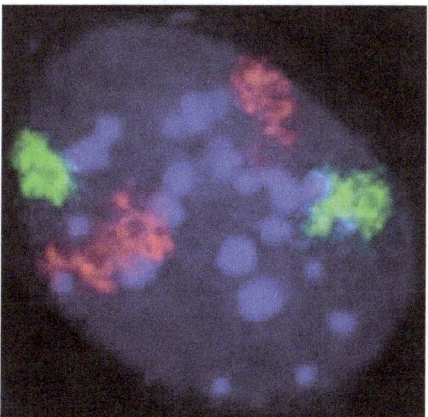

A mouse fibroblast nucleus in which DNA is stained blue. The distinct chromosome territories of chromosome 2 (red) and chromosome 9 (green) are stained with fluorescent in situ hybridization.

The cell nucleus contains the majority of the cell's genetic material in the form of multiple linear DNA molecules organized into structures called chromosomes. Each human cell contains roughly two meters of DNA. During most of the cell cycle these are organized in a DNA-protein complex known as chromatin, and during cell division the chromatin can be seen to form the well-defined chromosomes familiar from a karyotype. A small fraction of the cell's genes are located instead in the mitochondria.

There are two types of chromatin. Euchromatin is the less compact DNA form, and contains genes that are frequently expressed by the cell. The other type, heterochromatin, is the more compact form, and contains DNA that is infrequently transcribed. This structure is further categorized into *facultative* heterochromatin, consisting of genes that are organized as heterochromatin only in certain cell types or at certain stages of development, and *constitutive* heterochromatin that consists of chromosome structural components such as telomeres and centromeres. During interphase the chromatin organizes itself into discrete individual patches, called *chromosome territories*. Active genes, which are generally found in the euchromatic region of the chromosome, tend to be located towards the chromosome's territory boundary.

Antibodies to certain types of chromatin organization, in particular, nucleosomes, have been associated with a number of autoimmune diseases, such as systemic lupus erythematosus. These are known as anti-nuclear antibodies (ANA) and have also been observed in concert with multiple sclerosis as part of general immune system dysfunction. As in the case of progeria, the role played by the antibodies in inducing the symptoms of autoimmune diseases is not obvious.

Nucleolus

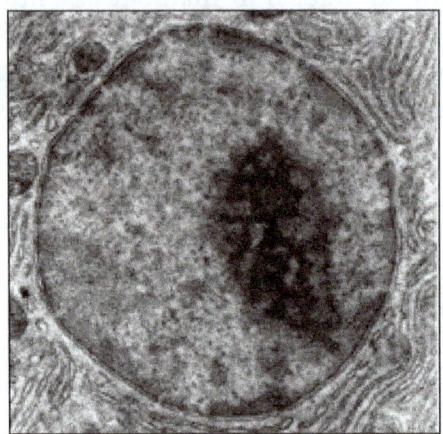

An electron micrograph of a cell nucleus,
showing the darkly stained nucleolus.

The nucleolus is the largest of the discrete densely stained, membraneless structures known as nuclear bodies found in the nucleus. It forms around tandem repeats of rDNA, DNA coding for ribosomal RNA (rRNA). These regions are called nucleolar organizer regions (NOR). The main roles of the nucleolus are to synthesize rRNA and assemble ribosomes. The structural cohesion of the nucleolus depends on its activity, as ribosomal assembly in the nucleolus results in the transient association of nucleolar components, facilitating further ribosomal assembly, and hence further association. This model is supported by observations that inactivation of rDNA results in intermingling of nucleolar structures.

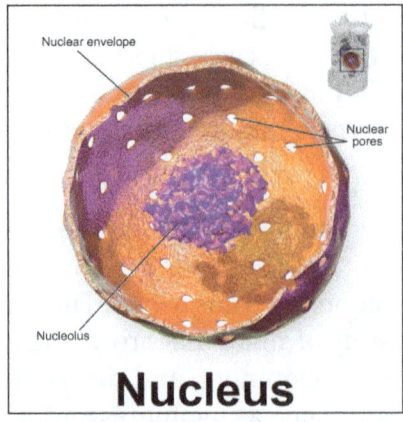

3D rendering of nucleus with location of nucleolus.

In the first step of ribosome assembly, a protein called RNA polymerase I transcribes rDNA, which forms a large pre-rRNA precursor. This is cleaved into the subunits 5.8S, 18S, and 28S rRNA. The transcription, post-transcriptional processing, and assembly of rRNA occurs in the nucleolus, aided by small nucleolar RNA (snoRNA) molecules, some of which are derived from spliced introns from messenger RNAs encoding genes related to ribosomal function. The assembled ribosomal subunits are the largest structures passed through the nuclear pores.

When observed under the electron microscope, the nucleolus can be seen to consist of three distinguishable regions: the innermost fibrillar centers (FCs), surrounded by the *dense* fibrillar component (DFC) (that contains fibrillarin and nucleolin), which in turn is bordered by the granular component (GC) (that contains the protein nucleophosmin). Transcription of the rDNA occurs either in the FC or at the FC-DFC boundary, and, therefore, when rDNA transcription in the cell is increased, more FCs are detected. Most of the cleavage and modification of rRNAs occurs in the DFC, while the latter steps involving protein assembly onto the ribosomal subunits occur in the GC.

Other Nuclear Bodies

Subnuclear structure sizes	
Structure name	Structure diameter
Cajal bodies	0.2–2.0 µm
Clastosomes	0.2-0.5 µm
PIKA	5 µm
PML bodies	0.2–1.0 µm
Paraspeckles	0.5–1.0 µm
Speckles	20–25 nm

Besides the nucleolus, the nucleus contains a number of other nuclear bodies. These include Cajal bodies, gemini of Cajal bodies, polymorphic interphase karyosomal association (PIKA), promyelocytic leukaemia (PML) bodies, paraspeckles, and splicing speckles. Although little is known about a number of these domains, they are significant in that they show that the nucleoplasm is not a uniform mixture, but rather contains organized functional subdomains.

Other subnuclear structures appear as part of abnormal disease processes. For example, the presence of small intranuclear rods has been reported in some cases of nemaline myopathy. This condition typically results from mutations in actin, and the rods themselves consist of mutant actin as well as other cytoskeletal proteins.

Cajal Bodies and Gems

A nucleus typically contains between 1 and 10 compact structures called Cajal bodies or coiled bodies (CB), whose diameter measures between 0.2 µm and 2.0 µm depending on the cell type and species. When seen under an electron microscope, they resemble balls of tangled thread and are dense foci of distribution for the protein coilin. CBs are involved in a number of different roles relating to RNA processing, specifically small nucleolar RNA (snoRNA) and small nuclear RNA (snRNA) maturation, and histone mRNA modification.

Similar to Cajal bodies are Gemini of Cajal bodies, or gems, whose name is derived from the Gemini constellation in reference to their close "twin" relationship with CBs. Gems are similar in size and shape to CBs, and in fact are virtually indistinguishable under the microscope. Unlike CBs, gems do not contain small nuclear ribonucleoproteins (snRNPs), but do contain a protein called survival of motor neuron (SMN) whose function relates to snRNP biogenesis. Gems are believed to assist CBs in snRNP biogenesis, though it has also been suggested from microscopy evidence that CBs and gems are different manifestations of the same structure. Later ultrastructural studies have shown gems to be twins of Cajal bodies with the difference being in the coilin component; Cajal bodies are SMN positive and coilin positive, and gems are SMN positive and coilin negative.

PIKA and PTF Domains

PIKA domains, or polymorphic interphase karyosomal associations, were first described in microscopy studies in 1991. Their function remains unclear, though they were not thought to be associated with active DNA replication, transcription, or RNA processing. They have been found to often associate with discrete domains defined by dense localization of the transcription factor PTF, which promotes transcription of small nuclear RNA (snRNA).

PML Bodies

Promyelocytic leukemia bodies (PML bodies) are spherical bodies found scattered throughout the nucleoplasm, measuring around 0.1–1.0 µm. They are known by a number of other names, including nuclear domain 10 (ND10), Kremer bodies, and PML oncogenic domains. PML bodies are named after one of their major components, the promyelocytic leukemia protein (PML). They are often seen in the nucleus in association with Cajal bodies and cleavage bodies. PML bodies belong to the nuclear matrix, an ill-defined super-structure of the nucleus proposed to anchor and regulate many nuclear functions, including DNA replication, transcription, or epigenetic silencing. The PML protein is the key organizer of these domains that recruits an ever-growing number of proteins, whose only common known feature to date is their ability to be SUMOylated. Yet, pml-/- mice (which have their PML gene deleted) cannot assemble nuclear bodies, develop normally and live well, demonstrating that PML bodies are dispensable for most basic biological functions.

Splicing Speckles

Speckles are subnuclear structures that are enriched in pre-messenger RNA splicing factors and are located in the interchromatin regions of the nucleoplasm of mammalian cells. At the fluorescence-microscope level they appear as irregular, punctate structures, which vary in size and shape, and when examined by electron microscopy they are seen as clusters of interchromatin granules. Speckles are dynamic structures, and both their protein and RNA-protein components can cycle continuously between speckles and other nuclear locations, including active transcription sites. Studies on the composition, structure and behaviour of speckles have provided a model for understanding the functional compartmentalization of the nucleus and the organization of the gene-expression machinery splicing snRNPs and other splicing proteins necessary for pre-mRNA processing. Because of a cell's changing requirements, the composition and location of these bodies changes according to mRNA transcription and regulation via phosphorylation of specific proteins. The splicing speckles are also known as nuclear speckles (nuclear specks), splicing factor compartments (SF compartments), interchromatin granule clusters (IGCs), and B snurposomes. B snurposomes are found in the amphibian oocyte nuclei and in *Drosophila melanogaster* embryos. B snurposomes appear alone or attached to the Cajal bodies in the electron micrographs of the amphibian nuclei. IGCs function as storage sites for the splicing factors.

Paraspeckles

Paraspeckles are irregularly shaped compartments in the interchromatin space of the nucleus. First documented in HeLa cells, where there are generally 10–30 per nucleus, paraspeckles are now known to also exist in all human primary cells, transformed cell lines, and tissue sections. Their name is derived from their distribution in the nucleus; the "para" is short for parallel and the "speckles" refers to the splicing speckles to which they are always in close proximity.

Paraspeckles are dynamic structures that are altered in response to changes in cellular metabolic activity. They are transcription dependent and in the absence of RNA Pol II transcription, the paraspeckle disappears and all of its associated protein components (PSP1, p54nrb, PSP2, CFI(m)68, and PSF) form a crescent shaped perinucleolar cap in the nucleolus. This phenomenon is demonstrated during the cell cycle. In the cell cycle, paraspeckles are present during interphase and during all of mitosis except for telophase. During telophase, when the two daughter nuclei are formed, there is no RNA Pol II transcription so the protein components instead form a perinucleolar cap.

Perichromatin Fibrils

Perichromatin fibrils are visible only under electron microscope. They are located next to the transcriptionally active chromatin and are hypothesized to be the sites of active pre-mRNA processing.

Clastosomes

Clastosomes are small nuclear bodies (0.2–0.5 μm) described as having a thick ring-shape due to the peripheral capsule around these bodies. Clastosomes are not typically present in normal cells, making them hard to detect. They form under high proteolytic conditions within the nucleus and degrade once there is a decrease in activity or if cells are treated with proteasome inhibitors. The scarcity of clastosomes in cells indicates that they are not required for proteasome function. Osmotic stress has also been shown to cause the formation of clastosomes. These nuclear bodies contain catalytic and regulatory sub-units of the proteasome and its substrates, indicating that clastosomes are sites for degrading proteins.

Function

The nucleus provides a site for genetic transcription that is segregated from the location of translation in the cytoplasm, allowing levels of gene regulation that are not available to prokaryotes. The main function of the cell nucleus is to control gene expression and mediate the replication of DNA during the cell cycle.

The nucleus is an organelle found in eukaryotic cells. Inside its fully enclosed nuclear membrane, it contains the majority of the cell's genetic material. This material is organized as DNA molecules, along with a variety of proteins, to form chromosomes.

Cell Compartmentalization

The nuclear envelope allows the nucleus to control its contents, and separate them from the rest of the cytoplasm where necessary. This is important for controlling processes on either side of the nuclear membrane. In most cases where a cytoplasmic process needs to be restricted, a key participant is removed to the nucleus, where it interacts with transcription factors to downregulate the production of certain enzymes in the pathway. This regulatory mechanism occurs in the case of glycolysis, a cellular pathway for breaking down glucose to produce energy. Hexokinase is an enzyme responsible for the first the step of glycolysis, forming glucose-6-phosphate from glucose. At high concentrations of fructose-6-phosphate, a molecule made later from glucose-6-phosphate, a regulator protein removes hexokinase to the nucleus, where it forms a transcriptional repressor complex with nuclear proteins to reduce the expression of genes involved in glycolysis.

In order to control which genes are being transcribed, the cell separates some transcription factor proteins responsible for regulating gene expression from physical access to the DNA until they are activated by other signaling pathways. This prevents even low levels of inappropriate gene expression. For example, in the case of NF-κB-controlled genes, which are involved in most inflammatory responses, transcription is induced in response to a signal pathway such as that initiated by the signaling molecule TNF-α, binds to a cell membrane receptor, resulting in the recruitment of signalling proteins, and eventually activating the transcription factor NF-κB. A nuclear localisation signal on the NF-κB protein allows it to be transported through the nuclear pore and into the nucleus, where it stimulates the transcription of the target genes.

The compartmentalization allows the cell to prevent translation of unspliced mRNA. Eukaryotic mRNA contains introns that must be removed before being translated to produce functional proteins. The splicing is done inside the nucleus before the mRNA can be accessed by ribosomes for translation. Without the nucleus, ribosomes would translate newly transcribed (unprocessed) mRNA, resulting in malformed and nonfunctional proteins.

Gene Expression

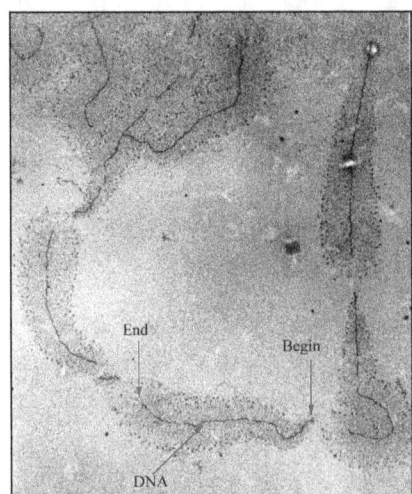

A micrograph of ongoing gene transcription of ribosomal RNA illustrating the growing primary transcripts. "Begin" indicates the 5' end of the DNA, where new RNA synthesis begins; "end" indicates the 3' end, where the primary transcripts are almost complete.

Gene expression first involves transcription, in which DNA is used as a template to produce RNA. In the case of genes encoding proteins, that RNA produced from this process is messenger RNA (mRNA), which then needs to be translated by ribosomes to form a protein. As ribosomes are located outside the nucleus, mRNA produced needs to be exported.

Since the nucleus is the site of transcription, it also contains a variety of proteins that either directly mediate transcription or are involved in regulating the process. These proteins include helicases, which unwind the double-stranded DNA molecule to facilitate access to it, RNA polymerases, which bind to the DNA promoter to synthesize the growing RNA molecule, topoisomerases, which change the amount of supercoiling in DNA, helping it wind and unwind, as well as a large variety of transcription factors that regulate expression.

Processing of Pre-mRNA

Newly synthesized mRNA molecules are known as primary transcripts or pre-mRNA. They must undergo post-transcriptional modification in the nucleus before being exported to the cytoplasm; mRNA that appears in the cytoplasm without these modifications is degraded rather than used for protein translation. The three main modifications are 5' capping, 3' polyadenylation, and RNA splicing. While in the nucleus, pre-mRNA is associated with a variety of proteins in complexes known as heterogeneous ribonucleoprotein particles (hnRNPs). Addition of the 5' cap occurs co-transcriptionally and is the first step in post-transcriptional modification. The 3' poly-adenine tail is only added after transcription is complete.

RNA splicing, carried out by a complex called the spliceosome, is the process by which introns, or regions of DNA that do not code for protein, are removed from the pre-mRNA and the remaining exons connected to re-form a single continuous molecule. This process normally occurs after 5' capping and 3' polyadenylation but can begin before synthesis is complete in transcripts with many exons. Many pre-mRNAs, including those encoding antibodies, can be spliced in multiple ways to produce different mature mRNAs that encode different protein sequences. This process is known as alternative splicing, and allows production of a large variety of proteins from a limited amount of DNA.

Dynamics and Regulation

Nuclear Transport

The entry and exit of large molecules from the nucleus is tightly controlled by the nuclear pore complexes. Although small molecules can enter the nucleus without regulation, macromolecules such as RNA and proteins require association karyopherins called importins to enter the nucleus and exportins to exit. "Cargo" proteins that must be translocated from the cytoplasm to the nucleus contain short amino acid sequences known as nuclear localization signals, which are bound by importins, while those transported from the nucleus to the cytoplasm carry nuclear export signals bound by exportins. The ability of importins and exportins to transport their cargo is regulated by GTPases, enzymes that hydrolyze the molecule guanosine triphosphate (GTP) to release energy. The key GTPase in nuclear transport is Ran, which can bind either GTP or GDP (guanosine diphosphate), depending on whether it is located in the nucleus or the cytoplasm. Whereas importins depend on RanGTP to dissociate from their cargo, exportins require RanGTP in order to bind to their cargo.

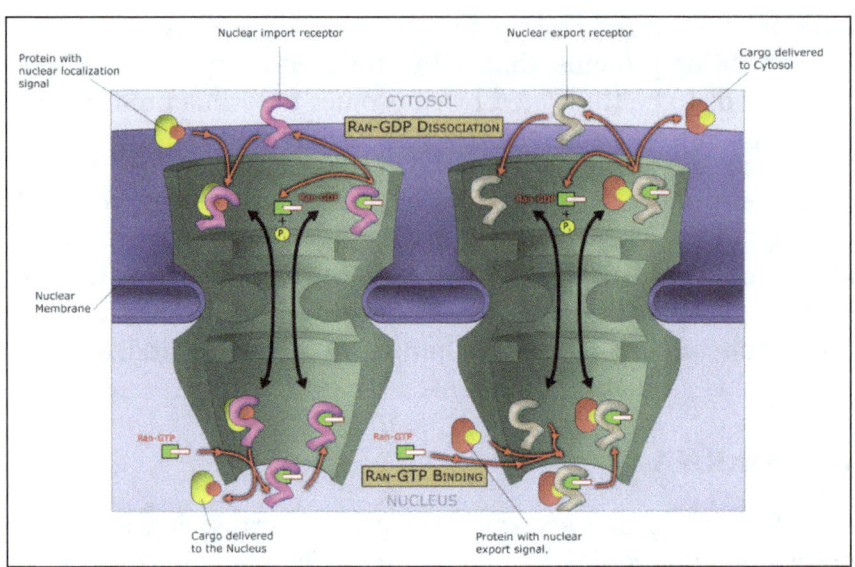

Macromolecules, such as RNA and proteins, are actively transported across the
nuclear membrane in a process called the Ran-GTP nuclear transport cycle.

Nuclear import depends on the importin binding its cargo in the cytoplasm and carrying it through
the nuclear pore into the nucleus. Inside the nucleus, RanGTP acts to separate the cargo from the
importin, allowing the importin to exit the nucleus and be reused. Nuclear export is similar, as the
exportin binds the cargo inside the nucleus in a process facilitated by RanGTP, exits through the
nuclear pore, and separates from its cargo in the cytoplasm.

Specialized export proteins exist for translocation of mature mRNA and tRNA to the cytoplasm
after post-transcriptional modification is complete. This quality-control mechanism is important
due to these molecules' central role in protein translation. Mis-expression of a protein due to incomplete excision of exons or mis-incorporation of amino acids could have negative consequences
for the cell; thus, incompletely modified RNA that reaches the cytoplasm is degraded rather than
used in translation.

Assembly and Disassembly

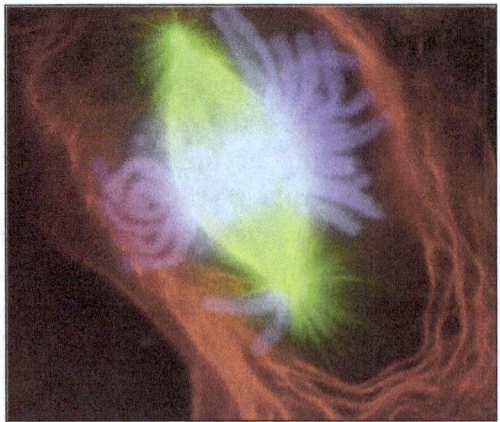

An image of a newt lung cell stained with fluorescent dyes during metaphase.
The mitotic spindle can be seen, stained green, attached to the two sets of chromosomes,
stained light blue. All chromosomes but one are already at the metaphase plate.

During its lifetime, a nucleus may be broken down or destroyed, either in the process of cell division or as a consequence of apoptosis (the process of programmed cell death). During these events, the structural components of the nucleus — the envelope and lamina — can be systematically degraded. In most cells, the disassembly of the nuclear envelope marks the end of the prophase of mitosis. However, this disassembly of the nucleus is not a universal feature of mitosis and does not occur in all cells. Some unicellular eukaryotes (e.g., yeasts) undergo so-called closed mitosis, in which the nuclear envelope remains intact. In closed mitosis, the daughter chromosomes migrate to opposite poles of the nucleus, which then divides in two. The cells of higher eukaryotes, however, usually undergo open mitosis, which is characterized by breakdown of the nuclear envelope. The daughter chromosomes then migrate to opposite poles of the mitotic spindle, and new nuclei reassemble around them.

At a certain point during the cell cycle in open mitosis, the cell divides to form two cells. In order for this process to be possible, each of the new daughter cells must have a full set of genes, a process requiring replication of the chromosomes as well as segregation of the separate sets. This occurs by the replicated chromosomes, the sister chromatids, attaching to microtubules, which in turn are attached to different centrosomes. The sister chromatids can then be pulled to separate locations in the cell. In many cells, the centrosome is located in the cytoplasm, outside the nucleus; the microtubules would be unable to attach to the chromatids in the presence of the nuclear envelope. Therefore, the early stages in the cell cycle, beginning in prophase and until around prometaphase, the nuclear membrane is dismantled. Likewise, during the same period, the nuclear lamina is also disassembled, a process regulated by phosphorylation of the lamins by protein kinases such as the CDC2 protein kinase. Towards the end of the cell cycle, the nuclear membrane is reformed, and around the same time, the nuclear lamina are reassembled by dephosphorylating the lamins.

However, in dinoflagellates, the nuclear envelope remains intact, the centrosomes are located in the cytoplasm, and the microtubules come in contact with chromosomes, whose centromeric regions are incorporated into the nuclear envelope (the so-called closed mitosis with extranuclear spindle). In many other protists (e.g., ciliates, sporozoans) and fungi, the centrosomes are intranuclear, and their nuclear envelope also does not disassemble during cell division.

Apoptosis is a controlled process in which the cell's structural components are destroyed, resulting in death of the cell. Changes associated with apoptosis directly affect the nucleus and its contents, for example, in the condensation of chromatin and the disintegration of the nuclear envelope and lamina. The destruction of the lamin networks is controlled by specialized apoptotic proteases called caspases, which cleave the lamin proteins and, thus, degrade the nucleus' structural integrity. Lamin cleavage is sometimes used as a laboratory indicator of caspase activity in assays for early apoptotic activity. Cells that express mutant caspase-resistant lamins are deficient in nuclear changes related to apoptosis, suggesting that lamins play a role in initiating the events that lead to apoptotic degradation of the nucleus. Inhibition of lamin assembly itself is an inducer of apoptosis.

The nuclear envelope acts as a barrier that prevents both DNA and RNA viruses from entering the nucleus. Some viruses require access to proteins inside the nucleus in order to replicate and/or assemble. DNA viruses, such as herpesvirus replicate and assemble in the cell nucleus, and exit by budding through the inner nuclear membrane. This process is accompanied by disassembly of the lamina on the nuclear face of the inner membrane.

Disease-related Dynamics

Initially, it has been suspected that immunoglobulins in general and autoantibodies in particular do not enter the nucleus. Now there is a body of evidence that under pathological conditions (e.g. lupus erythematosus) IgG can enter the nucleus.

Nuclei Per Cell

Most eukaryotic cell types usually have a single nucleus, but some have no nuclei, while others have several. This can result from normal development, as in the maturation of mammalian red blood cells, or from faulty cell division.

Anucleated Cells

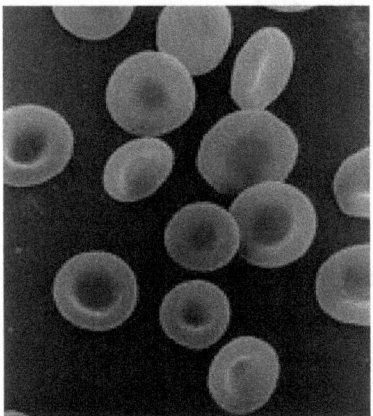

Human red blood cells, like those of other mammals, lack nuclei.
This occurs as a normal part of the cells' development.

An anucleated cell contains no nucleus and is, therefore, incapable of dividing to produce daughter cells. The best-known anucleated cell is the mammalian red blood cell, or erythrocyte, which also lacks other organelles such as mitochondria, and serves primarily as a transport vessel to ferry oxygen from the lungs to the body's tissues. Erythrocytes mature through erythropoiesis in the bone marrow, where they lose their nuclei, organelles, and ribosomes. The nucleus is expelled during the process of differentiation from an erythroblast to a reticulocyte, which is the immediate precursor of the mature erythrocyte. The presence of mutagens may induce the release of some immature "micronucleated" erythrocytes into the bloodstream. Anucleated cells can also arise from flawed cell division in which one daughter lacks a nucleus and the other has two nuclei.

In flowering plants, this condition occurs in sieve tube elements.

Multinucleated Cells

Multinucleated cells contain multiple nuclei. Most acantharean species of protozoa and some fungi in mycorrhizae have naturally multinucleated cells. Other examples include the intestinal parasites in the genus *Giardia*, which have two nuclei per cell. In humans, skeletal muscle cells, called myocytes and syncytium, become multinucleated during development; the resulting arrangement of nuclei near the periphery of the cells allows maximal intracellular space for myofibrils. Other multinucleate cells in the human are osteoclasts a type of bone cell. Multinucleated and binucleated cells can also be abnormal in humans; for example, cells arising from the fusion of monocytes and macrophages, known as giant multinucleated cells, sometimes accompany inflammation and are also implicated in tumor formation.

A number of dinoflagellates are known to have two nuclei. Unlike other multinucleated cells these nuclei contain two distinct lineages of DNA: one from the dinoflagellate and the other from a symbiotic diatom. The mitochondria and the plastids of the diatom somehow remain functional.

Evolution

As the major defining characteristic of the eukaryotic cell, the nucleus' evolutionary origin has been the subject of much speculation. Four major hypotheses have been proposed to explain the existence of the nucleus, although none have yet earned widespread support.

The first model known as the "syntrophic model" proposes that a symbiotic relationship between the archaea and bacteria created the nucleus-containing eukaryotic cell. (Organisms of the Archaea and Bacteria domain have no cell nucleus.) It is hypothesized that the symbiosis originated when ancient archaea, similar to modern methanogenic archaea, invaded and lived within bacteria similar to modern myxobacteria, eventually forming the early nucleus. This theory is analogous to the accepted theory for the origin of eukaryotic mitochondria and chloroplasts, which are thought to have developed from a similar endosymbiotic relationship between proto-eukaryotes and aerobic bacteria. The archaeal origin of the nucleus is supported by observations that archaea and eukarya have similar genes for certain proteins, including histones. Observations that myxobacteria are motile, can form multicellular complexes, and possess kinases and G proteins similar to eukarya, support a bacterial origin for the eukaryotic cell.

A second model proposes that proto-eukaryotic cells evolved from bacteria without an endosymbiotic stage. This model is based on the existence of modern planctomycetes bacteria that possess a nuclear structure with primitive pores and other compartmentalized membrane structures. A similar proposal states that a eukaryote-like cell, the chronocyte, evolved first and phagocytosed archaea and bacteria to generate the nucleus and the eukaryotic cell.

The most controversial model, known as *viral eukaryogenesis*, posits that the membrane-bound nucleus, along with other eukaryotic features, originated from the infection of a prokaryote by a virus. The suggestion is based on similarities between eukaryotes and viruses such as linear DNA strands, mRNA capping, and tight binding to proteins (analogizing histones to viral envelopes). One version of the proposal suggests that the nucleus evolved in concert with phagocytosis to form an early cellular "predator". Another variant proposes that eukaryotes originated from early archaea infected by poxviruses, on the basis o f observed similarity between the DNA polymerases in modern poxviruses and eukaryotes. It has been suggested that the unresolved question of the evolution of sex could be related to the viral eukaryogenesis hypothesis.

A more recent proposal, the *exomembrane hypothesis*, suggests that the nucleus instead originated from a single ancestral cell that evolved a second exterior cell membrane; the interior membrane enclosing the original cell then became the nuclear membrane and evolved increasingly elaborate pore structures for passage of internally synthesized cellular components such as ribosomal subunits.

MITOCHONDRION

The mitochondrion (plural mitochondria) is a double-membrane-bound organelle found in most eukaryotic organisms. Some cells in some multicellular organisms may, however, lack them (for example, mature mammalian red blood cells). A number of unicellular organisms, such as microsporidia, parabasalids, and diplomonads, have also reduced or transformed their mitochondria into other structures. To date, only one eukaryote, *Monocercomonoides*, is known to have completely lost its mitochondria. The word mitochondrion comes from the Greek μίτος, *mitos*, "thread", and χονδρίον, *chondrion*, "granule" or "grain-like". Mitochondria generate most of the cell's supply of adenosine triphosphate (ATP), used as a source of chemical energy. A mitochondrion is thus termed the *powerhouse* of the cell.

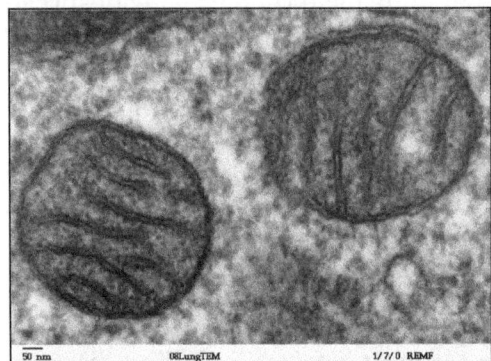

Two mitochondria from mammalian lung tissue displaying their
matrix and membranes as shown by electron microscopy

Mitochondria are commonly between 0.75 and 3 μm^2 in area but vary considerably in size and structure. Unless specifically stained, they are not visible. In addition to supplying cellular energy, mitochondria are involved in other tasks, such as signaling, cellular differentiation, and cell death, as well as maintaining control of the cell cycle and cell growth. Mitochondrial biogenesis is in turn temporally coordinated with these cellular processes. Mitochondria have been implicated in several human diseases, including mitochondrial disorders, cardiac dysfunction, heart failure and autism.

The number of mitochondria in a cell can vary widely by organism, tissue, and cell type. For instance, red blood cells have no mitochondria, whereas liver cells can have more than 2000. The organelle is composed of compartments that carry out specialized functions. These compartments or regions include the outer membrane, the intermembrane space, the inner membrane, and the cristae and matrix.

Although most of a cell's DNA is contained in the cell nucleus, the mitochondrion has its own independent genome that shows substantial similarity to bacterial genomes. Mitochondrial proteins (proteins transcribed from mitochondrial DNA) vary depending on the tissue and the species. In humans, 615 distinct types of protein have been identified from cardiac mitochondria, whereas in rats, 940 proteins have been reported. The mitochondrial proteome is thought to be dynamically regulated.

There are two hypotheses about the origin of mitochondria: endosymbiotic and autogenous. The endosymbiotic hypothesis suggests that mitochondria were originally prokaryotic cells, capable of implementing oxidative mechanisms that were not possible for eukaryotic cells; they became endosymbionts living inside the eukaryote. In the autogenous hypothesis, mitochondria were born by splitting off a portion of DNA from the nucleus of the eukaryotic cell at the time of divergence with the prokaryotes; this DNA portion would have been enclosed by membranes, which could not be crossed by proteins. Since mitochondria have many features in common with bacteria, the endosymbiotic hypothesis is more widely accepted.

A mitochondrion contains DNA, which is organized as several copies of a single, usually circular chromosome. This mitochondrial chromosome contains genes for redox proteins, such as those of the respiratory chain. The CoRR hypothesis proposes that this co-location is required for redox regulation. The mitochondrial genome codes for some RNAs of ribosomes, and the 22 tRNAs necessary for the translation of mRNAs into protein. The circular structure is also found in prokaryotes. The proto-mitochondrion was probably closely related to the *Rickettsia*. However, the exact relationship of the ancestor of mitochondria to the alphaproteobacteria and whether the mitochondrion was formed at the same time or after the nucleus, remains controversial. For example, it has been suggested that the SAR11 clade of bacteria shares a relatively recent common ancestor with the mitochondria, while phylogenomic analyses indicate that mitochondria evolved from a proteobacteria lineage that branched off before the divergence of all sampled alphaproteobacteria.

The ribosomes coded for by the mitochondrial DNA are similar to those from bacteria in size and structure. They closely resemble the bacterial 70S ribosome and not the 80S cytoplasmic ribosomes, which are coded for by nuclear DNA.

The endosymbiotic relationship of mitochondria with their host cells was popularized by Lynn Margulis. The endosymbiotic hypothesis suggests that mitochondria descended from bacteria that somehow survived endocytosis by another cell, and became incorporated into the cytoplasm. The ability of these bacteria to conduct respiration in host cells that had relied on glycolysis and fermentation would have provided a considerable evolutionary advantage. This symbiotic relationship probably developed 1.7 to 2 billion years ago. A few groups of unicellular eukaryotes have only vestigial mitochondria or derived structures: the microsporidians, metamonads, and archamoebae. These groups appear as the most primitive eukaryotes on phylogenetic trees constructed using rRNA information, which once suggested that they appeared before the origin of mitochondria. However, this is now known to be an artifact of long-branch attraction—they are derived groups and retain genes or organelles derived from mitochondria (e.g., mitosomes and hydrogenosomes).

Monocercomonoides appear to have lost their mitochondria completely and at least some of the mitochondrial functions seem to be carried out by cytoplasmic proteins now.

Structure

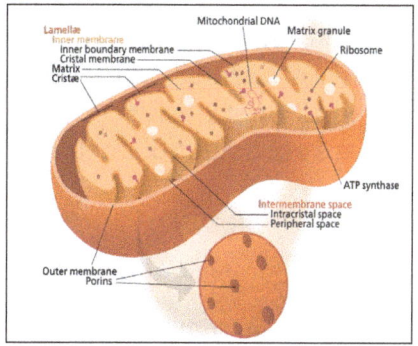

Mitochondrion ultrastructure *(interactive diagram)* A mitochondrion has a double membrane; the inner one contains its chemiosmotic apparatus and has deep grooves which increase its surface area. While commonly depicted as an "orange sausage with a blob inside of it" mitochondria can take many shapes and their intermembrane space is quite thin.

A mitochondrion contains outer and inner membranes composed of phospholipid bilayers and proteins. The two membranes have different properties. Because of this double-membraned organization, there are five distinct parts to a mitochondrion. They are:

1. The outer mitochondrial membrane,

2. The intermembrane space (the space between the outer and inner membranes),

3. The inner mitochondrial membrane,

4. The cristae space (formed by infoldings of the inner membrane), and

5. The matrix (space within the inner membrane).

Mitochondria stripped of their outer membrane are called mitoplasts.

Outer Membrane

The outer mitochondrial membrane, which encloses the entire organelle, is 60 to 75 angstroms (Å) thick. It has a protein-to-phospholipid ratio similar to that of the cell membrane (about 1:1 by weight). It contains large numbers of integral membrane proteins called porins. A major trafficking protein is the pore-forming voltage-dependent anion channel (VDAC). The VDAC is the primary transporter of nucleotides, ions and metabolites between the cytosol and the intermembrane space. It is formed as a beta barrel that spans the outer membrane, similar to that in the gram-negative bacterial membrane. Larger proteins can enter the mitochondrion if a signaling sequence at their N-terminus binds to a large multisubunit protein called translocase in the outer membrane, which then actively moves them across the membrane. Mitochondrial pro-proteins are imported through specialised translocation complexes.

The outer membrane also contains enzymes involved in such diverse activities as the elongation of fatty acids, oxidation of epinephrine, and the degradation of tryptophan. These enzymes include monoamine oxidase, rotenone-insensitive NADH-cytochrome c-reductase, kynurenine hydroxylase and fatty acid Co-A ligase. Disruption of the outer membrane permits proteins in the intermembrane space to leak into the cytosol, leading to certain cell death. The mitochondrial outer membrane can associate with the endoplasmic reticulum (ER) membrane, in a structure called MAM (mitochondria-associated ER-membrane). This is important in the ER-mitochondria calcium signaling and is involved in the transfer of lipids between the ER and mitochondria. Outside the outer membrane there are small (diameter: 60Å) particles named sub-units of Parson.

Intermembrane Space

The mitochondrial intermembrane space is the space between the outer membrane and the inner membrane. It is also known as perimitochondrial space. Because the outer membrane is freely permeable to small molecules, the concentrations of small molecules, such as ions and sugars, in the intermembrane space is the same as in the cytosol. However, large proteins must have a specific signaling sequence to be transported across the outer membrane, so the protein composition of this space is different from the protein composition of the cytosol. One protein that is localized to the intermembrane space in this way is cytochrome c.

Inner Membrane

The inner mitochondrial membrane contains proteins with five types of functions:

1. Those that perform the redox reactions of oxidative phosphorylation.

2. ATP synthase, which generates ATP in the matrix.

3. Specific transport proteins that regulate metabolite passage into and out of the mitochondrial matrix.

4. Protein import machinery.

5. Mitochondrial fusion and fission protein.

It contains more than 151 different polypeptides, and has a very high protein-to-phospholipid ratio (more than 3:1 by weight, which is about 1 protein for 15 phospholipids). The inner membrane is home to around 1/5 of the total protein in a mitochondrion. In addition, the inner membrane is rich in an unusual phospholipid, cardiolipin. This phospholipid was originally discovered in cow hearts in 1942, and is usually characteristic of mitochondrial and bacterial plasma membranes. Cardiolipin contains four fatty acids rather than two, and may help to make the inner membrane impermeable. Unlike the outer membrane, the inner membrane doesn't contain porins, and is highly impermeable to all molecules. Almost all ions and molecules require special membrane transporters to enter or exit the matrix. Proteins are ferried into the matrix via the translocase of the inner membrane (TIM) complex or via Oxa1. In addition, there is a membrane potential across the inner membrane, formed by the action of the enzymes of the electron transport chain.

Cristae

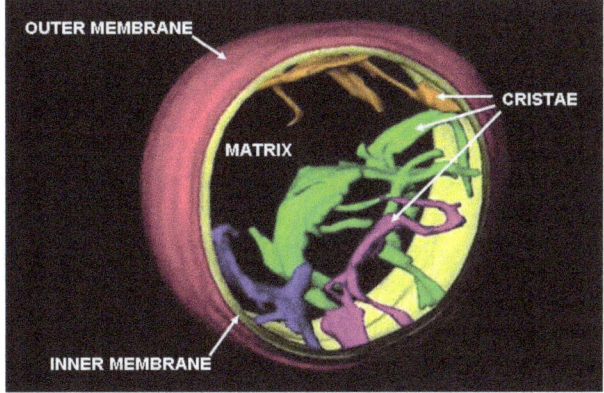

Cross-sectional image of cristae in rat liver mitochondrion to demonstrate the likely 3D structure and relationship to the inner membrane.

The inner mitochondrial membrane is compartmentalized into numerous cristae, which expand the surface area of the inner mitochondrial membrane, enhancing its ability to produce ATP. For typical liver mitochondria, the area of the inner membrane is about five times as large as the outer membrane. This ratio is variable and mitochondria from cells that have a greater demand for ATP, such as muscle cells, contain even more cristae. These folds are studded with small round bodies known as F_1 particles or oxysomes. These are not simple random folds but rather invaginations of the inner membrane, which can affect overall chemiosmotic function.

One recent mathematical modeling study has suggested that the optical properties of the cristae in filamentous mitochondria may affect the generation and propagation of light within the tissue.

Matrix

The matrix is the space enclosed by the inner membrane. It contains about 2/3 of the total protein in a mitochondrion. The matrix is important in the production of ATP with the aid of the ATP synthase contained in the inner membrane. The matrix contains a highly concentrated mixture of hundreds of enzymes, special mitochondrial ribosomes, tRNA, and several copies of the mitochondrial DNA genome. Of the enzymes, the major functions include oxidation of pyruvate and fatty acids, and the citric acid cycle. The DNA molecules are packaged into nucleoids by proteins, one of which is TFAM.

Mitochondria have their own genetic material, and the machinery to manufacture their own RNAs and proteins. A published human mitochondrial DNA sequence revealed 16,569 base pairs encoding 37 genes: 22 tRNA, 2 rRNA, and 13 peptide genes. The 13 mitochondrial peptides in humans are integrated into the inner mitochondrial membrane, along with proteins encoded by genes that reside in the host cell's nucleus.

Mitochondria-associated ER Membrane (MAM)

The mitochondria-associated ER membrane (MAM) is another structural element that is increasingly recognized for its critical role in cellular physiology and homeostasis. Once considered a technical snag in cell fractionation techniques, the alleged ER vesicle contaminants that invariably appeared in the mitochondrial fraction have been re-identified as membranous structures derived from the MAM—the interface between mitochondria and the ER. Physical coupling between these two organelles had previously been observed in electron micrographs and has more recently been probed with fluorescence microscopy. Such studies estimate that at the MAM, which may comprise up to 20% of the mitochondrial outer membrane, the ER and mitochondria are separated by a mere 10–25 nm and held together by protein tethering complexes.

Purified MAM from subcellular fractionation has been shown to be enriched in enzymes involved in phospholipid exchange, in addition to channels associated with Ca^{2+} signaling. These hints of a prominent role for the MAM in the regulation of cellular lipid stores and signal transduction have been borne out, with significant implications for mitochondrial-associated cellular phenomena, as discussed below. Not only has the MAM provided insight into the mechanistic basis underlying such physiological processes as intrinsic apoptosis and the propagation of calcium signaling, but it also favors a more refined view of the mitochondria. Though often seen as static, isolated 'powerhouses' hijacked for cellular metabolism through an ancient endosymbiotic event, the evolution of the MAM underscores the extent to which mitochondria have been integrated into overall cellular physiology, with intimate physical and functional coupling to the endomembrane system.

Phospholipid Transfer

The MAM is enriched in enzymes involved in lipid biosynthesis, such as phosphatidylserine synthase on the ER face and phosphatidylserine decarboxylase on the mitochondrial face. Because mitochondria are dynamic organelles constantly undergoing fission and fusion events, they require a constant and well-regulated supply of phospholipids for membrane integrity. But mitochondria are not only a destination for the phospholipids they finish synthesis of; rather, this organelle also plays a role in inter-organelle trafficking of the intermediates and products of phospholipid biosynthetic pathways, ceramide and cholesterol metabolism, and glycosphingolipid anabolism.

Such trafficking capacity depends on the MAM, which has been shown to facilitate transfer of lipid intermediates between organelles. In contrast to the standard vesicular mechanism of lipid transfer, evidence indicates that the physical proximity of the ER and mitochondrial membranes at the MAM allows for lipid flipping between opposed bilayers. Despite this unusual and seemingly energetically unfavorable mechanism, such transport does not require ATP. Instead, in yeast, it has been shown to be dependent on a multiprotein tethering structure termed the ER-mitochondria encounter structure, or ERMES, although it remains unclear whether this structure directly mediates lipid transfer or is required to keep the membranes in sufficiently close proximity to lower the energy barrier for lipid flipping.

The MAM may also be part of the secretory pathway, in addition to its role in intracellular lipid trafficking. In particular, the MAM appears to be an intermediate destination between the rough ER and the Golgi in the pathway that leads to very-low-density lipoprotein, or VLDL, assembly and secretion. The MAM thus serves as a critical metabolic and trafficking hub in lipid metabolism.

Calcium Signaling

A critical role for the ER in calcium signaling was acknowledged before such a role for the mitochondria was widely accepted, in part because the low affinity of Ca^{2+} channels localized to the outer mitochondrial membrane seemed to contradict this organelle's purported responsiveness to changes in intracellular Ca^{2+} flux. But the presence of the MAM resolves this apparent contradiction: the close physical association between the two organelles results in Ca^{2+} microdomains at contact points that facilitate efficient Ca^{2+} transmission from the ER to the mitochondria. Transmission occurs in response to so-called "Ca^{2+} puffs" generated by spontaneous clustering and activation of IP3R, a canonical ER membrane Ca^{2+} channel.

The fate of these puffs—in particular, whether they remain restricted to isolated locales or integrated into Ca^{2+} waves for propagation throughout the cell—is determined in large part by MAM dynamics. Although reuptake of Ca^{2+} by the ER (concomitant with its release) modulates the intensity of the puffs, thus insulating mitochondria to a certain degree from high Ca^{2+} exposure, the MAM often serves as a firewall that essentially buffers Ca^{2+} puffs by acting as a sink into which free ions released into the cytosol can be funneled. This Ca^{2+} tunneling occurs through the low-affinity Ca^{2+} receptor VDAC1, which recently has been shown to be physically tethered to the IP3R clusters on the ER membrane and enriched at the MAM. The ability of mitochondria to serve as a Ca^{2+} sink is a result of the electrochemical gradient generated during oxidative phosphorylation, which makes tunneling of the cation an exergonic process. Normal, mild calcium influx from cytosol into the mitochondrial matrix causes transient depolarization that is corrected by pumping out protons.

But transmission of Ca^{2+} is not unidirectional; rather, it is a two-way street. The properties of the Ca^{2+} pump SERCA and the channel IP3R present on the ER membrane facilitate feedback regulation coordinated by MAM function. In particular, the clearance of Ca^{2+} by the MAM allows for spatio-temporal patterning of Ca^{2+} signaling because Ca^{2+} alters IP3R activity in a biphasic manner. SERCA is likewise affected by mitochondrial feedback: uptake of Ca^{2+} by the MAM stimulates ATP production, thus providing energy that enables SERCA to reload the ER with Ca^{2+} for continued Ca^{2+} efflux at the MAM. Thus, the MAM is not a passive buffer for Ca^{2+} puffs; rather it helps modulate further Ca^{2+} signaling through feedback loops that affect ER dynamics.

Regulating ER release of Ca^{2+} at the MAM is especially critical because only a certain window of Ca^{2+} uptake sustains the mitochondria, and consequently the cell, at homeostasis. Sufficient intraorganelle Ca^{2+} signaling is required to stimulate metabolism by activating dehydrogenase enzymes critical to flux through the citric acid cycle. However, once Ca^{2+} signaling in the mitochondria passes a certain threshold, it stimulates the intrinsic pathway of apoptosis in part by collapsing the mitochondrial membrane potential required for metabolism. Studies examining the role of pro- and anti-apoptotic factors support this model; for example, the anti-apoptotic factor Bcl-2 has been shown to interact with IP3Rs to reduce Ca^{2+} filling of the ER, leading to reduced efflux at the MAM and preventing collapse of the mitochondrial membrane potential post-apoptotic stimuli. Given the need for such fine regulation of Ca^{2+} signaling, it is perhaps unsurprising that dysregulated mitochondrial Ca^{2+} has been implicated in several neurodegenerative diseases, while the catalogue of tumor suppressors includes a few that are enriched at the MAM.

Molecular Basis for Tethering

Recent advances in the identification of the tethers between the mitochondrial and ER membranes suggest that the scaffolding function of the molecular elements involved is secondary to other, non-structural functions. In yeast, ERMES, a multiprotein complex of interacting ER- and mitochondrial-resident membrane proteins, is required for lipid transfer at the MAM and exemplifies this principle. One of its components, for example, is also a constituent of the protein complex required for insertion of transmembrane beta-barrel proteins into the lipid bilayer. However, a homologue of the ERMES complex has not yet been identified in mammalian cells. Other proteins implicated in scaffolding likewise have functions independent of structural tethering at the MAM; for example, ER-resident and mitochondrial-resident mitofusins form heterocomplexes that regulate the number of inter-organelle contact sites, although mitofusins were first identified for their role in fission and fusion events between individual mitochondria. Glucose-related protein 75 (grp75) is another dual-function protein. In addition to the matrix pool of grp75, a portion serves as a chaperone that physically links the mitochondrial and ER Ca^{2+} channels VDAC and IP3R for efficient Ca^{2+} transmission at the MAM. Another potential tether is Sigma-1R, a non-opioid receptor whose stabilization of ER-resident IP3R may preserve communication at the MAM during the metabolic stress response.

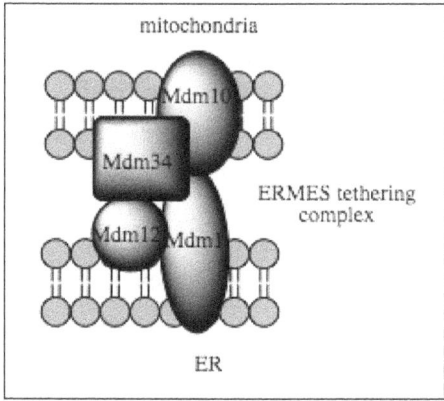

Model of the yeast multimeric tethering complex, ERMES.

Perspective

The MAM is a critical signaling, metabolic, and trafficking hub in the cell that allows for the integration of ER and mitochondrial physiology. Coupling between these organelles is not simply structural but functional as well and critical for overall cellular physiology and homeostasis. The MAM thus offers a perspective on mitochondria that diverges from the traditional view of this organelle as a static, isolated unit appropriated for its metabolic capacity by the cell. Instead, this mitochondrial-ER interface emphasizes the integration of the mitochondria, the product of an endosymbiotic event, into diverse cellular processes.

Organization and Distribution

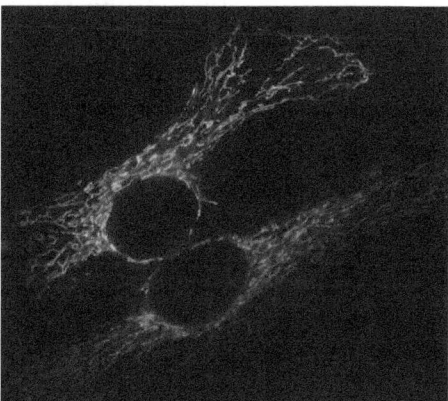

Typical mitochondrial network (green) in two human cells (HeLa cells).

Mitochondria (and related structures) are found in all eukaryotes (except one—the Oxymonad *Monocercomonoides* sp.). Although commonly depicted as bean-like structures they form a highly dynamic network in the majority of cells where they constantly undergo fission and fusion. The population of all the mitochondria of a given cell constitutes the chondriome. Mitochondria vary in number and location according to cell type. A single mitochondrion is often found in unicellular organisms. Conversely, the chondriome size of human liver cells is large, with about 1000–2000 mitochondria per cell, making up 1/5 of the cell volume. The mitochondrial content of otherwise similar cells can vary substantially in size and membrane potential, with differences arising from sources including uneven partitioning at cell divisions, leading to extrinsic differences in ATP levels and downstream cellular processes. The mitochondria can be found nestled between myofibrils of muscle or wrapped around the sperm flagellum. Often, they form a complex 3D branching network inside the cell with the cytoskeleton. The association with the cytoskeleton determines mitochondrial shape, which can affect the function as well: different structures of the mitochondrial network may afford the population a variety of physical, chemical, and signalling advantages or disadvantages. Mitochondria in cells are always distributed along microtubules and the distribution of these organelles is also correlated with the endoplasmic reticulum. Recent evidence suggests that vimentin, one of the components of the cytoskeleton, is also critical to the association with the cytoskeleton.

Function

The most prominent roles of mitochondria are to produce the energy currency of the cell, ATP (i.e., phosphorylation of ADP), through respiration, and to regulate cellular metabolism. The central set of reactions involved in ATP production are collectively known as the citric acid cycle, or the Krebs cycle. However, the mitochondrion has many other functions in addition to the production of ATP.

Energy Conversion

A dominant role for the mitochondria is the production of ATP, as reflected by the large number of proteins in the inner membrane for this task. This is done by oxidizing the major products of glucose: pyruvate, and NADH, which are produced in the cytosol. This type of cellular respiration known as aerobic respiration, is dependent on the presence of oxygen. When oxygen is limited, the glycolytic products will be metabolized by anaerobic fermentation, a process that is independent of the mitochondria. The production of ATP from glucose has an approximately 13-times higher yield during aerobic respiration compared to fermentation. Plant mitochondria can also produce a limited amount of ATP without oxygen by using the alternate substrate nitrite. ATP crosses out through the inner membrane with the help of a specific protein, and across the outer membrane via porins. ADP returns via the same route.

Pyruvate and the Citric Acid Cycle

Pyruvate molecules produced by glycolysis are actively transported across the inner mitochondrial membrane, and into the matrix where they can either be oxidized and combined with coenzyme A to form CO_2, acetyl-CoA, and NADH, or they can be carboxylated (by pyruvate carboxylase) to form oxaloacetate. This latter reaction "fills up" the amount of oxaloacetate in the citric acid cycle, and is therefore an anaplerotic reaction, increasing the cycle's capacity to metabolize acetyl-CoA when the tissue's energy needs (e.g. in muscle) are suddenly increased by activity.

In the citric acid cycle, all the intermediates (e.g. citrate, iso-citrate, alpha-ketoglutarate, succinate, fumarate, malate and oxaloacetate) are regenerated during each turn of the cycle. Adding more of any of these intermediates to the mitochondrion therefore means that the additional amount is retained within the cycle, increasing all the other intermediates as one is converted into the other. Hence, the addition of any one of them to the cycle has an anaplerotic effect, and its removal has a cataplerotic effect. These anaplerotic and cataplerotic reactions will, during the course of the cycle, increase or decrease the amount of oxaloacetate available to combine with acetyl-CoA to form citric acid. This in turn increases or decreases the rate of ATP production by the mitochondrion, and thus the availability of ATP to the cell.

Acetyl-CoA, on the other hand, derived from pyruvate oxidation, or from the beta-oxidation of fatty acids, is the only fuel to enter the citric acid cycle. With each turn of the cycle one molecule of acetyl-CoA is consumed for every molecule of oxaloacetate present in the mitochondrial matrix, and is never regenerated. It is the oxidation of the acetate portion of acetyl-CoA that produces CO_2 and water, with the energy thus released captured in the form of ATP.

In the liver, the carboxylation of cytosolic pyruvate into intra-mitochondrial oxaloacetate is an early step in the gluconeogenic pathway, which converts lactate and de-aminated alanine into glucose, under the influence of high levels of glucagon and epinephrine in the blood. Here, the addition of oxaloacetate to the mitochondrion does not have a net anaplerotic effect, as another citric acid cycle intermediate (malate) is immediately removed from the mitochondrion to be converted into cytosolic oxaloacetate, which is ultimately converted into glucose, in a process that is almost the reverse of glycolysis.

The enzymes of the citric acid cycle are located in the mitochondrial matrix, with the exception of succinate dehydrogenase, which is bound to the inner mitochondrial membrane as part of Complex II. The citric acid cycle oxidizes the acetyl-CoA to carbon dioxide, and, in the process, produces reduced cofactors (three molecules of NADH and one molecule of $FADH_2$) that are a source of electrons for the *electron transport chain*, and a molecule of GTP (that is readily converted to an ATP).

NADH and FADH2: The Electron Transport Chain

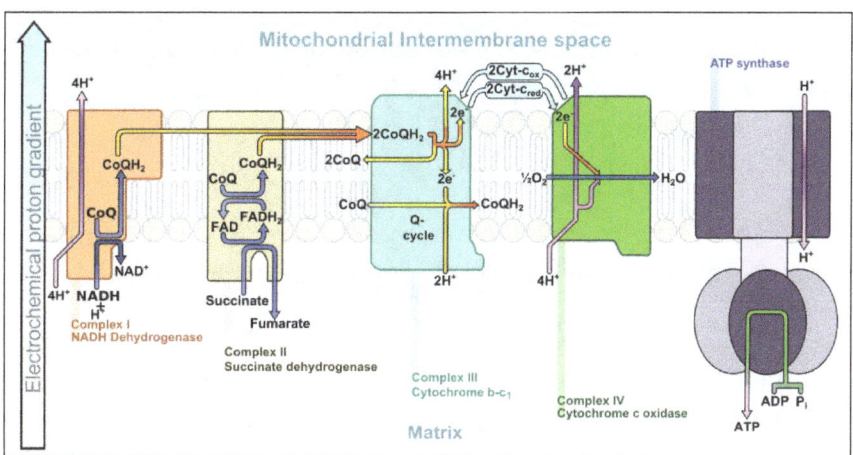

Electron transport chain in the mitochondrial intermembrane space.

The redox energy from NADH and $FADH_2$ is transferred to oxygen (O_2) in several steps via the electron transport chain. These energy-rich molecules are produced within the matrix via the citric acid cycle but are also produced in the cytoplasm by glycolysis. Reducing equivalents from the cytoplasm can be imported via the malate-aspartate shuttle system of antiporter proteins or feed into the electron transport chain using a glycerol phosphate shuttle. Protein complexes in the inner membrane (NADH dehydrogenase (ubiquinone), cytochrome c reductase, and cytochrome c oxidase) perform the transfer and the incremental release of energy is used to pump protons (H^+) into the intermembrane space. This process is efficient, but a small percentage of electrons may prematurely reduce oxygen, forming reactive oxygen species such as superoxide. This can cause oxidative stress in the mitochondria and may contribute to the decline in mitochondrial function associated with the aging process.

As the proton concentration increases in the intermembrane space, a strong electrochemical gradient is established across the inner membrane. The protons can return to the matrix through the ATP synthase complex, and their potential energy is used to synthesize ATP from ADP and inorganic phosphate (P_i). This process is called chemiosmosis, and was first described by Peter Mitchell who was awarded the 1978 Nobel Prize in Chemistry for his work. Later, part of the 1997 Nobel Prize in Chemistry was awarded to Paul D. Boyer and John E. Walker for their clarification of the working mechanism of ATP synthase.

Heat Production

Under certain conditions, protons can re-enter the mitochondrial matrix without contributing to ATP synthesis. This process is known as *proton leak* or mitochondrial uncoupling and is due to the facilitated diffusion of protons into the matrix. The process results in the unharnessed potential energy of the proton electrochemical gradient being released as heat. The process is mediated by a proton channel called thermogenin, or UCP1. Thermogenin is a 33 kDa protein first discovered in 1973. Thermogenin is primarily found in brown adipose tissue, or brown fat, and is responsible for non-shivering thermogenesis. Brown adipose tissue is found in mammals, and is at its highest levels in early life and in hibernating animals. In humans, brown adipose tissue is present at birth and decreases with age.

Storage of Calcium Ions

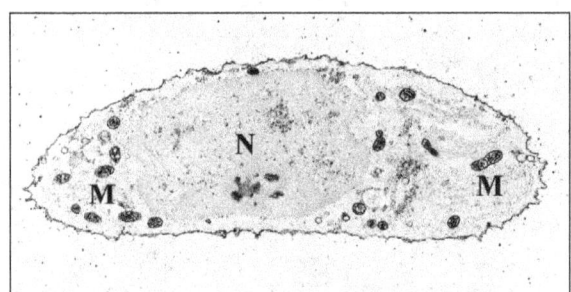

Transmission electron micrograph of a chondrocyte, stained for calcium, showing its nucleus (N) and mitochondria (M).

The concentrations of free calcium in the cell can regulate an array of reactions and is important for signal transduction in the cell. Mitochondria can transiently store calcium, a contributing process for the cell's homeostasis of calcium. In fact, their ability to rapidly take in calcium for later release makes them very good "cytosolic buffers" for calcium. The endoplasmic reticulum (ER) is the most significant storage site of calcium, and there is a significant interplay between the mitochondrion and ER with regard to calcium. The calcium is taken up into the matrix by the mitochondrial calcium uniporter on the inner mitochondrial membrane. It is primarily driven by the mitochondrial membrane potential. Release of this calcium back into the cell's interior can occur via a sodium-calcium exchange protein or via "calcium-induced-calcium-release" pathways. This can initiate calcium spikes or calcium waves with large changes in the membrane potential. These can activate a series of second messenger system proteins that can coordinate processes such as neurotransmitter release in nerve cells and release of hormones in endocrine cells.

Ca^{2+} influx to the mitochondrial matrix has recently been implicated as a mechanism to regulate respiratory bioenergetics by allowing the electrochemical potential across the membrane to transiently "pulse" from $\Delta\Psi$-dominated to pH-dominated, facilitating a reduction of oxidative stress. In neurons, concomitant increases in cytosolic and mitochondrial calcium act to synchronize neuronal activity with mitochondrial energy metabolism. Mitochondrial matrix calcium levels can reach the tens of micromolar levels, which is necessary for the activation of isocitrate dehydrogenase, one of the key regulatory enzymes of the Krebs cycle.

Additional Functions

Mitochondria play a central role in many other metabolic tasks, such as:

- Signaling through mitochondrial reactive oxygen species,

- Regulation of the membrane potential,

- Apoptosis-programmed cell death,

- Calcium signaling (including calcium-evoked apoptosis),

- Regulation of cellular metabolism,

- Certain heme synthesis reactions,

- Steroid synthesis,

- Hormonal signaling Mitochondria are sensitive and responsive to hormones, in part by the action of mitochondrial estrogen receptors (mtERs). These receptors have been found in various tissues and cell types, including brain and heart.

Some mitochondrial functions are performed only in specific types of cells. For example, mitochondria in liver cells contain enzymes that allow them to detoxify ammonia, a waste product of protein metabolism. A mutation in the genes regulating any of these functions can result in mitochondrial diseases.

Cellular Proliferation Regulation

The relationship between cellular proliferation and mitochondria has been investigated using cervical cancer HeLa cells. Tumor cells require an ample amount of ATP (Adenosine triphosphate) in order to synthesize bioactive compounds such as lipids, proteins, and nucleotides for rapid cell proliferation. The majority of ATP in tumor cells is generated via the oxidative phosphorylation pathway (OxPhos). Interference with OxPhos have shown to cause cell cycle arrest suggesting that mitochondria play a role in cell proliferation. Mitochondrial ATP production is also vital for cell division in addition to other basic functions in the cell including the regulation of cell volume, solute concentration, and cellular architecture. ATP levels differ at various stages of the cell cycle suggesting that there is a relationship between the abundance of ATP and the cell's ability to enter a new cell cycle. ATP's role in the basic functions of the cell make the cell cycle sensitive to changes in the availability of mitochondrial derived ATP. The variation in ATP levels at different stages of the cell cycle support the hypothesis that mitochondria play an important role in cell cycle regulation. Although the specific mechanisms between mitochondria and the cell cycle regulation is not well understood, studies have shown that low energy cell cycle checkpoints monitor the energy capability before committing to another round of cell division.

Genome

Mitochondria contain their own genome, an indication that they are derived from bacteria through endosymbiosis. However, the ancestral endosymbiont genome has lost most of its genes so that the mitochondrial genome (mitogenome) is one of the most reduced genomes across organisms.

The human mitochondrial genome is a circular DNA molecule of about 16 kilobases. It encodes 37 genes: 13 for subunits of respiratory complexes I, III, IV and V, 22 for mitochondrial tRNA (for the 20 standard amino acids, plus an extra gene for leucine and serine), and 2 for rRNA. One mitochondrion can contain two to ten copies of its DNA.

As in prokaryotes, there is a very high proportion of coding DNA and an absence of repeats. Mitochondrial genes are transcribed as multigenic transcripts, which are cleaved and polyadenylated to yield mature mRNAs. Not all proteins necessary for mitochondrial function are encoded by the mitochondrial genome; most are coded by genes in the cell nucleus and the corresponding proteins are imported into the mitochondrion. The exact number of genes encoded by the nucleus and the mitochondrial genome differs between species. Most mitochondrial genomes are circular, although exceptions have been reported. In general, mitochondrial DNA lacks introns, as is the case in the human mitochondrial genome; however, introns have been observed in some eukaryotic mitochondrial DNA, such as that of yeast and protists, including *Dictyostelium discoideum*. Between protein-coding regions, tRNAs are present. During transcription, the tRNAs acquire their characteristic L-shape that gets recognized and cleaved by specific enzymes. Mitochondrial tRNA genes have different sequences from the nuclear tRNAs but lookalikes of mitochondrial tRNAs have been found in the nuclear chromosomes with high sequence similarity.

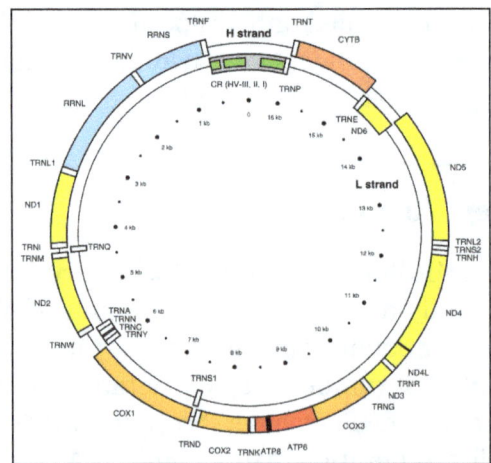

The circular 16,569 bp human mitochondrial genome
encoding 37 genes, *i.e.*, 28 on the H-strand and 9 on the L-strand.

In animals, the mitochondrial genome is typically a single circular chromosome that is approximately 16 kb long and has 37 genes. The genes, while highly conserved, may vary in location. Curiously, this pattern is not found in the human body louse (*Pediculus humanus*). Instead, this mitochondrial genome is arranged in 18 minicircular chromosomes, each of which is 3–4 kb long and has one to three genes. This pattern is also found in other sucking lice, but not in chewing lice. Recombination has been shown to occur between the minichromosomes. The reason for this difference is not known.

Alternative Genetic Code

While slight variations on the standard genetic code had been predicted earlier, none was discovered until 1979, when researchers studying human mitochondrial genes determined that they used an alternative code. However, the mitochondria of many other eukaryotes, including most plants, use the standard code. Many slight variants have been discovered since, including various alternative mitochondrial codes. Further, the AUA, AUC, and AUU codons are all allowable start codons.

Exceptions to the standard genetic code in mitochondria			
Organism	Codon	Standard	Mitochondria
Mammals	AGA, AGG	Arginine	Stop codon
Invertebrates	AGA, AGG	Arginine	Serine
Fungi	CUA	Leucine	Threonine
All of the above	AUA	Isoleucine	Methionine
	UGA	Stop codon	Tryptophan

Some of these differences should be regarded as pseudo-changes in the genetic code due to the phenomenon of RNA editing, which is common in mitochondria. In higher plants, it was thought that CGG encoded for tryptophan and not arginine; however, the codon in the processed RNA was discovered to be the UGG codon, consistent with the standard genetic code for tryptophan. Of note, the arthropod mitochondrial genetic code has undergone parallel evolution within a phylum, with some organisms uniquely translating AGG to lysine.

Evolution and Diversity

Mitochondrial genomes have far fewer genes than the bacteria from which they are thought to be descended. Although some have been lost altogether, many have been transferred to the nucleus, such as the respiratory complex II protein subunits. This is thought to be relatively common over evolutionary time. A few organisms, such as the *Cryptosporidium*, actually have mitochondria that lack any DNA, presumably because all their genes have been lost or transferred. In *Cryptosporidium*, the mitochondria have an altered ATP generation system that renders the parasite resistant to many classical mitochondrial inhibitors such as cyanide, azide, and atovaquone.

Replication and Inheritance

Mitochondria divide by binary fission, similar to bacterial cell division. The regulation of this division differs between eukaryotes. In many single-celled eukaryotes, their growth and division is linked to the cell cycle. For example, a single mitochondrion may divide synchronously with the nucleus. This division and segregation process must be tightly controlled so that each daughter cell receives at least one mitochondrion. In other eukaryotes (in mammals for example), mitochondria may replicate their DNA and divide mainly in response to the energy needs of the cell, rather than in phase with the cell cycle. When the energy needs of a cell are high, mitochondria grow and divide. When the energy use is low, mitochondria are destroyed or become inactive. In such examples, and in contrast to the situation in many single celled eukaryotes, mitochondria are apparently randomly distributed to the daughter cells during the division of the cytoplasm. Understanding of mitochondrial dynamics, which is described as the balance between mitochondrial fusion and fission, has revealed that functional and structural alterations in mitochondrial morphology are important factors in pathologies associated with several disease conditions.

The hypothesis of mitochondrial binary fission has relied on the visualization by fluorescence microscopy and conventional transmission electron microscopy (TEM). The resolution of fluorescence microscopy(~200 nm) is insufficient to distinguish structural details, such as double mitochondrial membrane in mitochondrial division or even to distinguish individual mitochondria when several are close together. Conventional TEM has also some technical limitations in verifying mitochondrial division. Cryo-electron tomography was recently used to visualize mitochondrial division in frozen hydrated intact cells. It revealed that mitochondria divide by budding.

An individual's mitochondrial genes are not inherited by the same mechanism as nuclear genes. Typically, the mitochondria are inherited from one parent only. In humans, when an egg cell is fertilized by a sperm, the egg nucleus and sperm nucleus each contribute equally to the genetic makeup of the zygote nucleus. In contrast, the mitochondria, and therefore the mitochondrial DNA, usually come from the egg only. The sperm's mitochondria enter the egg, but do not contribute genetic information to the embryo. Instead, paternal mitochondria are marked with ubiquitin to select them for later destruction inside the embryo. The egg cell contains relatively few mitochondria, but it is these mitochondria that survive and divide to populate the cells of the adult organism. Mitochondria are, therefore, in most cases inherited only from mothers, a pattern known as maternal inheritance. This mode is seen in most organisms, including the majority of animals. However, mitochondria in some species can sometimes be inherited paternally. This is the norm among certain coniferous plants, although not in pine trees and yews. For Mytilids, paternal inheritance only occurs within males of the species. It has been suggested that it occurs at a very low level in humans. There is a recent suggestion that mitochondria that shorten male lifespan stay in the system because they are inherited only through the mother. By contrast, natural selection weeds out mitochondria that reduce female survival as such mitochondria are less likely to be passed on to the next generation. Therefore, it is suggested that human females and female animals tend to live longer than males. The authors claim that this is a partial explanation.

Uniparental inheritance leads to little opportunity for genetic recombination between different lineages of mitochondria, although a single mitochondrion can contain 2–10 copies of its DNA. For this reason, mitochondrial DNA is usually thought to reproduce by binary fission. What recombination does take place maintains genetic integrity rather than maintaining diversity. However, there are studies showing evidence of recombination in mitochondrial DNA. It is clear that the enzymes necessary for recombination are present in mammalian cells. Further, evidence suggests that animal mitochondria can undergo recombination. The data are a bit more controversial in humans, although indirect evidence of recombination exists. If recombination does not occur, the whole mitochondrial DNA sequence represents a single haplotype, which makes it useful for studying the evolutionary history of populations.

Entities undergoing uniparental inheritance and with little to no recombination may be expected to be subject to Muller's ratchet, the inexorable accumulation of deleterious mutations until functionality is lost. Animal populations of mitochondria avoid this buildup through a developmental process known as the mtDNA bottleneck. The bottleneck exploits stochastic processes in the cell to increase in the cell-to-cell variability in mutant load as an organism develops: a single egg cell with some proportion of mutant mtDNA thus produces an embryo where different cells have different mutant loads. Cell-level selection may then act to remove those cells with more mutant mtDNA, leading to a stabilisation or reduction in mutant load between generations. The mechanism underlying the bottleneck is debated, with a recent mathematical and experimental metastudy providing evidence for a combination of random partitioning of mtDNAs at cell divisions and random turnover of mtDNA molecules within the cell.

DNA Repair

Mitochondria can repair oxidative DNA damage by mechanisms that are analogous to those occurring in the cell nucleus. The proteins that are employed in mtDNA repair are encoded by nuclear genes, and are translocated to the mitochondria. The DNA repair pathways in mammalian mitochondria include base excision repair, double-strand break repair, direct reversal and mismatch repair. Also DNA damages may be bypassed, rather than repaired, by translesion synthesis.

Of the several DNA repair process in mitochondria, the base excision repair pathway is the one that has been most comprehensively studied. Base excision repair is carried out by a sequence of enzymatic catalyzed steps that include recognition and excision of a damaged DNA base, removal of the resulting abasic site, end processing, gap filling and ligation. A common damage in mtDNA that is repaired by base excision repair is 8-oxoguanine produced by the oxidation of guanine.

Double-strand breaks can be repaired by homologous recombinational repair in both mammalian mtDNA and plant mtDNA. Double-strand breaks in mtDNA can also be repaired by microhomology-mediated end joining. Although there is evidence for the repair processes of direct reversal and mismatch repair in mtDNA, these processes are still not well characterized.

Lack of Mitochondrial DNA

Eukaryotic cells typically have mitochondrial DNA; however, mitochondria that lack their own DNA have been found in a marine parasitic dinoflagellate from the genus *Amoebophyra*. This microorganism, *A. cerati*, has functional mitochondria that lack a genome. In related species, the mitochondrial genome still has three genes, but in *A. cerati* only a single mitochondrial gene — the cytochrome c oxidase I gene (*cox1*) — is found, and it has migrated to the genome of the nucleus.

Population Genetic Studies

The near-absence of genetic recombination in mitochondrial DNA makes it a useful source of information for scientists involved in population genetics and evolutionary biology. Because all the mitochondrial DNA is inherited as a single unit, or haplotype, the relationships between mitochondrial DNA from different individuals can be represented as a gene tree. Patterns in these gene trees can be used to infer the evolutionary history of populations. The classic example of this is in human evolutionary genetics, where the molecular clock can be used to provide a recent date for mitochondrial Eve. This is often interpreted as strong support for a recent modern human expansion out of Africa. Another human example is the sequencing of mitochondrial DNA from Neanderthal bones. The relatively large evolutionary distance between the mitochondrial DNA sequences of Neanderthals and living humans has been interpreted as evidence for the lack of interbreeding between Neanderthals and anatomically modern humans.

However, mitochondrial DNA reflects only the history of the females in a population and so may not represent the history of the population as a whole. This can be partially overcome by the use of paternal genetic sequences, such as the non-recombining region of the Y-chromosome. In a broader sense, only studies that also include nuclear DNA can provide a comprehensive evolutionary history of a population.

Recent measurements of the molecular clock for mitochondrial DNA reported a value of 1 mutation every 7884 years dating back to the most recent common ancestor of humans and apes, which is consistent with estimates of mutation rates of autosomal DNA (10^{-8} per base per generation.

Dysfunction and Disease

Mitochondrial Diseases

Damage and subsequent dysfunction in mitochondria is an important factor in a range of human diseases due to their influence in cell metabolism. Mitochondrial disorders often present themselves as neurological disorders, including autism. They can also manifest as myopathy, diabetes, multiple endocrinopathy, and a variety of other systemic disorders. Diseases caused by mutation in the mtDNA include Kearns-Sayre syndrome, MELAS syndrome and Leber's hereditary optic neuropathy. In the vast majority of cases, these diseases are transmitted by a female to her children, as the zygote derives its mitochondria and hence its mtDNA from the ovum. Diseases such as Kearns-Sayre syndrome, Pearson syndrome, and progressive external ophthalmoplegia are thought to be due to large-scale mtDNA rearrangements, whereas other diseases such as MELAS syndrome, Leber's hereditary optic neuropathy, myoclonic epilepsy with ragged red fibers (MERRF), and others are due to point mutations in mtDNA.

In other diseases, defects in nuclear genes lead to dysfunction of mitochondrial proteins. This is the case in Friedreich's ataxia, hereditary spastic paraplegia, and Wilson's disease. These diseases are inherited in a dominance relationship, as applies to most other genetic diseases. A variety of disorders can be caused by nuclear mutations of oxidative phosphorylation enzymes, such as co-enzyme Q10 deficiency and Barth syndrome. Environmental influences may interact with hereditary predispositions and cause mitochondrial disease. For example, there may be a link between pesticide exposure and the later onset of Parkinson's disease. Other pathologies with etiology involving mitochondrial dysfunction include schizophrenia, bipolar disorder, dementia, Alzheimer's disease, Parkinson's disease, epilepsy, stroke, cardiovascular disease, chronic fatigue syndrome, retinitis pigmentosa, and diabetes mellitus.

Mitochondria-mediated oxidative stress plays a role in cardiomyopathy in Type 2 diabetics. Increased fatty acid delivery to the heart increases fatty acid uptake by cardiomyocytes, resulting in increased fatty acid oxidation in these cells. This process increases the reducing equivalents available to the electron transport chain of the mitochondria, ultimately increasing reactive oxygen species (ROS) production. ROS increases uncoupling proteins (UCPs) and potentiate proton leakage through the adenine nucleotide translocator (ANT), the combination of which uncouples the mitochondria. Uncoupling then increases oxygen consumption by the mitochondria, compounding the increase in fatty acid oxidation. This creates a vicious cycle of uncoupling; furthermore, even though oxygen consumption increases, ATP synthesis does not increase proportionally because the mitochondria is uncoupled. Less ATP availability ultimately results in an energy deficit presenting as reduced cardiac efficiency and contractile dysfunction. To compound the problem, impaired sarcoplasmic reticulum calcium release and reduced mitochondrial reuptake limits peak cytosolic levels of the important signaling ion during muscle contraction. The decreased intra-mitochondrial calcium concentration increases dehydrogenase activation and ATP synthesis. So in addition to lower ATP synthesis due to fatty acid oxidation, ATP synthesis is impaired by poor calcium signaling as well, causing cardiac problems for diabetics.

Possible Relationships to Aging

Given the role of mitochondria as the cell's powerhouse, there may be some leakage of the high-energy electrons in the respiratory chain to form reactive oxygen species. This was thought to result in significant oxidative stress in the mitochondria with high mutation rates of mitochondrial DNA (mtDNA). Hypothesized links between aging and oxidative stress are not new and were proposed in 1956, which was later refined into the mitochondrial free radical theory of aging. A vicious cycle was thought to occur, as oxidative stress leads to mitochondrial DNA mutations, which can lead to enzymatic abnormalities and further oxidative stress.

A number of changes can occur to mitochondria during the aging process. Tissues from elderly patients show a decrease in enzymatic activity of the proteins of the respiratory chain. However, mutated mtDNA can only be found in about 0.2% of very old cells. Large deletions in the mitochondrial genome have been hypothesized to lead to high levels of oxidative stress and neuronal death in Parkinson's disease.

CHLOROPLAST

Chloroplasts are organelles that conduct photosynthesis, where the photosynthetic pigment chlorophyll captures the energy from sunlight, converts it, and stores it in the energy-storage molecules ATP and NADPH while freeing oxygen from water in plant and algal cells. They then use the ATP and NADPH to make organic molecules from carbon dioxide in a process known as the Calvin cycle. Chloroplasts carry out a number of other functions, including fatty acid synthesis, much amino acid synthesis, and the immune response in plants. The number of chloroplasts per cell varies from one, in unicellular algae, up to 100 in plants like *Arabidopsis* and wheat.

Chloroplasts visible in the cells of *Bryum capillare*, a type of moss.

A chloroplast is a type of organelle known as a plastid, characterized by its two membranes and a high concentration of chlorophyll. Other plastid types, such as the leucoplast and the chromoplast, contain little chlorophyll and do not carry out photosynthesis.

Chloroplasts are highly dynamic—they circulate and are moved around within plant cells, and occasionally pinch in two to reproduce. Their behavior is strongly influenced by environmental factors like light color and intensity. Chloroplasts, like mitochondria, contain their own DNA, which is thought to be inherited from their ancestor—a photosynthetic cyanobacterium that was engulfed by an early eukaryotic cell. Chloroplasts cannot be made by the plant cell and must be inherited by each daughter cell during cell division.

With one exception (the amoeboid *Paulinella chromatophora*), all chloroplasts can probably be traced back to a single endosymbiotic event, when a cyanobacterium was engulfed by the eukaryote. Despite this, chloroplasts can be found in an extremely wide set of organisms, some not even directly related to each other—a consequence of many secondary and even tertiary endosymbiotic events.

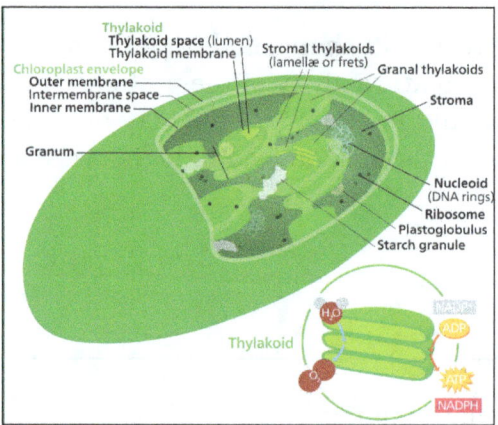

Structure of a typical higher-plant chloroplast.

Chloroplast Lineages and Evolution

Chloroplasts are one of many types of organelles in the plant cell. They are considered to have originated from cyanobacteria through endosymbiosis—when a eukaryotic cell engulfed a photosynthesizing cyanobacterium that became a permanent resident in the cell. Mitochondria are thought to have come from a similar event, where an aerobic prokaryote was engulfed. This origin of chloroplasts was first suggested by the Russian biologist Konstantin Mereschkowski in 1905 after Andreas Schimper observed in 1883 that chloroplasts closely resemble cyanobacteria. Chloroplasts are only found in plants, algae, and the amoeboid *Paulinella chromatophora*.

Cyanobacterial Ancestor

Cyanobacteria are considered the ancestors of chloroplasts. They are sometimes called blue-green algae even though they are prokaryotes. They are a diverse phylum of bacteria capable of carrying out photosynthesis, and are gram-negative, meaning that they have two cell membranes. Cyanobacteria also contain a peptidoglycan cell wall, which is thicker than in other gram-negative bacteria, and which is located between their two cell membranes. Like chloroplasts, they have thylakoids within. On the thylakoid membranes are photosynthetic pigments, including chlorophyll *a.* Phycobilins are also common cyanobacterial pigments, usually organized into hemispherical phycobilisomes attached to the outside of the thylakoid membranes (phycobilins are not shared with all chloroplasts though).

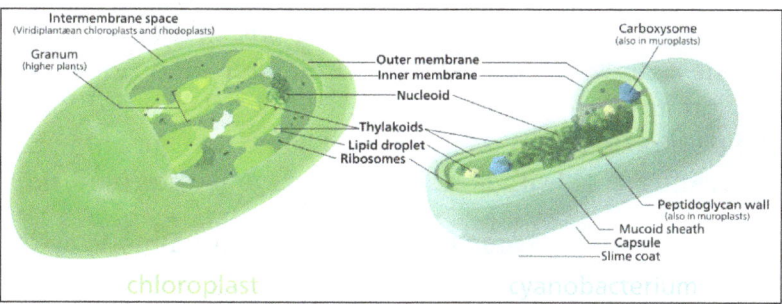

Both chloroplasts and cyanobacteria have a double membrane, DNA, ribosomes, and thylakoids. Both the chloroplast and cyanobacterium depicted are idealized versions (the chloroplast is that of a higher plant)—a lot of diversity exists among chloroplasts and cyanobacteria.

Primary Endosymbiosis

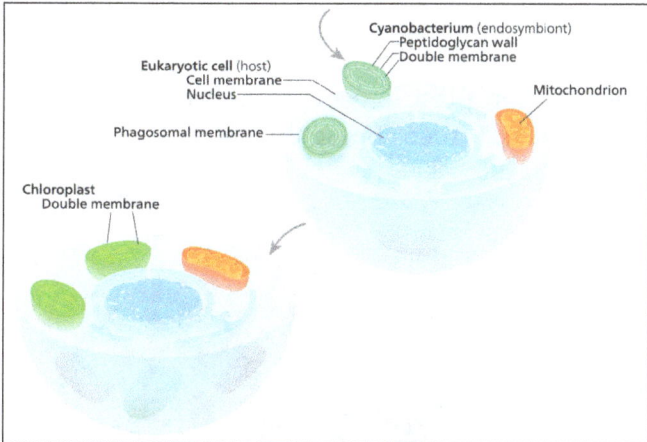

Primary endosymbiosis.

A eukaryote with mitochondria engulfed a cyanobacterium in an event of serial primary endosymbiosis, creating a lineage of cells with both organelles. It is important to note that the cyanobacterial endosymbiont already had a double membrane—the phagosomal vacuole-derived membrane was lost.

Somewhere around 1 to 2 billion years ago, a free-living cyanobacterium entered an early eukaryotic cell, either as food or as an internal parasite, but managed to escape the phagocytic vacuole it was contained in. The two innermost lipid-bilayer membranes that surround all chloroplasts correspond to the outer and inner membranes of the ancestral cyanobacterium's gram negative cell wall, and not the phagosomal membrane from the host, which was probably lost. The new cellular resident quickly became an advantage, providing food for the eukaryotic host, which allowed it to live within it. Over time, the cyanobacterium was assimilated, and many of its genes were lost or transferred to the nucleus of the host. From genomes that probably originally contained over 3000 genes only about 130 genes remain in the chloroplasts of contemporary plants. Some of its proteins were then synthesized in the cytoplasm of the host cell, and imported back into the chloroplast (formerly the cyanobacterium). Separately, somewhere around 500 million years ago, it happened again and led to the amoeboid *Paulinella chromatophora*.

This event is called *endosymbiosis*, or "cell living inside another cell with a mutual benefit for both". The external cell is commonly referred to as the *host* while the internal cell is called the *endosymbiont*. Chloroplasts are believed to have arisen after mitochondria, since all eukaryotes contain mitochondria, but not all have chloroplasts. This is called *serial endosymbiosis*—an early eukaryote engulfing the mitochondrion ancestor, and some descendants of it then engulfing the chloroplast ancestor, creating a cell with both chloroplasts and mitochondria.

Whether or not primary chloroplasts came from a single endosymbiotic event, or many independent engulfments across various eukaryotic lineages, has long been debated. It is now generally held that organisms with primary chloroplasts share a single ancestor that took in a cyanobacterium 600–2000 million years ago. It has been proposed this bacterium was *Gloeomargarita lithophora*. The exception is the amoeboid *Paulinella chromatophora*, which descends from an ancestor that took in a *Prochlorococcus* cyanobacterium 90–500 million years ago.

These chloroplasts, which can be traced back directly to a cyanobacterial ancestor, are known as *primary plastids* ("*plastid*" in this context means almost the same thing as chloroplast). All primary chloroplasts belong to one of four chloroplast lineages—the glaucophyte chloroplast lineage, the amoeboid *Paulinella chromatophora* lineage, the rhodophyte (red algal) chloroplast lineage, or the chloroplastidan (green) chloroplast lineage. The rhodophyte and chloroplastidan lineages are the largest, with chloroplastidan (green) being the one that contains the land plants.

Glaucophyta

Diversity of red algae Clockwise from top left: *Bornetia secundiflora*, *Peyssonnelia squamaria*, *Cyanidium*, *Laurencia*, *Callophyllis laciniata*. Red algal chloroplasts are characterized by phycobilin pigments which often give them their reddish color.

The alga *Cyanophora*, a glaucophyte, is thought to be one of the first organisms to contain a chloroplast. The glaucophyte chloroplast group is the smallest of the three primary chloroplast lineages, being found in only 13 species, and is thought to be the one that branched off the earliest. Glauco-phytes have chloroplasts that retain a peptidoglycan wall between their double membranes, like their cyanobacterial parent. For this reason, glaucophyte chloroplasts are also known as 'muro-plasts' (besides 'cyanoplasts' or 'cyanelles'). Glaucophyte chloroplasts also contain concentric un-stacked thylakoids, which surround a carboxysome – an icosahedral structure that glaucophyte chloroplasts and cyanobacteria keep their carbon fixation enzyme RuBisCO in. The starch that they synthesize collects outside the chloroplast. Like cyanobacteria, glaucophyte and rhodophyte chlo-roplast thylakoids are studded with light collecting structures called phycobilisomes. For these rea-sons, glaucophyte chloroplasts are considered a primitive intermediate between cyanobacteria and the more evolved chloroplasts in red algae and plants.

Rhodophyceae (Red Algae)

The rhodophyte, or red algal chloroplast group is another large and diverse chloroplast lineage. Rhodophyte chloroplasts are also called *rhodoplasts*, literally "red chloroplasts".

Rhodoplasts have a double membrane with an intermembrane space and phycobilin pigments or-ganized into phycobilisomes on the thylakoid membranes, preventing their thylakoids from stack-ing. Some contain pyrenoids. Rhodoplasts have chlorophyll *a* and phycobilins for photosynthetic pigments; the phycobilin phycoerytherin is responsible for giving many red algae their distinctive red color. However, since they also contain the blue-green chlorophyll *a* and other pigments, many are reddish to purple from the combination. The red phycoerytherin pigment is an adaptation to help red algae catch more sunlight in deep water—as such, some red algae that live in shallow water have less phycoerytherin in their rhodoplasts, and can appear more greenish. Rhodoplasts synthesize a form of starch called floridean starch, which collects into granules outside the rhodo-plast, in the cytoplasm of the red alga.

Chloroplastida (Green Algae and Plants)

Diversity of green algae Clockwise from top left: *Scenedesmus, Micrasterias, Hydrodictyon, Volvox, Stigeoclonium*. Green algal chloroplasts are characterized by their pigments chlorophyll *a* and chlorophyll *b* which give them their green color.

The chloroplastidan chloroplasts, or green chloroplasts, are another large, highly diverse primary chloroplast lineage. Their host organisms are commonly known as the green algae and land plants. They differ from glaucophyte and red algal chloroplasts in that they have lost their phycobilisomes, and contain chlorophyll *b* instead. Most green chloroplasts are (obviously) green, though some aren't, like some forms of *Hæmatococcus pluvialis*, due to accessory pigments that override the chlorophylls' green colors. Chloroplastidan chloroplasts have lost the peptidoglycan wall between their double membrane, leaving an intermembrane space. Some plants seem to have kept the genes for the synthesis of the peptidoglycan layer, though they've been repurposed for use in chloroplast division instead.

Green algae and plants keep their starch *inside* their chloroplasts, and in plants and some algae, the chloroplast thylakoids are arranged in grana stacks. Some green algal chloroplasts contain a structure called a pyrenoid, which is functionally similar to the glaucophyte carboxysome in that it is where RuBisCO and CO_2 are concentrated in the chloroplast.

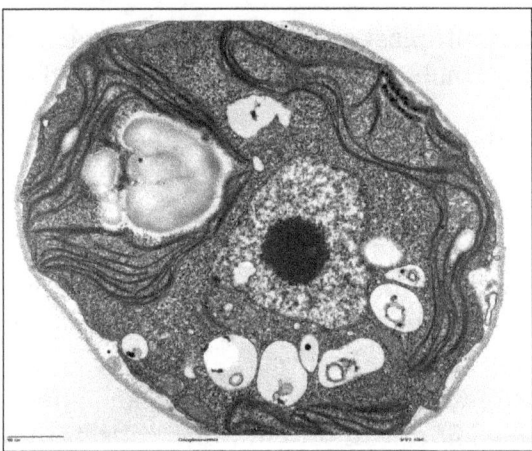

Transmission electron micrograph of *Chlamydomonas reinhardtii*,
a green alga that contains a pyrenoid surrounded by starch.

Helicosporidium is a genus of nonphotosynthetic parasitic green algae that is thought to contain a vestigial chloroplast. Genes from a chloroplast and nuclear genes indicating the presence of a chloroplast have been found in *Helicosporidium* even if nobody's seen the chloroplast itself.

Paulinella Chromatophora

While most chloroplasts originate from that first set of endosymbiotic events, *Paulinella chromatophora* is an exception that acquired a photosynthetic cyanobacterial endosymbiont more recently. It is not clear whether that symbiont is closely related to the ancestral chloroplast of other eukaryotes. Being in the early stages of endosymbiosis, *Paulinella chromatophora* can offer some insights into how chloroplasts evolved. *Paulinella* cells contain one or two sausage shaped blue-green photosynthesizing structures called chromatophores, descended from the cyanobacterium *Synechococcus*. Chromatophores cannot survive outside their host. Chromatophore DNA is about a million base pairs long, containing around 850 protein encoding genes—far less than the three million base pair *Synechococcus* genome, but much larger than the approximately 150,000 base pair genome of the more assimilated chloroplast. Chromatophores have transferred much less of their DNA to the nucleus of their host. About 0.3–0.8% of the nuclear DNA in *Paulinella* is from the chromatophore, compared with 11–14% from the chloroplast in plants.

Secondary and Tertiary Endosymbiosis

Many other organisms obtained chloroplasts from the primary chloroplast lineages through secondary endosymbiosis—engulfing a red or green alga that contained a chloroplast. These chloroplasts are known as secondary plastids.

While primary chloroplasts have a double membrane from their cyanobacterial ancestor, secondary chloroplasts have additional membranes outside of the original two, as a result of the secondary endosymbiotic event, when a nonphotosynthetic eukaryote engulfed a chloroplast-containing alga but failed to digest it—much like the cyanobacterium at the beginning of this story. The engulfed alga was broken down, leaving only its chloroplast, and sometimes its cell membrane and nucleus, forming a chloroplast with three or four membranes—the two cyanobacterial membranes, sometimes the eaten alga's cell membrane, and the phagosomal vacuole from the host's cell membrane.

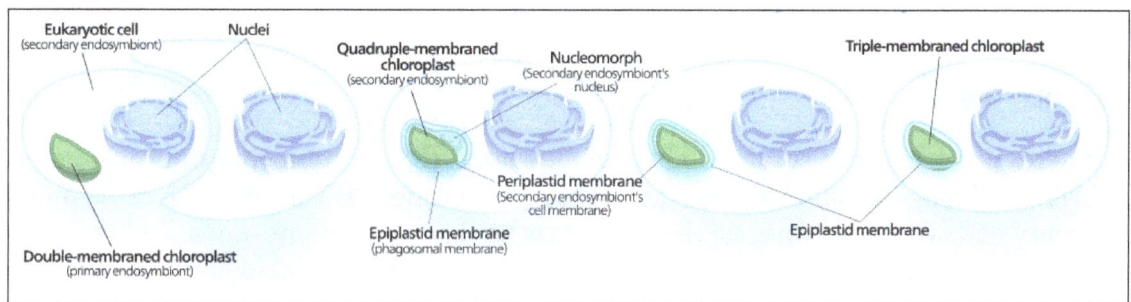

Secondary endosymbiosis consisted of a eukaryotic alga being engulfed by another eukaryote, forming a chloroplast with three or four membranes.

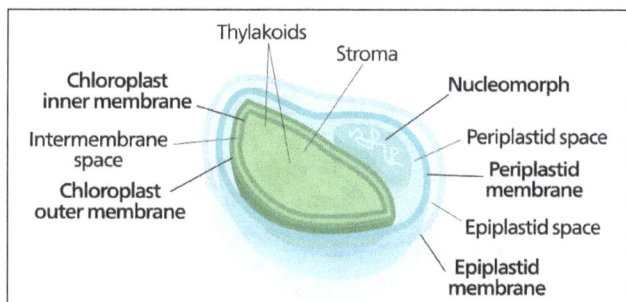

Diagram of a four membraned chloroplast containing a nucleomorph.

The genes in the phagocytosed eukaryote's nucleus are often transferred to the secondary host's nucleus. Cryptomonads and chlorarachniophytes retain the phagocytosed eukaryote's nucleus, an object called a nucleomorph, located between the second and third membranes of the chloroplast.

All secondary chloroplasts come from green and red algae—no secondary chloroplasts from glaucophytes have been observed, probably because glaucophytes are relatively rare in nature, making them less likely to have been taken up by another eukaryote.

Green Algal Derived Chloroplasts

Green algae have been taken up by the euglenids, chlorarachniophytes, a lineage of dinoflagellates, and possibly the ancestor of the CASH lineage (cryptomonads, alveolates, stramenopiles and haptophytes) in three or four separate engulfments. Many green algal derived chloroplasts contain pyrenoids, but unlike chloroplasts in their green algal ancestors, storage product collects in granules outside the chloroplast.

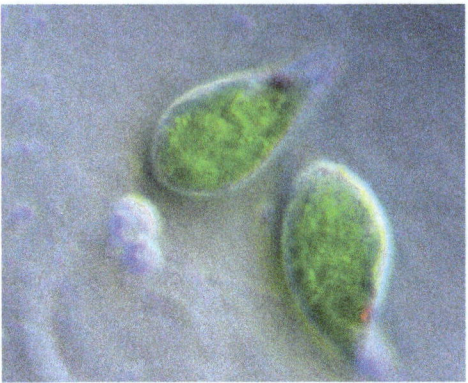

Euglena, a euglenophyte, contains secondary chloroplasts from green algae.

Euglenophytes

Euglenophytes are a group of common flagellated protists that contain chloroplasts derived from a green alga. Euglenophyte chloroplasts have three membranes—it is thought that the membrane of the primary endosymbiont was lost, leaving the cyanobacterial membranes, and the secondary host's phagosomal membrane. Euglenophyte chloroplasts have a pyrenoid and thylakoids stacked in groups of three. Photosynthetic product is stored in the form of paramylon, which is contained in membrane-bound granules in the cytoplasm of the euglenophyte.

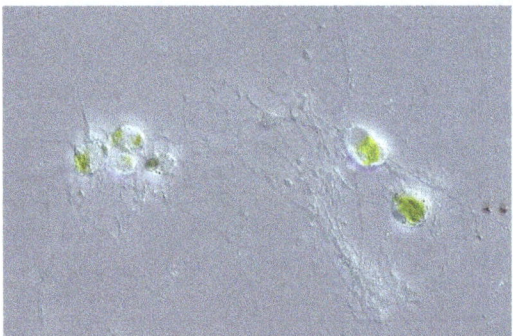

Chlorarachnion reptans is a chlorarachniophyte. Chlorarachniophytes replaced their original red algal endosymbiont with a green alga.

Chlorarachniophytes

Chlorarachniophytes are a rare group of organisms that also contain chloroplasts derived from green algae, though their story is more complicated than that of the euglenophytes. The ancestor of chlorarachniophytes is thought to have been a eukaryote with a *red* algal derived chloroplast. It is then thought to have lost its first red algal chloroplast, and later engulfed a green alga, giving it its second, green algal derived chloroplast.

Chlorarachniophyte chloroplasts are bounded by four membranes, except near the cell membrane, where the chloroplast membranes fuse into a double membrane. Their thylakoids are arranged in loose stacks of three. Chlorarachniophytes have a form of polysaccharide called chrysolaminarin, which they store in the cytoplasm, often collected around the chloroplast pyrenoid, which bulges into the cytoplasm. Chlorarachniophyte chloroplasts are notable because the green alga they are derived from has not been completely broken down—its nucleus still persists as a nucleomorph found between the second and third chloroplast membranes—the periplastid space, which corresponds to the green alga's cytoplasm.

Prasinophyte-derived Dinophyte Chloroplast

Lepidodinium viride and its close relatives are dinophytes that lost their original peridinin chloroplast and replaced it with a green algal derived chloroplast (more specifically, a prasinophyte). *Lepidodinium* is the only dinophyte that has a chloroplast that's not from the rhodoplast lineage. The chloroplast is surrounded by two membranes and has no nucleomorph—all the nucleomorph genes have been transferred to the dinophyte nucleus. The endosymbiotic event that led to this chloroplast was serial secondary endosymbiosis rather than tertiary endosymbiosis—the endosymbiont was a green alga containing a primary chloroplast (making a secondary chloroplast).

Red Algal Derived Chloroplasts

Cryptophytes

Cryptophytes, or cryptomonads are a group of algae that contain a red-algal derived chloroplast. Cryptophyte chloroplasts contain a nucleomorph that superficially resembles that of the chlorarachniophytes. Cryptophyte chloroplasts have four membranes, the outermost of which is continuous with the rough endoplasmic reticulum. They synthesize ordinary starch, which is stored in granules found in the periplastid space—outside the original double membrane, in the place that corresponds to the red alga's cytoplasm. Inside cryptophyte chloroplasts is a pyrenoid and thylakoids in stacks of two.

Their chloroplasts do not have phycobilisomes, but they do have phycobilin pigments which they keep in their thylakoid space, rather than anchored on the outside of their thylakoid membranes. Cryptophytes may have played a key role in the spreading of red algal based chloroplasts.

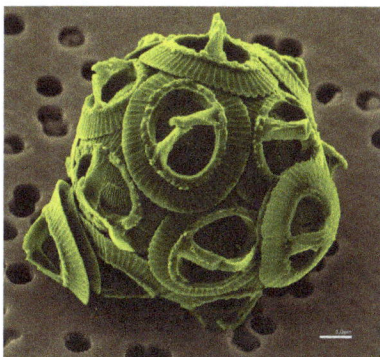

Scanning electron micrograph of *Gephyrocapsa oceanica*, a haptophyte.

Haptophytes

Haptophytes are similar and closely related to cryptophytes or heterokontophytes. Their chloroplasts lack a nucleomorph, their thylakoids are in stacks of three, and they synthesize chrysolaminarin sugar, which they store completely outside of the chloroplast, in the cytoplasm of the haptophyte.

Heterokontophytes (Stramenopiles)

The photosynthetic pigments present in their chloroplasts
give diatoms a greenish-brown color.

The heterokontophytes, also known as the stramenopiles, are a very large and diverse group of eukaryotes. The photoautotrophic lineage, Ochrophyta, including the diatoms and the brown algae, golden algae, and yellow-green algae, also contains red algal derived chloroplasts.

Heterokont chloroplasts are very similar to haptophyte chloroplasts, containing a pyrenoid, triplet thylakoids, and with some exceptions, having four layer plastidic envelope, the outermost epiplastid membrane connected to the endoplasmic reticulum. Like haptophytes, heterokontophytes store sugar in chrysolaminarin granules in the cytoplasm. Heterokontophyte chloroplasts contain chlorophyll *a* and with a few exceptions chlorophyll *c*, but also have carotenoids which give them their many colors.

Apicomplexans, Chromerids and Dinophytes

The alveolates are a major clade of unicellular eukaryotes of both autotrophic and heterotrophic members. The most notable shared characteristic is the presence of cortical (outer-region) alveoli (sacs). These are flattened vesicles (sacs) packed into a continuous layer just under the membrane and supporting it, typically forming a flexible pellicle (thin skin). In dinoflagellates they often form armor plates. Many members contain a red-algal derived plastid. One notable characteristic of this diverse group is the frequent loss of photosynthesis. However, a majority of these heterotrophs continue to process a non-photosynthetic plastid.

Apicomplexans

Apicomplexans are a group of alveolates. Like the helicosproidia, they're parasitic, and have a non-photosynthetic chloroplast. They were once thought to be related to the helicosproidia, but it is now known that the helicosproida are green algae rather than part of the CASH lineage. The apicomplexans include *Plasmodium*, the malaria parasite. Many apicomplexans keep a vestigial red algal derived chloroplast called an apicoplast, which they inherited from their ancestors. Other apicomplexans like *Cryptosporidium* have lost the chloroplast completely. Apicomplexans store their energy in amylopectin granules that are located in their cytoplasm, even though they are nonphotosynthetic.

Apicoplasts have lost all photosynthetic function, and contain no photosynthetic pigments or true thylakoids. They are bounded by four membranes, but the membranes are not connected to the endoplasmic reticulum. The fact that apicomplexans still keep their nonphotosynthetic chloroplast around demonstrates how the chloroplast carries out important functions other than photosynthesis. Plant chloroplasts provide plant cells with many important things besides sugar, and apicoplasts are no different—they synthesize fatty acids, isopentenyl pyrophosphate, iron-sulfur clusters, and carry out part of the heme pathway. This makes the apicoplast an attractive target for drugs to cure apicomplexan-related diseases. The most important apicoplast function is isopentenyl pyrophosphate synthesis—in fact, apicomplexans die when something interferes with this apicoplast function, and when apicomplexans are grown in an isopentenyl pyrophosphate-rich medium, they dump the organelle.

Chromerids

The Chromerida is a newly discovered group of algae from Australian corals which comprises some close photosynthetic relatives of the apicomplexans. The first member, *Chromera velia*, was discovered and first isolated in 2001. The discovery of *Chromera velia* with similar structure to the apicomplexanss, provides an important link in the evolutionary history of the apicomplexans and dinophytes. Their plastids have four membranes, lack chlorophyll c and use the type II form of RuBisCO obtained from a horizontal transfer event.

Dinophytes

The dinoflagellates are yet another very large and diverse group of protists, around half of which are (at least partially) photosynthetic.

Most dinophyte chloroplasts are secondary red algal derived chloroplasts. Many other dinophytes have lost the chloroplast (becoming the nonphotosynthetic kind of dinoflagellate), or replaced it though *tertiary* endosymbiosis—the engulfment of another eukaryotic algae containing a red algal derived chloroplast. Others replaced their original chloroplast with a green algal derived one.

Most dinophyte chloroplasts contain form II RuBisCO, at least the photosynthetic pigments chlorophyll *a*, chlorophyll c_2, *beta*-carotene, and at least one dinophyte-unique xanthophyll (peridinin, dinoxanthin, or diadinoxanthin), giving many a golden-brown color. All dinophytes store starch in their cytoplasm, and most have chloroplasts with thylakoids arranged in stacks of three.

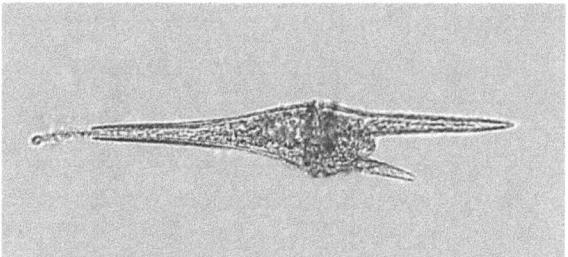

Ceratium furca, a peridinin-containing dinophyte.

The most common dinophyte chloroplast is the peridinin-type chloroplast, characterized by the carotenoid pigment peridinin in their chloroplasts, along with chlorophyll *a* and chlorophyll c_2. Peridinin is not found in any other group of chloroplasts. The peridinin chloroplast is bounded by three membranes (occasionally two), having lost the red algal endosymbiont's original cell membrane. The outermost membrane is not connected to the endoplasmic reticulum. They contain a pyrenoid, and have triplet-stacked thylakoids. Starch is found outside the chloroplast. An important feature of these chloroplasts is that their chloroplast DNA is highly reduced and fragmented into many small circles. Most of the genome has migrated to the nucleus, and only critical photosynthesis-related genes remain in the chloroplast.

The peridinin chloroplast is thought to be the dinophytes' "original" chloroplast, which has been lost, reduced, replaced, or has company in several other dinophyte lineages.

Fucoxanthin-containing (Haptophyte-derived) Dinophyte Chloroplasts

The fucoxanthin dinophyte lineages (including *Karlodinium* and *Karenia*) lost their original red algal derived chloroplast, and replaced it with a new chloroplast derived from a haptophyte endosymbiont. *Karlodinium* and *Karenia* probably took up different heterokontophytes. Because the haptophyte chloroplast has four membranes, tertiary endosymbiosis would be expected to create a six membraned chloroplast, adding the haptophyte's cell membrane and the dinophyte's phagosomal vacuole. However, the haptophyte was heavily reduced, stripped of a few membranes and its nucleus, leaving only its chloroplast (with its original double membrane), and possibly one or two additional membranes around it.

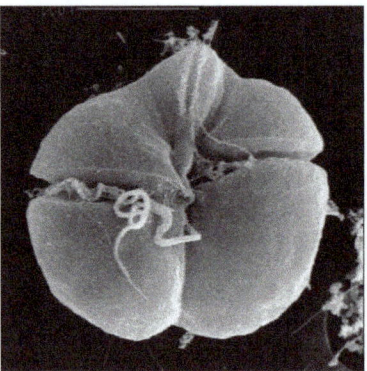

Karenia brevis is a fucoxanthin-containing dynophyte
responsible for algal blooms called "red tides".

Fucoxanthin-containing chloroplasts are characterized by having the pigment fucoxanthin (actually 19′-hexanoyloxy-fucoxanthin 19′-butanoyloxy-fucoxanthin) and no peridinin. Fucoxanthin is also found in haptophyte chloroplasts, providing evidence of ancestry.

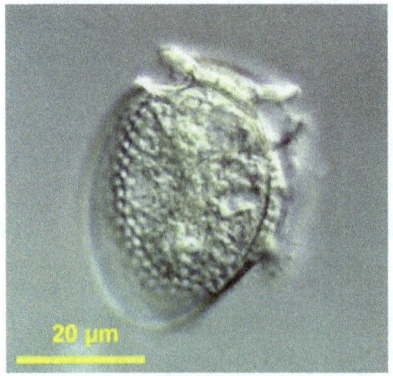

Dinophysis acuminata has chloroplasts
taken from a cryptophyte.

Diatom-derived Dinophyte Chloroplasts

Some dinophytes, like *Kryptoperidinium* and *Durinskia* have a diatom (heterokontophyte) derived chloroplast. These chloroplasts are bounded by up to *five* membranes, (depending on whether you count the entire diatom endosymbiont as the chloroplast, or just the red algal derived chloroplast inside it). The diatom endosymbiont has been reduced relatively little—it still retains its original mitochondria, and has endoplasmic reticulum, ribosomes, a nucleus, and of course, red algal derived chloroplasts—practically a complete cell, all inside the host's endoplasmic reticulum lumen. However the diatom endosymbiont can't store its own food—its storage polysaccharide is found in granules in the dinophyte host's cytoplasm instead. The diatom endosymbiont's nucleus is present, but it probably can't be called a nucleomorph because it shows no sign of genome reduction, and might have even been *expanded*. Diatoms have been engulfed by dinoflagellates at least three times.

The diatom endosymbiont is bounded by a single membrane, inside it are chloroplasts with four membranes. Like the diatom endosymbiont's diatom ancestor, the chloroplasts have triplet thylakoids and pyrenoids. In some of these genera, the diatom endosymbiont's chloroplasts aren't the only chloroplasts in the dinophyte. The original three-membraned peridinin chloroplast is still around, converted to an eyespot.

Kleptoplastidy

In some groups of mixotrophic protists, like some dinoflagellates (e.g. *Dinophysis*), chloroplasts are separated from a captured alga and used temporarily. These klepto chloroplasts may only have a lifetime of a few days and are then replaced.

Cryptophyte-derived Dinophyte Chloroplast

Members of the genus *Dinophysis* have a phycobilin-containing chloroplast taken from a cryptophyte. However, the cryptophyte is not an endosymbiont—only the chloroplast seems to have been taken, and the chloroplast has been stripped of its nucleomorph and outermost two membranes, leaving just a two-membraned chloroplast. Cryptophyte chloroplasts require their nucleomorph to maintain themselves, and *Dinophysis* species grown in cell culture alone cannot survive, so it is possible (but not confirmed) that the *Dinophysis* chloroplast is a kleptoplast—if so, *Dinophysis* chloroplasts wear out and *Dinophysis* species must continually engulf cryptophytes to obtain new chloroplasts to replace the old ones.

Chloroplast DNA

Chloroplasts have their own DNA, often abbreviated as ctDNA, or cpDNA. It is also known as the plastome. Its existence was first proved in 1962, and first sequenced in 1986—when two Japanese research teams sequenced the chloroplast DNA of liverwort and tobacco. Since then, hundreds of chloroplast DNAs from various species have been sequenced, but they are mostly those of land plants and green algae—glaucophytes, red algae, and other algal groups are extremely underrepresented, potentially introducing some bias in views of "typical" chloroplast DNA structure and content.

Molecular Structure

With few exceptions, most chloroplasts have their entire chloroplast genome combined into a single large circular DNA molecule, typically 120,000–170,000 base pairs long. They can have a contour length of around 30–60 micrometers, and have a mass of about 80–130 million daltons.

While usually thought of as a circular molecule, there is some evidence that chloroplast DNA molecules more often take on a linear shape.

Inverted Repeats

Many chloroplast DNAs contain two *inverted repeats*, which separate a long single copy section (LSC) from a short single copy section (SSC). While a given pair of inverted repeats are rarely completely identical, they are always very similar to each other, apparently resulting from concerted evolution.

The inverted repeats vary wildly in length, ranging from 4,000 to 25,000 base pairs long each and containing as few as four or as many as over 150 genes. Inverted repeats in plants tend to be at the upper end of this range, each being 20,000–25,000 base pairs long.

The inverted repeat regions are highly conserved among land plants, and accumulate few mutations. Similar inverted repeats exist in the genomes of cyanobacteria and the other two chloroplast lineages (glaucophyta and rhodophyceae), suggesting that they predate the chloroplast, though some chloroplast DNAs have since lost or flipped the inverted repeats (making them direct repeats). It is possible that the inverted repeats help stabilize the rest of the chloroplast genome, as chloroplast DNAs which have lost some of the inverted repeat segments tend to get rearranged more.

Nucleoids

New chloroplasts may contain up to 100 copies of their DNA, though the number of chloroplast DNA copies decreases to about 15–20 as the chloroplasts age. They are usually packed into nucleoids, which can contain several identical chloroplast DNA rings. Many nucleoids can be found in each chloroplast. In primitive red algae, the chloroplast DNA nucleoids are clustered in the center of the chloroplast, while in green plants and green algae, the nucleoids are dispersed throughout the stroma.

Though chloroplast DNA is not associated with true histones, in red algae, similar proteins that tightly pack each chloroplast DNA ring into a nucleoid have been found.

DNA Repair

In chloroplasts of the moss *Physcomitrella patens*, the DNA mismatch repair protein Msh1 interacts with the recombinational repair proteins RecA and RecG to maintain chloroplast genome stability. In chloroplasts of the plant *Arabidopsis thaliana* the RecA protein maintains the integrity of the chloroplast's DNA by a process that likely involves the recombinational repair of DNA damage.

DNA Replication

Leading Model of cpDNA Replication

The mechanism for chloroplast DNA (cpDNA) replication has not been conclusively determined, but two main models have been proposed. Scientists have attempted to observe chloroplast replication via electron microscopy since the 1970s. The results of the microscopy experiments led to the idea that chloroplast DNA replicates using a double displacement loop (D-loop). As the D-loop moves through the circular DNA, it adopts a theta intermediary form, also known as a Cairns replication intermediate, and completes replication with a rolling circle mechanism. Transcription starts at specific points of origin. Multiple replication forks open up, allowing replication machinery to transcribe the DNA. As replication continues, the forks grow and eventually converge. The new cpDNA structures separate, creating daughter cpDNA chromosomes.

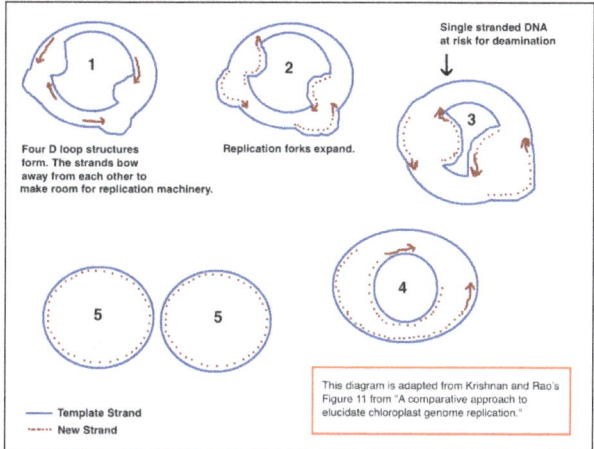

Chloroplast DNA replication via multiple D loop mechanisms.
Adapted from Krishnan NM, Rao BJ's paper "A comparative
approach to elucidate chloroplast genome replication."

In addition to the early microscopy experiments, this model is also supported by the amounts of deamination seen in cpDNA. Deamination occurs when an amino group is lost and is a mutation that often results in base changes. When adenine is deaminated, it becomes hypoxanthine. Hypoxanthine can bind to cytosine, and when the XC base pair is replicated, it becomes a GC (thus, an A → G base change).

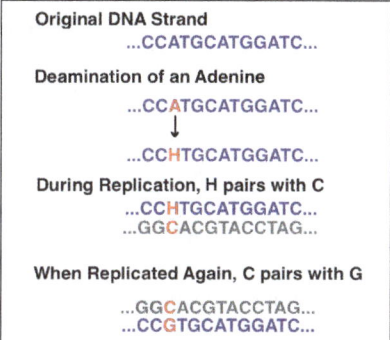

Over time, base changes in the DNA sequence can arise from deamination mutations.
When adenine is deaminated, it becomes hypoxanthine, which can pair with cytosine.
During replication, the cytosine will pair with guanine, causing an A --> G base change.

Deamination

In cpDNA, there are several A → G deamination gradients. DNA becomes susceptible to deamination events when it is single stranded. When replication forks form, the strand not being copied is single stranded, and thus at risk for A → G deamination. Therefore, gradients in deamination indicate that replication forks were most likely present and the direction that they initially opened (the highest gradient is most likely nearest the start site because it was single stranded for the longest amount of time). This mechanism is still the leading theory today; however, a second theory suggests that most cpDNA is actually linear and replicates through homologous recombination. It further contends that only a minority of the genetic material is kept in circular chromosomes while the rest is in branched, linear, or other complex structures.

Alternative Model of Replication

One of competing model for cpDNA replication asserts that most cpDNA is linear and participates in homologous recombination and replication structures similar to the linear and circular DNA structures of bacteriophage T4. It has been established that some plants have linear cpDNA, such as maize, and that more species still contain complex structures that scientists do not yet understand. When the original experiments on cpDNA were performed, scientists did notice linear structures; however, they attributed these linear forms to broken circles. If the branched and complex structures seen in cpDNA experiments are real and not artifacts of concatenated circular DNA or broken circles, then a D-loop mechanism of replication is insufficient to explain how those structures would replicate. At the same time, homologous recombination does not expand the multiple A --> G gradients seen in plastomes. Because of the failure to explain the deamination gradient as well as the numerous plant species that have been shown to have circular cpDNA, the predominant theory continues to hold that most cpDNA is circular and most likely replicates via a D loop mechanism.

Gene Content and Protein Synthesis

The chloroplast genome most commonly includes around 100 genes that code for a variety of things, mostly to do with the protein pipeline and photosynthesis. As in prokaryotes, genes in chloroplast DNA are organized into operons. Unlike prokaryotic DNA molecules, chloroplast DNA molecules contain introns (plant mitochondrial DNAs do too, but not human mtDNAs). Among land plants, the contents of the chloroplast genome are fairly similar.

Chloroplast genome Reduction and Gene Transfer

Over time, many parts of the chloroplast genome were transferred to the nuclear genome of the host, a process called *endosymbiotic gene transfer*. As a result, the chloroplast genome is heavily reduced compared to that of free-living cyanobacteria. Chloroplasts may contain 60–100 genes whereas cyanobacteria often have more than 1500 genes in their genome. Recently, a plastid without a genome was found, demonstrating chloroplasts can lose their genome during endosymbiotic the gene transfer process.

Endosymbiotic gene transfer is how we know about the lost chloroplasts in many CASH lineages. Even if a chloroplast is eventually lost, the genes it donated to the former host's nucleus persist, providing evidence for the lost chloroplast's existence. For example, while diatoms (a heterokontophyte) now have a red algal derived chloroplast, the presence of many green algal genes in the diatom nucleus provide evidence that the diatom ancestor had a green algal de-rived chloroplast at some point, which was subsequently replaced by the red chloroplast.

In land plants, some 11–14% of the DNA in their nuclei can be traced back to the chloroplast, up to 18% in *Arabidopsis*, corresponding to about 4,500 protein-coding genes. There have been a few recent transfers of genes from the chloroplast DNA to the nuclear genome in land plants.

Of the approximately 3000 proteins found in chloroplasts, some 95% of them are encoded by nucle-ar genes. Many of the chloroplast's protein complexes consist of subunits from both the chloroplast genome and the host's nuclear genome. As a result, protein synthesis must be coordinated between the chloroplast and the nucleus. The chloroplast is mostly under nuclear control, though chloroplasts can also give out signals regulating gene expression in the nucleus, called *retrograde signaling*.

Protein Synthesis

Protein synthesis within chloroplasts relies on two RNA polymerases. One is coded by the chlo-roplast DNA, the other is of nuclear origin. The two RNA polymerases may recognize and bind to different kinds of promoters within the chloroplast genome. The ribosomes in chloroplasts are similar to bacterial ribosomes.

Protein Targeting and Import

Because so many chloroplast genes have been moved to the nucleus, many proteins that would orig-inally have been translated in the chloroplast are now synthesized in the cytoplasm of the plant cell. These proteins must be directed back to the chloroplast, and imported through at least two chloro-plast membranes.

Curiously, around half of the protein products of transferred genes aren't even targeted back to the chloroplast. Many became exaptations, taking on new functions like participating in cell division, protein routing, and even disease resistance. A few chloroplast genes found new homes in the mito-chondrial genome—most became nonfunctional pseudogenes, though a few tRNA genes still work in the mitochondrion. Some transferred chloroplast DNA protein products get directed to the secretory pathway though many secondary plastids are bounded by an outermost membrane derived from the host's cell membrane, and therefore topologically outside of the cell, because to reach the chloroplast from the cytosol, you have to cross the cell membrane, just like if you were headed for the extracellu-lar space. In those cases, chloroplast-targeted proteins do initially travel along the secretory pathway.

Because the cell acquiring a chloroplast already had mitochondria (and peroxisomes, and a cell membrane for secretion), the new chloroplast host had to develop a unique protein targeting sys-tem to avoid having chloroplast proteins being sent to the wrong organelle.

The two ends of a polypeptide are called the N-terminus, or *amino end,* and the C-terminus, or *carboxyl end*. This polypeptide has four amino acids linked together. At the left is the N-terminus, with its amino (H_2N) group in green. The blue C-terminus, with its carboxyl group (CO_2H) is at the right.

In most, but not all cases, nuclear-encoded chloroplast proteins are translated with a *cleavable transit peptide* that's added to the N-terminus of the protein precursor. Sometimes the transit sequence is found on the C-terminus of the protein, or within the functional part of the protein.

Transport Proteins and Membrane Translocons

After a chloroplast polypeptide is synthesized on a ribosome in the cytosol, an enzyme specific to chloroplast proteins phosphorylates, or adds a phosphate group to many (but not all) of them in their transit sequences. Phosphorylation helps many proteins bind the polypeptide, keeping it from folding prematurely. This is important because it prevents chloroplast proteins from assuming their active form and carrying out their chloroplast functions in the wrong place—the cytosol. At the same time, they have to keep just enough shape so that they can be recognized by the chloroplast. These proteins also help the polypeptide get imported into the chloroplast.

From here, chloroplast proteins bound for the stroma must pass through two protein complexes—the TOC complex, or *translocon on the outer chloroplast membrane*, and the TIC translocon, or *translocon on the inner chloroplast membrane translocon*. Chloroplast polypeptide chains probably often travel through the two complexes at the same time, but the TIC complex can also retrieve preproteins lost in the intermembrane space.

Structure

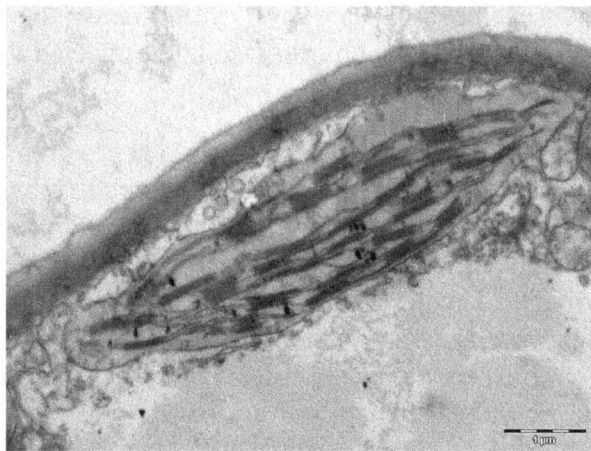

Transmission electron microscope image of a chloroplast. Grana of
thylakoids and their connecting lamellae are clearly visible.

In land plants, chloroplasts are generally lens-shaped, 3–10 μm in diameter and 1–3 μm thick. Corn seedling chloroplasts are ≈20 μm³ in volume. Greater diversity in chloroplast shapes exists among the algae, which often contain a single chloroplast that can be shaped like a net (e.g., *Oedogonium*), a cup (e.g., *Chlamydomonas*), a ribbon-like spiral around the edges of the cell (e.g., *Spirogyra*), or slightly twisted bands at the cell edges (e.g., *Sirogonium*). Some algae have two chloroplasts in each cell; they are star-shaped in *Zygnema*, or may follow the shape of half the cell in order Desmidiales. In some algae, the chloroplast takes up most of the cell, with pockets for the nucleus and other organelles, for example, some species of *Chlorella* have a cup-shaped chloroplast that occupies much of the cell.

All chloroplasts have at least three membrane systems—the outer chloroplast membrane, the inner chloroplast membrane, and the thylakoid system. Chloroplasts that are the product of secondary endosymbiosis may have additional membranes surrounding these three. Inside the outer and inner chloroplast membranes is the chloroplast stroma, a semi-gel-like fluid that makes up much of a chloroplast's volume, and in which the thylakoid system floats.

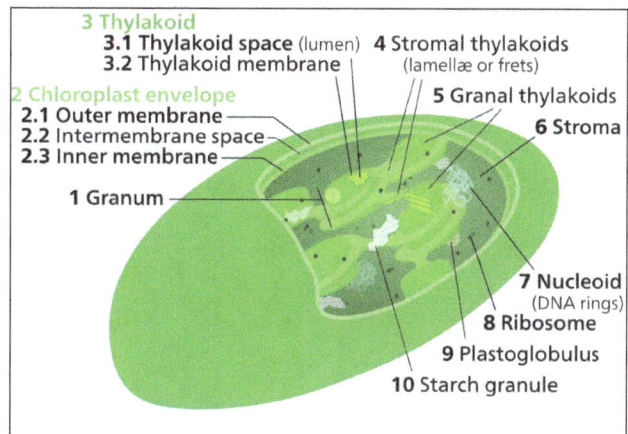

Chloroplast ultrastructure *(interactive diagram)* Chloroplasts have at least three distinct membrane systems, and a variety of things can be found in their stroma.

There are some common misconceptions about the outer and inner chloroplast membranes. The fact that chloroplasts are surrounded by a double membrane is often cited as evidence that they are the descendants of endosymbiotic cyanobacteria. This is often interpreted as meaning the outer chloroplast membrane is the product of the host's cell membrane infolding to form a vesicle to surround the ancestral cyanobacterium—which is not true—both chloroplast membranes are homologous to the cyanobacterium's original double membranes.

The chloroplast double membrane is also often compared to the mitochondrial double membrane. This is not a valid comparison—the inner mitochondria membrane is used to run proton pumps and carry out oxidative phosphorylation across to generate ATP energy. The only chloroplast structure that can considered analogous to it is the internal thylakoid system. Even so, in terms of "in-out", the direction of chloroplast H^+ ion flow is in the opposite direction compared to oxidative phosphorylation in mitochondria. In addition, in terms of function, the inner chloroplast membrane, which regulates metabolite passage and synthesizes some materials, has no counterpart in the mitochondrion.

Outer Chloroplast Membrane

The outer chloroplast membrane is a semi-porous membrane that small molecules and ions can easily diffuse across. However, it is not permeable to larger proteins, so chloroplast polypeptides being synthesized in the cell cytoplasm must be transported across the outer chloroplast membrane by the TOC complex, or *translocon on the outer chloroplast* membrane.

The chloroplast membranes sometimes protrude out into the cytoplasm, forming a stromule, or stroma-containing tubule. Stromules are very rare in chloroplasts, and are much more common in other plastids like chromoplasts and amyloplasts in petals and roots, respectively. They may exist to increase the chloroplast's surface area for cross-membrane transport, because they are often branched and tangled with the endoplasmic reticulum. When they were first observed in 1962, some plant biologists dismissed the structures as artifactual, claiming that stromules were just oddly shaped chloroplasts with constricted regions or dividing chloroplasts. However, there is a growing body of evidence that stromules are functional, integral features of plant cell plastids, not merely artifacts.

Intermembrane Space and Peptidoglycan Wall

Instead of an intermembrane space, glaucophyte algae have a peptidoglycan
wall between their inner and outer chloroplast membranes.

Usually, a thin intermembrane space about 10–20 nanometers thick exists between the outer and inner chloroplast membranes.

Glaucophyte algal chloroplasts have a peptidoglycan layer between the chloroplast membranes. It corresponds to the peptidoglycan cell wall of their cyanobacterial ancestors, which is located between their two cell membranes. These chloroplasts are called *muroplasts* (from Latin *"mura"*, meaning "wall"). Other chloroplasts have lost the cyanobacterial wall, leaving an intermembrane space between the two chloroplast envelope membranes.

Inner Chloroplast Membrane

The inner chloroplast membrane borders the stroma and regulates passage of materials in and out of the chloroplast. After passing through the TOC complex in the outer chloroplast membrane, polypeptides must pass through the TIC complex translocon on the inner chloroplast membrane which is located in the inner chloroplast membrane.

In addition to regulating the passage of materials, the inner chloroplast membrane is where fatty acids, lipids, and carotenoids are synthesized.

Peripheral Reticulum

Some chloroplasts contain a structure called the chloroplast peripheral reticulum. It is often found in the chloroplasts of C_4 plants, though it has also been found in some C_3 angiosperms, and even some gymnosperms. The chloroplast peripheral reticulum consists of a maze of membranous tubes and vesicles continuous with the inner chloroplast membrane that extends into the internal stromal fluid of the chloroplast. Its purpose is thought to be to increase the chloroplast's surface area for cross-membrane transport between its stroma and the cell cytoplasm. The small vesicles sometimes observed may serve as transport vesicles to shuttle stuff between the thylakoids and intermembrane space.

Stroma

The protein-rich, alkaline, aqueous fluid within the inner chloroplast membrane and outside of the thylakoid space is called the stroma, which corresponds to the cytosol of the original cyanobacterium. Nucleoids of chloroplast DNA, chloroplast ribosomes, the thylakoid system with plastoglobuli, starch granules, and many proteins can be found floating around in it. The Calvin cycle, which fixes CO_2 into G3P takes place in the stroma.

Chloroplast Ribosomes

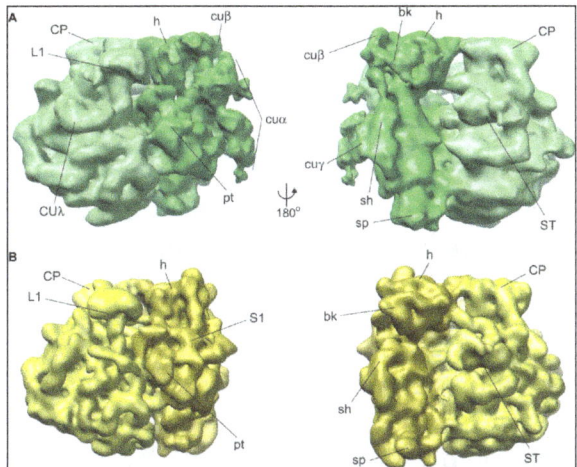

Chloroplast ribosomes Comparison of a chloroplast ribosome (green) and a bacterial ribosome (yellow). Important features common to both ribosomes and chloroplast-unique features are labeled.

Chloroplasts have their own ribosomes, which they use to synthesize a small fraction of their proteins. Chloroplast ribosomes are about two-thirds the size of cytoplasmic ribosomes (around 17 nm vs 25 nm). They take mRNAs transcribed from the chloroplast DNA and translate them into protein. While similar to bacterial ribosomes, chloroplast translation is more complex than in bacteria, so chloroplast ribosomes include some chloroplast-unique features. Small subunit ribosomal RNAs in several Chlorophyta and euglenid chloroplasts lack motifs for shine-dalgarno sequence recognition, which is considered essential for translation initiation in most chloroplasts and prokaryotes. Such loss is also rarely observed in other plastids and prokaryotes.

Plastoglobuli

Plastoglobuli (singular *plastoglobulus*, sometimes spelled *plastoglobules*), are spherical bubbles of lipids and proteins about 45–60 nanometers across. They are surrounded by a lipid monolayer. Plastoglobuli are found in all chloroplasts, but become more common when the chloroplast is under oxidative stress, or when it ages and transitions into a gerontoplast. Plastoglobuli also exhibit a greater size variation under these conditions. They are also common in etioplasts, but decrease in number as the etioplasts mature into chloroplasts.

Plastoglubuli contain both structural proteins and enzymes involved in lipid synthesis and metabolism. They contain many types of lipids including plastoquinone, vitamin E, carotenoids and chlorophylls.

Plastoglobuli were once thought to be free-floating in the stroma, but it is now thought that they are permanently attached either to a thylakoid or to another plastoglobulus attached to a thylakoid, a configuration that allows a plastoglobulus to exchange its contents with the thylakoid network. In normal green chloroplasts, the vast majority of plastoglobuli occur singularly, attached directly to their parent thylakoid. In old or stressed chloroplasts, plastoglobuli tend to occur in linked groups or chains, still always anchored to a thylakoid.

Plastoglobuli form when a bubble appears between the layers of the lipid bilayer of the thylakoid membrane, or bud from existing plastoglubuli—though they never detach and float off into the stroma. Practically all plastoglobuli form on or near the highly curved edges of the thylakoid disks or sheets. They are also more common on stromal thylakoids than on granal ones.

Starch Granules

Starch granules are very common in chloroplasts, typically taking up 15% of the organelle's volume, though in some other plastids like amyloplasts, they can be big enough to distort the shape of the organelle. Starch granules are simply accumulations of starch in the stroma, and are not bounded by a membrane.

Starch granules appear and grow throughout the day, as the chloroplast synthesizes sugars, and are consumed at night to fuel respiration and continue sugar export into the phloem, though in mature chloroplasts, it is rare for a starch granule to be completely consumed or for a new granule to accumulate. Starch granules vary in composition and location across different chloroplast lineages. In red algae, starch granules are found in the cytoplasm rather than in the chloroplast. In C_4 plants, mesophyll chloroplasts, which do not synthesize sugars, lack starch granules.

RuBisCO

RuBisCO, shown here in a space-filling model, is the main enzyme responsible for carbon fixation in chloroplasts.

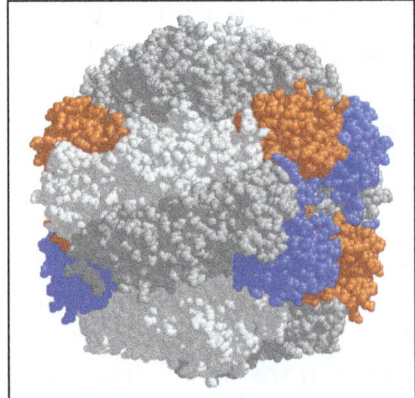

The chloroplast stroma contains many proteins, though the most common and important is RuBisCO, which is probably also the most abundant protein on the planet. RuBisCO is the enzyme that fixes CO_2 into sugar molecules. In C_3 plants, RuBisCO is abundant in all chloroplasts, though in C_4 plants, it is confined to the bundle sheath chloroplasts, where the Calvin cycle is carried out in C_4 plants.

Pyrenoids

The chloroplasts of some hornworts and algae contain structures called pyrenoids. They are not found in higher plants. Pyrenoids are roughly spherical and highly refractive bodies which are a site of starch accumulation in plants that contain them. They consist of a matrix opaque to electrons, surrounded by two hemispherical starch plates. The starch is accumulated as the pyrenoids mature. In algae with carbon concentrating mechanisms, the enzyme RuBisCO is found in the pyrenoids. Starch can also accumulate around the pyrenoids when CO_2 is scarce. Pyrenoids can divide to form new pyrenoids, or be produced "de novo".

Thylakoid System

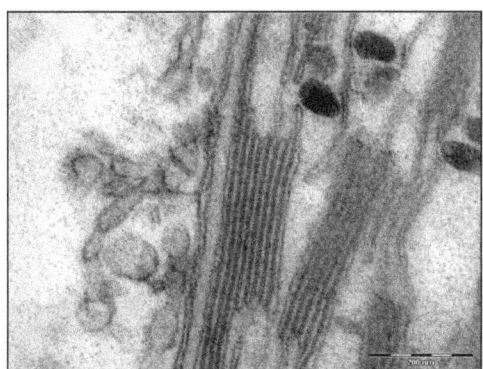

Transmission electron microscope image of some thylakoids arranged in grana stacks and lamellæ. Plastoglobuli (dark blobs) are also present.

Suspended within the chloroplast stroma is the thylakoid system, a highly dynamic collection of membranous sacks called thylakoids where chlorophyll is found and the light reactions of photosynthesis happen. In most vascular plant chloroplasts, the thylakoids are arranged in stacks called grana, though in certain C_4 plant chloroplasts and some algal chloroplasts, the thylakoids are free floating.

Granal Structure

Using a light microscope, it is just barely possible to see tiny green granules—which were named grana. With electron microscopy, it became possible to see the thylakoid system in more detail, revealing it to consist of stacks of flat thylakoids which made up the grana, and long interconnecting stromal thylakoids which linked different grana. In the transmission electron microscope, thylakoid membranes appear as alternating light-and-dark bands, 8.5 nanometers thick.

For a long time, the three-dimensional structure of the thylakoid system has been unknown or disputed. One model has the granum as a stack of thylakoids linked by helical stromal thylakoids; the other has the granum as a single folded thylakoid connected in a "hub and spoke" way to other grana by stromal thylakoids. While the thylakoid system is still commonly depicted according to the folded thylakoid model, it was determined in 2011 that the stacked and helical thylakoids model is correct.

In the helical thylakoid model, grana consist of a stack of flattened circular granal thylakoids that resemble pancakes. Each granum can contain anywhere from two to a hundred thylakoids, though grana with 10–20 thylakoids are most common. Wrapped around the grana are helicoid stromal thylakoids, also known as frets or lamellar thylakoids. The helices ascend at an angle of 20–25°, connecting to each granal thylakoid at a bridge-like slit junction. The helicoids may extend as large sheets that link multiple grana, or narrow to tube-like bridges between grana. While different parts of the thylakoid system contain different membrane proteins, the thylakoid membranes are continuous and the thylakoid space they enclose form a single continuous labyrinth.

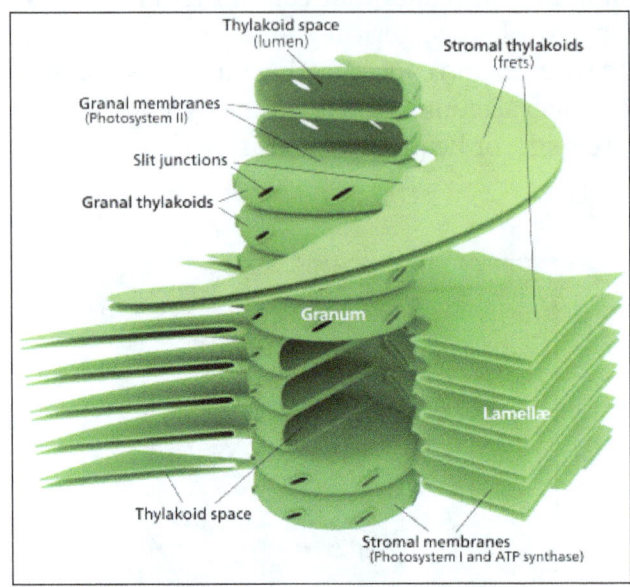

Granum structure: The prevailing model for granal structure is a stack of granal thylakoids linked by helical stromal thylakoids that wrap around the grana stacks and form large sheets that connect different grana.

Thylakoids

Thylakoids (sometimes spelled *thylakoïds*), are small interconnected sacks which contain the membranes that the light reactions of photosynthesis take place on. The word *thylakoid* comes from the Greek word *thylakos* which means "sack".

Embedded in the thylakoid membranes are important protein complexes which carry out the light reactions of photosynthesis. Photosystem II and photosystem I contain light-harvesting complexes with chlorophyll and carotenoids that absorb light energy and use it to energize electrons. Molecules in the thylakoid membrane use the energized electrons to pump hydrogen ions into the thylakoid space, decreasing the pH and turning it acidic. ATP synthase is a large protein complex that harnesses the concentration gradient of the hydrogen ions in the thylakoid space to generate ATP energy as the hydrogen ions flow back out into the stroma—much like a dam turbine.

There are two types of thylakoids—granal thylakoids, which are arranged in grana, and stromal thylakoids, which are in contact with the stroma. Granal thylakoids are pancake-shaped circular disks about 300–600 nanometers in diameter. Stromal thylakoids are helicoid sheets that spiral around grana. The flat tops and bottoms of granal thylakoids contain only the relatively flat photosystem II protein complex. This allows them to stack tightly, forming grana with many layers of tightly appressed membrane, called granal membrane, increasing stability and surface area for light capture.

In contrast, photosystem I and ATP synthase are large protein complexes which jut out into the stroma. They can't fit in the appressed granal membranes, and so are found in the stromal thylakoid membrane—the edges of the granal thylakoid disks and the stromal thylakoids. These large protein complexes may act as spacers between the sheets of stromal thylakoids.

The number of thylakoids and the total thylakoid area of a chloroplast is influenced by light exposure. Shaded chloroplasts contain larger and more grana with more thylakoid membrane area than chloroplasts exposed to bright light, which have smaller and fewer grana and less thylakoid area. Thylakoid extent can change within minutes of light exposure or removal.

Pigments and Chloroplast Colors

Inside the photosystems embedded in chloroplast thylakoid membranes are various photosynthetic pigments, which absorb and transfer light energy. The types of pigments found are different in various groups of chloroplasts, and are responsible for a wide variety of chloroplast colorations.

Chlorophylls

Chlorophyll a is found in all chloroplasts, as well as their cyanobacterial ancestors. Chlorophyll a is a blue-green pigment partially responsible for giving most cyanobacteria and chloroplasts their color. Other forms of chlorophyll exist, such as the accessory pigments chlorophyll b, chlorophyll c, chlorophyll d, and chlorophyll f.

Chlorophyll b is an olive green pigment found only in the chloroplasts of plants, green algae, any secondary chloroplasts obtained through the secondary endosymbiosis of a green alga, and a few cyanobacteria. It is the chlorophylls a and b together that make most plant and green algal chloroplasts green.

Chlorophyll c is mainly found in secondary endosymbiotic chloroplasts that originated from a red alga, although it is not found in chloroplasts of red algae themselves. Chlorophyll c is also found in some green algae and cyanobacteria.

Chlorophylls d and f are pigments found only in some cyanobacteria.

Carotenoids

Delesseria sanguinea, a red alga, has chloroplasts that contain red pigments like phycoerytherin that mask their blue-green chlorophyll a.

In addition to chlorophylls, another group of yellow–orange pigments called carotenoids are also found in the photosystems. There are about thirty photosynthetic carotenoids. They help transfer and dissipate excess energy, and their bright colors sometimes override the chlorophyll green, like during the fall, when the leaves of some land plants change color. β-carotene is a bright red-orange carotenoid found in nearly all chloroplasts, like chlorophyll *a*. Xanthophylls, especially the orange-red zeaxanthin, are also common. Many other forms of carotenoids exist that are only found in certain groups of chloroplasts.

Phycobilins

Phycobilins are a third group of pigments found in cyanobacteria, and glaucophyte, red algal, and cryptophyte chloroplasts. Phycobilins come in all colors, though phycoerytherin is one of the pigments that makes many red algae red. Phycobilins often organize into relatively large protein complexes about 40 nanometers across called phycobilisomes. Like photosystem I and ATP synthase, phycobilisomes jut into the stroma, preventing thylakoid stacking in red algal chloroplasts. Cryptophyte chloroplasts and some cyanobacteria don't have their phycobilin pigments organized into phycobilisomes, and keep them in their thylakoid space instead.

Photosynthetic pigments table of the presence of various pigments across chloroplast groups. Colored cells represent pigment presence.

	Chloro-phyll a	Chloro-phyll b	Chloro-phyll c	Chloro-phyll d and f	Xantho-phylls	α-carotene	β-carotene	Phycobilins
Land plants	■	■			■	■	■	
Green algae	■	■	■		■		■	
Euglenophytes and Chlorarachniophytes	■	■			■		■	
Multicellular red algae	■				■	■	■	■
Unicellular red algae	■				■		■	■
Haptophytes and Dinophytes	■		■		■		■	
Cryptophytes	■		■					■
Glaucophytes	■				■		■	■
Cyanobacteria	■	■	■	■	■		■	■

Specialized Chloroplasts in C4 Plants

To fix carbon dioxide into sugar molecules in the process of photosynthesis, chloroplasts use an enzyme called RuBisCO. RuBisCO has a problem—it has trouble distinguishing between carbon dioxide and oxygen, so at high oxygen concentrations, RuBisCO starts accidentally adding oxygen to sugar precursors. This has the end result of ATP energy being wasted and CO_2 being released, all with no sugar being produced. This is a big problem, since O_2 is produced by the initial light reactions of photosynthesis, causing issues down the line in the Calvin cycle which uses RuBisCO.

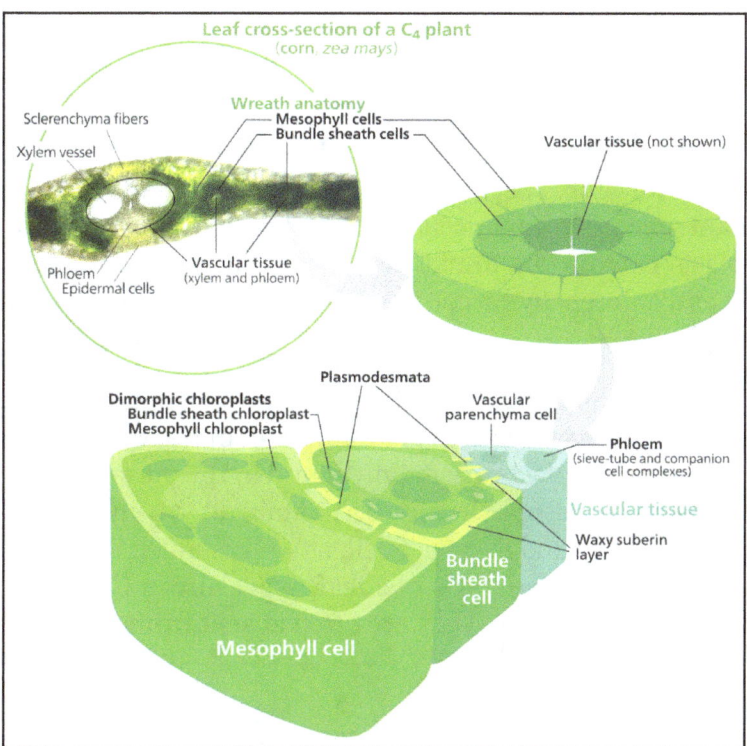

Many C_4 plants have their mesophyll cells and bundle sheath cells arranged radially around their leaf veins. The two types of cells contain different types of chloroplasts specialized for a particular part of photosynthesis.

C_4 plants evolved a way to solve this—by spatially separating the light reactions and the Calvin cycle. The light reactions, which store light energy in ATP and NADPH, are done in the mesophyll cells of a C_4 leaf. The Calvin cycle, which uses the stored energy to make sugar using RuBisCO, is done in the bundle sheath cells, a layer of cells surrounding a vein in a leaf.

As a result, chloroplasts in C_4 mesophyll cells and bundle sheath cells are specialized for each stage of photosynthesis. In mesophyll cells, chloroplasts are specialized for the light reactions, so they lack RuBisCO, and have normal grana and thylakoids, which they use to make ATP and NADPH, as well as oxygen. They store CO_2 in a four-carbon compound, which is why the process is called C_4 *photosynthesis*. The four-carbon compound is then transported to the bundle sheath chloroplasts, where it drops off CO_2 and returns to the mesophyll. Bundle sheath chloroplasts do not carry out the light reactions, preventing oxygen from building up in them and disrupting RuBisCO activity. Because of this, they lack thylakoids organized into grana stacks—though bundle sheath chloroplasts still have free-floating thylakoids in the stroma where they still carry out cyclic electron flow, a light-driven method of synthesizing ATP to power the Calvin cycle without generating oxygen. They lack photosystem II, and only have photosystem I—the only protein complex needed for cyclic electron flow. Because the job of bundle sheath chloroplasts is to carry out the Calvin cycle and make sugar, they often contain large starch grains.

Both types of chloroplast contain large amounts of chloroplast peripheral reticulum, which they use to get more surface area to transport stuff in and out of them. Mesophyll chloroplasts have a little more peripheral reticulum than bundle sheath chloroplasts.

Location

Distribution in a Plant

Not all cells in a multicellular plant contain chloroplasts. All green parts of a plant contain chloroplasts—the chloroplasts, or more specifically, the chlorophyll in them are what make the photosynthetic parts of a plant green. The plant cells which contain chloroplasts are usually parenchyma cells, though chloroplasts can also be found in collenchyma tissue. A plant cell which contains chloroplasts is known as a chlorenchyma cell. A typical chlorenchyma cell of a land plant contains about 10 to 100 chloroplasts.

In some plants such as cacti, chloroplasts are found in the stems, though in most plants, chloroplasts are concentrated in the leaves. One square millimeter of leaf tissue can contain half a million chloroplasts. Within a leaf, chloroplasts are mainly found in the mesophyll layers of a leaf, and the guard cells of stomata. Palisade mesophyll cells can contain 30–70 chloroplasts per cell, while stomatal guard cells contain only around 8–15 per cell, as well as much less chlorophyll. Chloroplasts can also be found in the bundle sheath cells of a leaf, especially in C_4 plants, which carry out the Calvin cycle in their bundle sheath cells. They are often absent from the epidermis of a leaf.

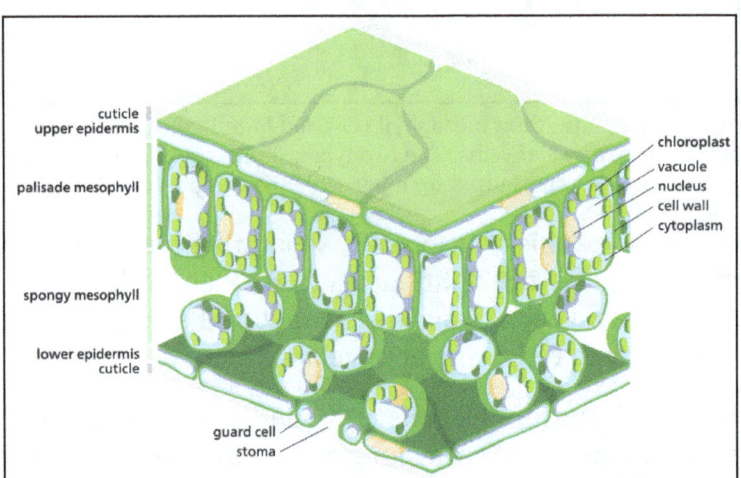

A cross section of a leaf, showing chloroplasts in its mesophyll cells. Stomal guard cells also have chloroplasts, though much fewer than mesophyll cells.

Cellular Location

Chloroplast Movement

The chloroplasts of plant and algal cells can orient themselves to best suit the available light. In low-light conditions, they will spread out in a sheet—maximizing the surface area to absorb light. Under intense light, they will seek shelter by aligning in vertical columns along the plant cell's cell wall or turning sideways so that light strikes them edge-on. This reduces exposure and protects them from photooxidative damage. This ability to distribute chloroplasts so that they can take shelter behind each other or spread out may be the reason why land plants evolved to have many small chloroplasts instead of a few big ones. Chloroplast movement is considered one of the most closely regulated stimulus-response systems that can be found in plants. Mitochondria have also been observed to follow chloroplasts as they move.

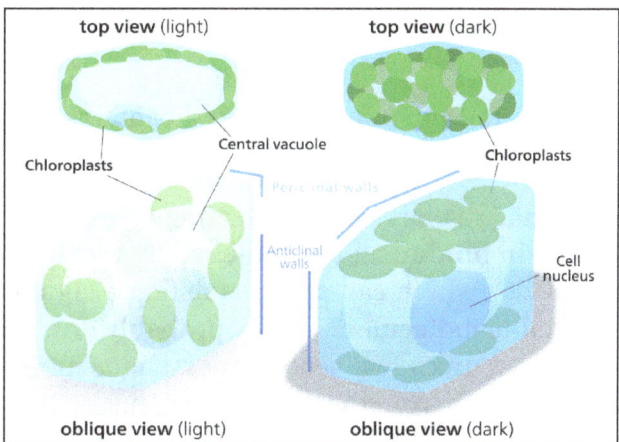

When chloroplasts are exposed to direct sunlight, they stack along the anticlinal cell walls to minimize exposure. In the dark they spread out in sheets along the periclinal walls to maximize light absorption.

In higher plants, chloroplast movement is run by phototropins, blue light photoreceptors also responsible for plant phototropism. In some algae, mosses, ferns, and flowering plants, chloroplast movement is influenced by red light in addition to blue light, though very long red wavelengths inhibit movement rather than speeding it up. Blue light generally causes chloroplasts to seek shelter, while red light draws them out to maximize light absorption.

Studies of *Vallisneria gigantea*, an aquatic flowering plant, have shown that chloroplasts can get moving within five minutes of light exposure, though they don't initially show any net directionality. They may move along microfilament tracks, and the fact that the microfilament mesh changes shape to form a honeycomb structure surrounding the chloroplasts after they have moved suggests that microfilaments may help to anchor chloroplasts in place.

Function and Chemistry

Guard Cell Chloroplasts

Unlike most epidermal cells, the guard cells of plant stomata contain relatively well-developed chloroplasts. However, exactly what they do is controversial.

Plant Innate Immunity

Plants lack specialized immune cells—all plant cells participate in the plant immune response. Chloroplasts, along with the nucleus, cell membrane, and endoplasmic reticulum, are key players in pathogen defense. Due to its role in a plant cell's immune response, pathogens frequently target the chloroplast.

Plants have two main immune responses—the hypersensitive response, in which infected cells seal themselves off and undergo programmed cell death, and systemic acquired resistance, where infected cells release signals warning the rest of the plant of a pathogen's presence. Chloroplasts stimulate both responses by purposely damaging their photosynthetic system, producing reactive oxygen species. High levels of reactive oxygen species will cause the hypersensitive response. The reactive oxygen species also directly kill any pathogens within the cell. Lower levels of reactive oxygen species initiate systemic acquired resistance, triggering defense-molecule production in the rest of the plant.

In some plants, chloroplasts are known to move closer to the infection site and the nucleus during an infection.

Chloroplasts can serve as cellular sensors. After detecting stress in a cell, which might be due to a pathogen, chloroplasts begin producing molecules like salicylic acid, jasmonic acid, nitric oxide and reactive oxygen species which can serve as defense-signals. As cellular signals, reactive oxygen species are unstable molecules, so they probably don't leave the chloroplast, but instead pass on their signal to an unknown second messenger molecule. All these molecules initiate retrograde signaling—signals from the chloroplast that regulate gene expression in the nucleus.

In addition to defense signaling, chloroplasts, with the help of the peroxisomes, help synthesize an important defense molecule, jasmonate. Chloroplasts synthesize all the fatty acids in a plant cell—linoleic acid, a fatty acid, is a precursor to jasmonate.

Photosynthesis

One of the main functions of the chloroplast is its role in photosynthesis, the process by which light is transformed into chemical energy, to subsequently produce food in the form of sugars. Water (H_2O) and carbon dioxide (CO_2) are used in photosynthesis, and sugar and oxygen (O_2) is made, using light energy. Photosynthesis is divided into two stages—the light reactions, where water is split to produce oxygen, and the dark reactions, or Calvin cycle, which builds sugar molecules from carbon dioxide. The two phases are linked by the energy carriers adenosine triphosphate (ATP) and nicotinamide adenine dinucleotide phosphate ($NADP^+$).

Light Reactions

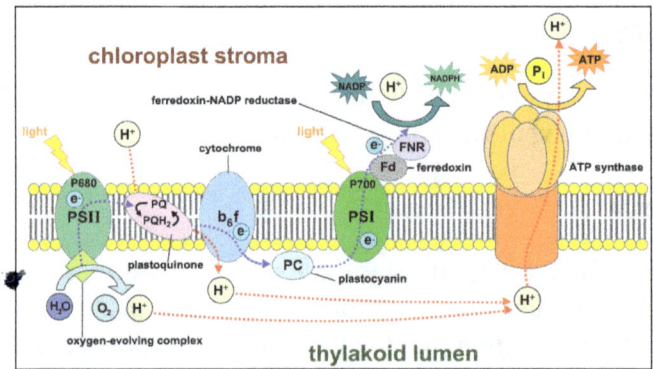

The light reactions of photosynthesis take place across the thylakoid membranes.

The light reactions take place on the thylakoid membranes. They take light energy and store it in NADPH, a form of $NADP^+$, and ATP to fuel the dark reactions.

Energy Carriers

ATP is the phosphorylated version of adenosine diphosphate (ADP), which stores energy in a cell and powers most cellular activities. ATP is the energized form, while ADP is the (partially) depleted form. $NADP^+$ is an electron carrier which ferries high energy electrons. In the light reactions, it gets reduced, meaning it picks up electrons, becoming NADPH.

Photophosphorylation

Like mitochondria, chloroplasts use the potential energy stored in an H^+, or hydrogen ion gradient to generate ATP energy. The two photosystems capture light energy to energize electrons taken from water, and release them down an electron transport chain. The molecules between the photosystems harness the electrons' energy to pump hydrogen ions into the thylakoid space, creating a concentration gradient, with more hydrogen ions (up to a thousand times as many) inside the thylakoid system than in the stroma. The hydrogen ions in the thylakoid space then diffuse back down their concentration gradient, flowing back out into the stroma through ATP synthase. ATP synthase uses the energy from the flowing hydrogen ions to phosphorylate adenosine diphosphate into adenosine triphosphate, or ATP. Because chloroplast ATP synthase projects out into the stroma, the ATP is synthesized there, in position to be used in the dark reactions.

$NADP^+$ Reduction

Electrons are often removed from the electron transport chains to charge $NADP^+$ with electrons, reducing it to NADPH. Like ATP synthase, ferredoxin-$NADP^+$ reductase, the enzyme that reduces $NADP^+$, releases the NADPH it makes into the stroma, right where it is needed for the dark reactions.

Because $NADP^+$ reduction removes electrons from the electron transport chains, they must be replaced—the job of photosystem II, which splits water molecules (H_2O) to obtain the electrons from its hydrogen atoms.

Cyclic Photophosphorylation

While photosystem II photolyzes water to obtain and energize new electrons, photosystem I simply reenergizes depleted electrons at the end of an electron transport chain. Normally, the reenergized electrons are taken by $NADP^+$, though sometimes they can flow back down more H^+-pumping electron transport chains to transport more hydrogen ions into the thylakoid space to generate more ATP. This is termed cyclic photophosphorylation because the electrons are recycled. Cyclic photophosphorylation is common in C_4 plants, which need more ATP than NADPH.

Dark Reactions

The Calvin cycle, also known as the dark reactions, is a series of biochemical reactions that fixes CO_2 into G3P sugar molecules and uses the energy and electrons from the ATP and NADPH made in the light reactions. The Calvin cycle takes place in the stroma of the chloroplast.

While named *"the dark reactions"*, in most plants, they take place in the light, since the dark reactions are dependent on the products of the light reactions.

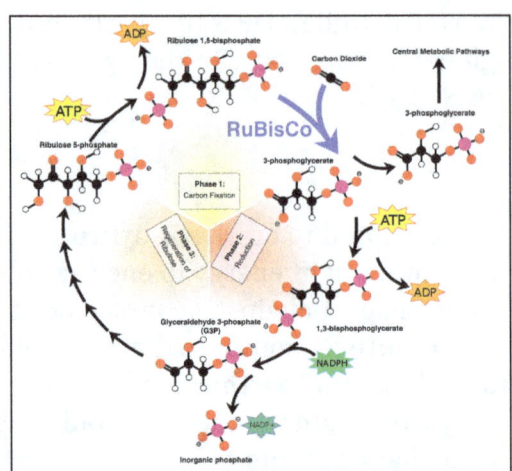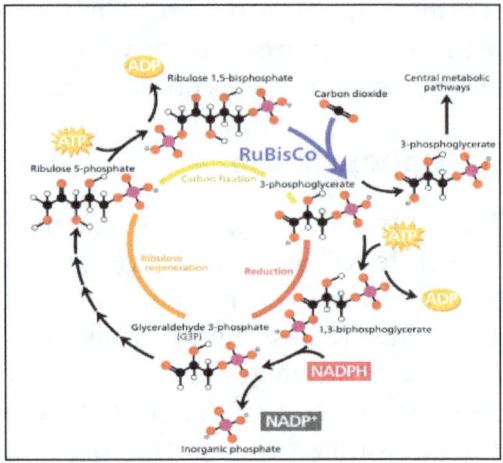

The Calvin cycle *(Interactive diagram)* The Calvin cycle
incorporates carbon dioxide into sugar molecules.

Carbon Fixation and G3P Synthesis

The Calvin cycle starts by using the enzyme RuBisCO to fix CO_2 into five-carbon Ribulose bi-sphosphate (RuBP) molecules. The result is unstable six-carbon molecules that immediately break down into three-carbon molecules called 3-phosphoglyceric acid, or 3-PGA. The ATP and NADPH made in the light reactions is used to convert the 3-PGA into glyceraldehyde-3-phosphate, or G3P sugar molecules. Most of the G3P molecules are recycled back into RuBP using energy from more ATP, but one out of every six produced leaves the cycle—the end product of the dark reactions.

Sugars and Starches

Glyceraldehyde-3-phosphate can double up to form larger sugar molecules like glucose and fructose. These molecules are processed, and from them, the still larger sucrose, a disaccharide commonly known as table sugar, is made, though this process takes place outside of the chloroplast, in the cytoplasm.

Sucrose is made up of a glucose monomer (left), and a fructose monomer (right).

Alternatively, glucose monomers in the chloroplast can be linked together to make starch, which accumulates into the starch grains found in the chloroplast. Under conditions such as high atmospheric CO_2 concentrations, these starch grains may grow very large, distorting the grana and thylakoids. The starch granules displace the thylakoids, but leave them intact. Waterlogged roots can also cause starch buildup in the chloroplasts, possibly due to less sucrose being exported out of the chloroplast (or more accurately, the plant cell). This depletes a plant's free phosphate supply, which indirectly stimulates chloroplast starch synthesis. While linked to low photosynthesis rates, the starch grains themselves may not necessarily interfere significantly with the efficiency of photosynthesis, and might simply be a side effect of another photosynthesis-depressing factor.

Photorespiration

Photorespiration can occur when the oxygen concentration is too high. RuBisCO cannot distinguish between oxygen and carbon dioxide very well, so it can accidentally add O_2 instead of CO_2 to RuBP. This process reduces the efficiency of photosynthesis—it consumes ATP and oxygen, releases CO_2, and produces no sugar. It can waste up to half the carbon fixed by the Calvin cycle. Several mechanisms have evolved in different lineages that raise the carbon dioxide concentration relative to oxygen within the chloroplast, increasing the efficiency of photosynthesis. These mechanisms are called carbon dioxide concentrating mechanisms, or CCMs. These include Crassulacean acid metabolism, C_4 carbon fixation, and pyrenoids. Chloroplasts in C_4 plants are notable as they exhibit a distinct chloroplast dimorphism.

pH

Because of the H^+ gradient across the thylakoid membrane, the interior of the thylakoid is acidic, with a pH around 4, while the stroma is slightly basic, with a pH of around 8. The optimal stroma pH for the Calvin cycle is 8.1, with the reaction nearly stopping when the pH falls below 7.3. CO_2 in water can form carbonic acid, which can disturb the pH of isolated chloroplasts, interfering with photosynthesis, even though CO_2 is used in photosynthesis. However, chloroplasts in living plant cells are not affected by this as much.

Chloroplasts can pump K^+ and H^+ ions in and out of themselves using a poorly understood light-driven transport system. In the presence of light, the pH of the thylakoid lumen can drop up to 1.5 pH units, while the pH of the stroma can rise by nearly one pH unit.

Amino Acid Synthesis

Chloroplasts alone make almost all of a plant cell's amino acids in their stroma except the sulfur-containing ones like cysteine and methionine. Cysteine is made in the chloroplast (the proplastid too) but it is also synthesized in the cytosol and mitochondria, probably because it has trouble crossing membranes to get to where it is needed. The chloroplast is known to make the precursors to methionine but it is unclear whether the organelle carries out the last leg of the pathway or if it happens in the cytosol.

Other Nitrogen Compounds

Chloroplasts make all of a cell's purines and pyrimidines—the nitrogenous bases found in DNA and RNA. They also convert nitrite (NO_2^-) into ammonia (NH_3) which supplies the plant with nitrogen to make its amino acids and nucleotides.

Other Chemical Products

The plastid is the site of diverse and complex lipid synthesis in plants. The carbon used to form the majority of the lipid is from acetyl-CoA, which is the decarboxylation product of pyruvate.

Pyruvate may enter the plastid from the cytosol by passive diffusion through the membrane after production in glycolysis. Pyruvate is also made in the plastid from phosphoenolpyruvate, a metabolite made in the cytosol from pyruvate or PGA. Acetate in the cytosol is unavailable for lipid biosynthesis in the plastid. The typical length of fatty acids produced in the plastid are 16 or 18 carbons, with 0-3 cis double bonds.

The biosynthesis of fatty acids from acetyl-CoA primarily requires two enzymes. Acetyl-CoA carboxylase creates malonyl-CoA, used in both the first step and the extension steps of synthesis. Fatty acid synthase (FAS) is a large complex of enzymes and cofactors including acyl carrier protein (ACP) which holds the acyl chain as it is synthesized. The initiation of synthesis begins with the condensation of malonyl-ACP with acetyl-CoA to produce ketobutyryl-ACP. 2 reductions involving the use of NADPH and one dehydration creates butyryl-ACP. Extension of the fatty acid comes from repeated cycles of malonyl-ACP condensation, reduction, and dehydration.

Other lipids are derived from the methyl-erythritol phosphate (MEP) pathway and consist of gibberelins, sterols, abscisic acid, phytol, and innumerable secondary metabolites.

Differentiation, Replication, and Inheritance

Chloroplasts are a special type of a plant cell organelle called a plastid, though the two terms are sometimes used interchangeably. There are many other types of plastids, which carry out various functions. All chloroplasts in a plant are descended from undifferentiated proplastids found in the zygote, or fertilized egg. Proplastids are commonly found in an adult plant's apical meristems. Chloroplasts do not normally develop from proplastids in root tip meristems—instead, the formation of starch-storing amyloplasts is more common.

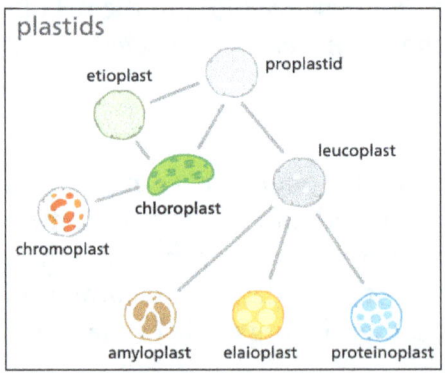

Plastid types (*Interactive diagram*) Plants contain many different kinds of plastids in their cells.

In shoots, proplastids from shoot apical meristems can gradually develop into chloroplasts in photosynthetic leaf tissues as the leaf matures, if exposed to the required light. This process involves invaginations of the inner plastid membrane, forming sheets of membrane that project into the internal stroma. These membrane sheets then fold to form thylakoids and grana.

If angiosperm shoots are not exposed to the required light for chloroplast formation, proplastids may develop into an etioplast stage before becoming chloroplasts. An etioplast is a plastid that lacks chlorophyll, and has inner membrane invaginations that form a lattice of tubes in their stroma, called a prolamellar body. While etioplasts lack chlorophyll, they have a yellow chlorophyll precursor stocked. Within a few minutes of light exposure, the prolamellar body begins to reorganize into stacks of thylakoids, and chlorophyll starts to be produced. This process, where the etioplast becomes a chloroplast, takes several hours. Gymnosperms do not require light to form chloroplasts.

Light, however, does not guarantee that a proplastid will develop into a chloroplast. Whether a proplastid develops into a chloroplast some other kind of plastid is mostly controlled by the nucleus and is largely influenced by the kind of cell it resides in.

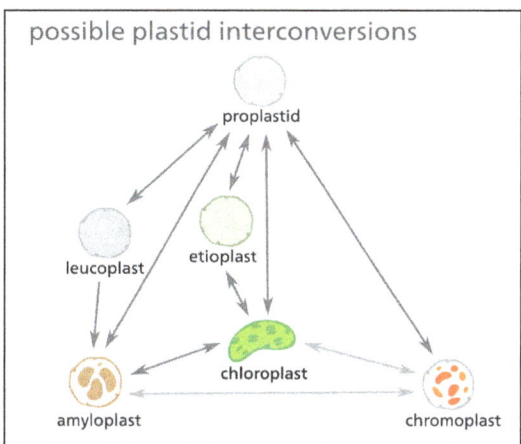

Many plastid interconversions are possible.

Plastid Interconversion

Plastid differentiation is not permanent, in fact many interconversions are possible. Chloroplasts may be converted to chromoplasts, which are pigment-filled plastids responsible for the bright colors seen in flowers and ripe fruit. Starch storing amyloplasts can also be converted to chromoplasts, and it is possible for proplastids to develop straight into chromoplasts. Chromoplasts and amyloplasts can also become chloroplasts, like what happens when a carrot or a potato is illuminated. If a plant is injured, or something else causes a plant cell to revert to a meristematic state, chloroplasts and other plastids can turn back into proplastids. Chloroplast, amyloplast, chromoplast, proplast, etc., are not absolute states—intermediate forms are common.

Chloroplast Division

Most chloroplasts in a photosynthetic cell do not develop directly from proplastids or etioplasts. In fact, a typical shoot meristematic plant cell contains only 7–20 proplastids. These proplastids differentiate into chloroplasts, which divide to create the 30–70 chloroplasts found in a mature photosynthetic plant cell. If the cell divides, chloroplast division provides the additional chloroplasts to partition between the two daughter cells.

In single-celled algae, chloroplast division is the only way new chloroplasts are formed. There is no proplastid differentiation—when an algal cell divides, its chloroplast divides along with it, and each daughter cell receives a mature chloroplast.

Almost all chloroplasts in a cell divide, rather than a small group of rapidly dividing chloroplasts. Chloroplasts have no definite S-phase—their DNA replication is not synchronized or limited to that of their host cells. Much of what we know about chloroplast division comes from studying organisms like *Arabidopsis* and the red alga *Cyanidioschyzon merolæ*.

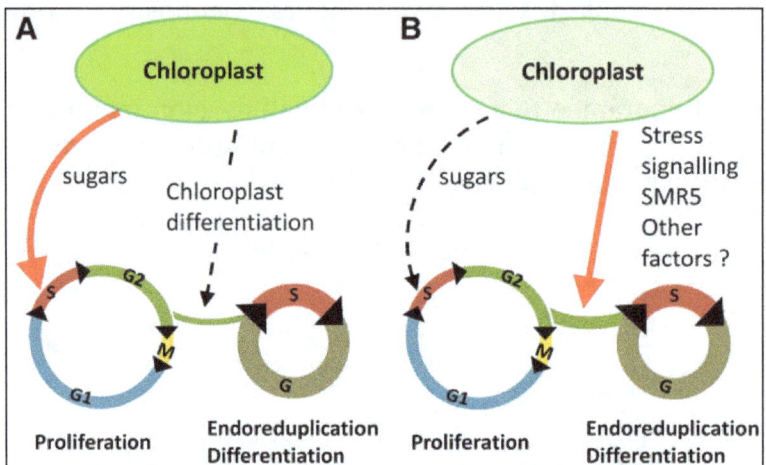

Most chloroplasts in plant cells, and all chloroplasts in algae arise from chloroplast division.

The division process starts when the proteins FtsZ1 and FtsZ2 assemble into filaments, and with the help of a protein ARC6, form a structure called a Z-ring within the chloroplast's stroma. The Min system manages the placement of the Z-ring, ensuring that the chloroplast is cleaved more or less evenly. The protein MinD prevents FtsZ from linking up and forming filaments. Another protein ARC3 may also be involved, but it is not very well understood. These proteins are active at the poles of the chloroplast, preventing Z-ring formation there, but near the center of the chloroplast, MinE inhibits them, allowing the Z-ring to form.

Next, the two plastid-dividing rings, or PD rings form. The inner plastid-dividing ring is located in the inner side of the chloroplast's inner membrane, and is formed first. The outer plastid-dividing ring is found wrapped around the outer chloroplast membrane. It consists of filaments about 5 nanometers across, arranged in rows 6.4 nanometers apart, and shrinks to squeeze the chloroplast. This is when chloroplast constriction begins. In a few species like *Cyanidioschyzon merolæ*, chloroplasts have a third plastid-dividing ring located in the chloroplast's intermembrane space.

Late into the constriction phase, dynamin proteins assemble around the outer plastid-dividing ring, helping provide force to squeeze the chloroplast. Meanwhile, the Z-ring and the inner plastid-dividing ring break down. During this stage, the many chloroplast DNA plasmids floating around in the stroma are partitioned and distributed to the two forming daughter chloroplasts.

Later, the dynamins migrate under the outer plastid dividing ring, into direct contact with the chloroplast's outer membrane, to cleave the chloroplast in two daughter chloroplasts.

A remnant of the outer plastid dividing ring remains floating between the two daughter chloroplasts, and a remnant of the dynamin ring remains attached to one of the daughter chloroplasts.

Of the five or six rings involved in chloroplast division, only the outer plastid-dividing ring is present for the entire constriction and division phase—while the Z-ring forms first, constriction does not begin until the outer plastid-dividing ring forms.

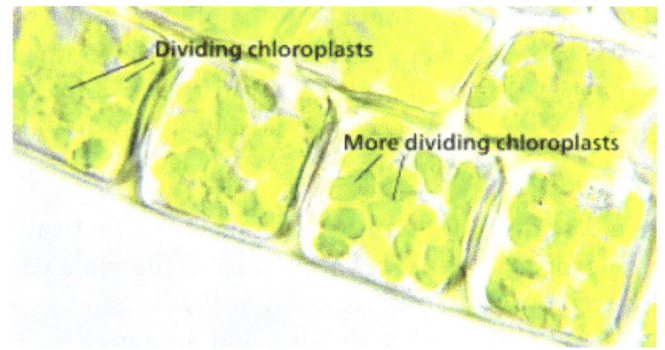

Chloroplast division In this light micrograph of some moss chloroplasts,
many dumbbell-shaped chloroplasts can be seen dividing.
Grana are also just barely visible as small granules.

Regulation

In species of algae that contain a single chloroplast, regulation of chloroplast division is extremely important to ensure that each daughter cell receives a chloroplast—chloroplasts can't be made from scratch. In organisms like plants, whose cells contain multiple chloroplasts, coordination is looser and less important. It is likely that chloroplast and cell division are somewhat synchronized, though the mechanisms for it are mostly unknown.

Light has been shown to be a requirement for chloroplast division. Chloroplasts can grow and progress through some of the constriction stages under poor quality green light, but are slow to complete division—they require exposure to bright white light to complete division. Spinach leaves grown under green light have been observed to contain many large dumbbell-shaped chloroplasts. Exposure to white light can stimulate these chloroplasts to divide and reduce the population of dumbbell-shaped chloroplasts.

Chloroplast Inheritance

Like mitochondria, chloroplasts are usually inherited from a single parent. Biparental chloroplast inheritance—where plastid genes are inherited from both parent plants—occurs in very low levels in some flowering plants.

Many mechanisms prevent biparental chloroplast DNA inheritance, including selective destruction of chloroplasts or their genes within the gamete or zygote, and chloroplasts from one parent being excluded from the embryo. Parental chloroplasts can be sorted so that only one type is present in each offspring.

Gymnosperms, such as pine trees, mostly pass on chloroplasts paternally, while flowering plants often inherit chloroplasts maternally. Flowering plants were once thought to only inherit chloroplasts maternally. However, there are now many documented cases of angiosperms inheriting chloroplasts paternally.

Angiosperms, which pass on chloroplasts maternally, have many ways to prevent paternal inheritance. Most of them produce sperm cells that do not contain any plastids. There are many other documented mechanisms that prevent paternal inheritance in these flowering plants, such as different rates of chloroplast replication within the embryo.

Among angiosperms, paternal chloroplast inheritance is observed more often in hybrids than in offspring from parents of the same species. This suggests that incompatible hybrid genes might interfere with the mechanisms that prevent paternal inheritance.

Transplastomic Plants

Recently, chloroplasts have caught attention by developers of genetically modified crops. Since, in most flowering plants, chloroplasts are not inherited from the male parent, transgenes in these plastids cannot be disseminated by pollen. This makes plastid transformation a valuable tool for the creation and cultivation of genetically modified plants that are biologically contained, thus posing significantly lower environmental risks. This biological containment strategy is therefore suitable for establishing the coexistence of conventional and organic agriculture. While the reliability of this mechanism has not yet been studied for all relevant crop species, recent results in tobacco plants are promising, showing a failed containment rate of transplastomic plants at 3 in 1,000,000.

CENTROSOME

Centrosomes are organelles which serve as the main microtubule organizing centers for animal cells. Centrosomes are made of from arrangement of two barrel-shaped clusters of microtubules, called "centrioles," and a complex of proteins that help additional microtubules to form.

These proteins allow the centrosomes to start and stop the formation of microtubule proteins. This allows them to control the formation of mitotic spindle fibers and other structures that play important roles in cellular development.

Centrosomes assist with several important functions, including:

- Organizing changes to the shape of the cell membrane that allow the membrane to "pinch" in two during cell division.

- Ensuring that chromosomes are properly distributed to daughter cells by creating and shortening mitotic spindle fibers.

- Overseeing other important changes to cell membrane shape, such as those seen in phago-cytosis.

In animal cells, centrosomes are treated very much the same way as DNA.

Each daughter cell gets one centrosome from the parent cell during cell division. The centrosome is then copied during the cell cycle, so that the cell can give one to each daughter cell when it divides.

During cell division, when chromosomes are lined up and then pulled toward opposite ends of the cell, it is the centrosomes that are responsible.

The centrosomes, which migrate to opposite "poles" of the cell as the cell prepares for division, direct the mitotic spindle fibers. These spindle fibers pull the sister chromatids apart and ensure that one copy of each chromosome ends up in each daughter cell.

The graphic below shows a cell midway through telophase of cell division. You can see that its DNA has already been pulled by mitotic spindle fibers to opposite sides of the parent cell, and that the cytoskeleton is now beginning to "pinch" the cell in two.

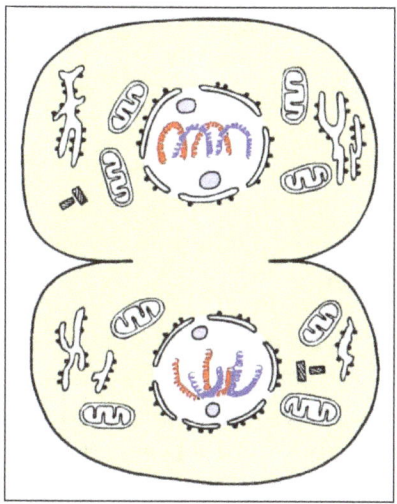

Schtelophase.

In the absence of centrosomes, some animal cells can still complete this assortment of DNA, but the process is less reliable. A few animal species can develop normally without centrosomes, but in most species, cells may begin dividing incorrectly or stop dividing at all if centrosomes are destroyed.

Mutations that harm centrosome function are associated with rates of cancer in some species, which is consistent with failures to correctly sort DNA. Biologists think that some cases of cancer are caused in by errors in copying and distribution of chromosomes.

Centrosomes are not necessary in plant and fungi cells, because these cells do not change the shape of their cell membranes during cell division. These cells have stuff, inflexible cell walls which prevent them from changing their membrane shape to "pinch" in two during mitosis.

Function of Centrosomes

Centrosomes are sometimes referred to as the "MTOC," or "microtubule organizing center" of the cell. They serve to direct the movements of microtubules and other cytoskeletal structures and proteins, ultimately allowing large changes to the shapes of animal cell membranes.

Animal cells are unique among cell types because they are highly flexible, giving animals their soft tissues and highly versatile bodies. But they also have the ability to have structure and change their shape, which permits movement and many other functions. When animals cells want to change their shapes, complexes of proteins move the cell's membranes along a network of microtubules – stiff "skeletal" fibers which can bend and change shape in response to intra- and extra-cellular signals.

The largest changes to a cell's membrane shape occurs during mitosis, when the entire cell splits in two to form daughter cells. Mitosis is also when centrosomes play a starring role as the organizers of the microtubules that pull sister chromatids apart, ensuring that each daughter cell gets a full compliment of the parent cells' DNA.

Centrosomes can also orchestrate large changes to cell membrane shape under other circumstances, such as phagocytosis.

This process, which comes from the Greek words for "cell eating," occurs when the cell changes shape to completely wrap itself around and "swallow" another cell or item in its environment.

Controversy Over Necessity

For many years, it was believed that animal cells could not divide successfully without centrosomes coordinating the separation of sister chromatids, the changes to the cytoskeleton, etc.. Upholding this theory, some cells in the lab were observed to stop dividing altogether, or to divide incorrectly, when their centrosomes were destroyed.

But in recent years, it has been discovered that some species of animals can develop normally, even if they are genetic mutants who have no centrosomes at all. Fruit flies and flatworms are among those that accomplish successful cell division without centrosomes. This has raised questions about the real utility of centrosomes, and whether the cell can "make up for" their absence through other mechanisms. Some scientists propose that centrosomes might assist the processes described here, but not be vital to them.

More data is needed before scientists can say for sure whether centrosomes are essential to cell division, and what they can do that cells don't have other ways to accomplish. But in the meantime, it's better to assume that they are important than that they aren't!

RIBOSOME

One of the essential cell organelles are ribosomes, which are in charge of protein synthesis. The ribosome is a complex made of protein and RNA and which adds up to numerous million Daltons in size and assumes an important part in the course of decoding the genetic message reserved in the genome into protein.

The essential chemical step of protein synthesis is peptidyl transfer, that the developing or nascent peptide is moved from one tRNA molecule to the amino acid together with another tRNA. Amino acids are included in the developing polypeptide in line with the arrangement of codons of a mRNA. The ribosome, therefore, has necessary sites for one mRNA and no less than two tRNAs.

Made of two subunits, the big and the little subunit which comprises a couple of ribosomal RNA (rRNA) molecules and an irregular number of ribosomal proteins. Numerous protein factors catalyze distinct impression of protein synthesis. The translation of the genetic code is of essential significance for the manufacturing of useful proteins and for the growth of the cell.

Structure

Ribosomes are made of proteins and ribonucleic acid (abbreviated as RNA), in almost equal amounts. It comprises of two sections, known as subunits. The tinier subunit is the place the mRNA binds and it decodes, whereas the bigger subunit is the place the amino acids are included.

Both subunits comprise of both ribonucleic acid and protein components and are linked to each other by interactions between the proteins in one subunit and the rRNAs in the other subunit. The ribonucleic acid is obtained from the nucleolus, at the point where ribosomes are arranged in a cell.

The structures of ribosomes include:

- Situated in two areas of the cytoplasm.

- They are seen scattered in the cytoplasm and a few are connected to the endoplasmic reticulum.

- Whenever joined to the ER they are called the rough endoplasmic reticulum.

- The free and the bound ribosomes are very much alike in structure and are associated with protein synthesis.

- Around 37 to 62% of RNA is comprised of RNA and the rest is proteins.

- Prokaryotes have 70S ribosomes respectively subunits comprising the little subunit of 30S and the bigger subunit of 50S. Eukaryotes have 80S ribosomes respectively comprising of little (40S) and substantial (60S) subunits.

- The ribosomes seen in the chloroplasts of mitochondria of eukaryotes are comprised of big and little subunits composed of proteins inside a 70S particle.

- Share a center structure which is very much alike to all ribosomes in spite of changes in its size.

- The RNA is arranged in different tertiary structures. The RNA in the bigger ribosomes is into numerous continuous infusions as they create loops out of the center of the structure without disturbing or altering it.

- The contrast between those of eukaryotic and bacteria are utilized to make antibiotics that can crush bacterial disease without damaging human cells.

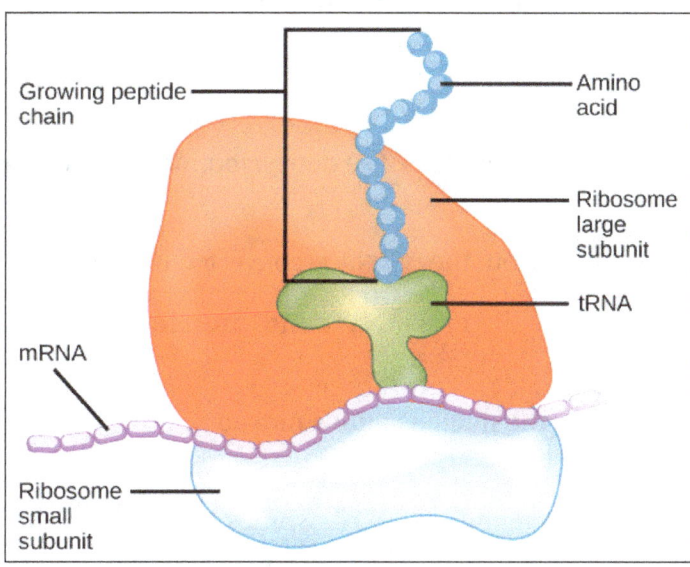

By CNX OpenStax.

Ribosomes Size

Ribosomes comprise of two subunits that are suitably composed and function as one to translate the mRNA into a polypeptide chain amid protein synthesis. Due to the fact that they are made from two subunits of differing size, they are a little longer in the hinge than in diameter. They vary in size between prokaryotic cells and eukaryotic cells.

The prokaryotic is comprised of a 30s (Svedberg) subunit and a 50s (Svedberg) subunit meaning 70s for the entire organelle equal to the molecular weight of 2.7×106 Daltons. Prokaryotic ribosomes are about 20 nm (200 Å) in diameter and are made of 35% ribosomal proteins and 65% rRNA.

Notwithstanding, the eukaryotic are amidst 25 and 30 nm (250–300 Å) in diameter. They comprise of a 40s (Svedberg) subunit and a 60s (Svedberg) subunit which means 80s (Svedberg) for the entire organelle which is equal to the molecular weight of 4×106 Daltons.

Location

Ribosomes are organelles located inside the animal, human cell, and plant cells. They are situated in the cytosol, some bound and free-floating to the membrane of the coarse endoplasmic reticulum.

They are utilized in decoding DNA (deoxyribonucleic acid) to proteins and no rRNA is forever bound to the RER, they release or bind as directed by the kind of protein they proceed to combine. In an animal or human cell, there could be up to 10 million ribosomes and numerous ribosomes can be connected to the equivalent mRNA strand, this structure is known as a POLYSOME.

Function

When it comes to the main functions of ribosomes, they assume the role of bringing together amino acids to form particular proteins, which are important for completing the cell's activities.

Protein is required for numerous cell functions, for example, directing chemical processes or fixing the damage. Ribosomes can yet be discovered floating inside the cytoplasm or joined to the endoplasmic reticulum.

The other functions include:

1. The procedure of creation of proteins, the deoxyribonucleic acid makes mRNA by the step of DNA transcription.

2. The hereditary information from the mRNA is converted into proteins amid DNA translation.

3. The arrangements of protein assembly amid protein synthesis are indicated in the mRNA.

4. The mRNA is arranged in the nucleus and is moved to the cytoplasm for an additional operation of protein synthesis.

5. The proteins which are arranged by the ribosomes currently in the cytoplasm are utilized inside the cytoplasm by itself. The proteins created by the bound ribosomes are moved outside the cell.

Taking into consideration their main function in developing proteins, it's clear that a cell can't function in the absence of ribosomes.

Those that live inside bacteria, parasites and different creatures, for example, lower and microscopic level creatures are the ones which are called prokaryotic ribosomes. While those that live inside humans and others such as higher level creatures are those ones we call the eukaryotic ribosome.

The other major differences include:

1. Prokaryotes have 70S ribosomes, singly made of a 30S and a 50S subunit. While the Eukaryotes have 80S ribosomes, singly made of a 40S and 60S subunit.

2. 70S Ribosomes are relatively smaller than 80S while the 80S Ribosomes are relatively bigger than 70S ribosomes.

3. Prokaryotes have 30S subunit with a 16S RNA subunit and comprise of 1540 nucleotides bound to 21 proteins. The 50S subunit gets produced from a 5S RNA subunit that involves 120 nucleotides, a 23S RNA subunit that contains 2900 nucleotides and 31 proteins.

4. Eukaryotes have 40S subunit with 18S RNA and also 33 proteins and 1900 nucleotides. The big subunit contains 5S RNA and also 120 nucleotides, 4700 nucleotides and also 28S RNA, 5.8S RNA as well as 160 nucleotides subunits and 46 proteins.

5. Eukaryotic cells have mitochondria and chloroplasts as organelles and those organelles additionally have ribosomes 70S. Hence, eukaryotic cells have different kinds of ribosomes (70S and 80S), while prokaryotic cells just have 70S ribosomes.

VACUOLE

A vacuole is an organelle in cells which functions to hold various solutions or materials. This includes solutions that have been created and are being stored or excreted, and those that have been phagocytized, or engulfed, by the cell. A vacuole is simply a chamber surrounded by a membrane, which keeps the cytosol from being exposed to the contents inside. Because vacuoles are surrounded by semi-permeable membranes, they only let certain molecules through.

Vacuole Structure

A vacuole has a broad definition, and includes a variety of membrane-bound sacs. The membranes are composed of phospholipids, but each organism may use slightly different phospholipids. Embedded in the membranes are proteins, which can function to transport molecules across the membrane or give it structure. Various combinations of these proteins allow different vacuoles to handle and hold different materials.

In each organism, different genetics cause different proteins to be embedded in the membrane of the vacuole, which allow different molecules through, and gives the vacuoles different properties.

Most plant cells have evolved to use vacuoles as water storage organelles, which provide a variety of functions to the cell. Animals don't rely on this water storage for the rigidity of their form, and use their vacuoles for the storage of various products, and for exocytosis and endocytosis.

Functions of a Vacuole

Water Storage

In plants, a large vacuole occupies the majority of the cell. This vacuole is surrounded by the tonoplast, a type of cytoplasmic membrane that can stretch and fills itself with a solution known as cell sap. The vacuole also fill itself with protons from the cytosol, creating an acid environment inside of the cell. The vacuole can then use the chemical gradient created to transport materials in and out of the vacuole, a type of movement called proton motive force. This includes the movement of water and other molecules. The following is a picture of a plant cell, and the vacuole inside.

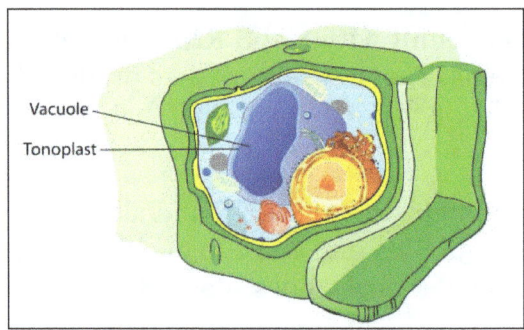

Plant cell structure vacuole.

Turgor Pressure

Plants use their vacuoles for a second function, which is of utmost importance to all plants. The vacuole, when completely filled with water, can become pressurized and exert a force on the cell walls. Although the force in each cell is small, this turgor pressure allows the cells to create a form, and stand up to wind, rain and even hail. Although woody plants create additional proteins and fibers that help them stand tall, many non-woody plants can reach a height of several feet on turgor pressure alone.

While this is an efficient way for plants to structure themselves, the plant will droop when the pH balance is off, or they don't have enough water. If you see your houseplants drooping, make sure they have water. Oftentimes a quick drink will have them from wilted to turgid in a matter of hours. If that doesn't work, check the pH of the soil. Without the right conditions, the roots can't take up water or nutrients to store in the vacuoles of the cells, and the plant wilts. Change the pH, and the plant will stiffen right up.

Endocytosis and Exocytosis

A vacuole is used whenever a large amount of substance is taken in through endocytosis, or excreted through exocytosis. Many cells, plant and animal, take in substances and must store them separate from the cytosol. This could be because the substances are reactive, in which case they will cause unwanted reactions. It could also be because the substances would interfere with cellular processes because they are two large. Lysosomes are vesicles used to intake substances to be digested. Sometimes these lysosomes can fuse to form a large, digestive vesicle that can digest nutrients in an acid environment, then transfer them to the cytosol or other organelles to be used. This process is endocytosis, and varies among different types of cells.

Going the opposite direction, many cells function as secretory cells, and must produce and excrete large amounts of different substances. The substances are produces in the endoplasmic reticulum, travel to the Golgi apparatus to be modified and labeled for distribution. The substances can then be put into vesicles. The vesicles travel into the cytoplasm and can merge into a larger vacuole before being excreted. This is known as exocytosis. The vacuoles that carry different substances to and fro vary in structure in different cells, and even within cells when they have different functions. An animal cell may contain many vacuoles that preform many functions. An example of an animal cell and its vacuoles can be seen below, the smaller unlabeled sphere would be vesicles. Once they fuse together, they would also be considered vacuoles.

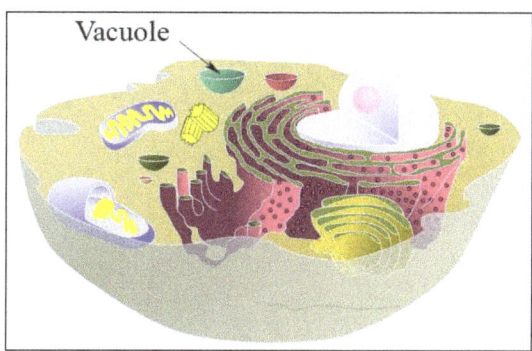

Biological cell vacuole.

Other Storage Functions

Vacuoles are able to store many different types of molecules. Fat cells, for instance, store huge amounts of lipids in specialized vacuoles. This allows single cells to store a large amount of fat, which organisms can use when resources are low. The expandability of the vacuole means an organism can gain or lose weight without too many cells being created or lost. Other times, vacuoles of organisms are used to create entire ecosystems, in which symbiotic organisms can live. Coral polyps often eat algae through endocytosis, but the algae are allowed to live in vacuoles within the coral. This allows the coral to gain the oxygen and nutrients given off by the algae.

References

- Nuclear-membrane: biologydictionary.net, Retrieved 29 March, 2019

- Underwood, emily (2018). "when the brain's waste disposal system fails". Knowable magazine. Doi:10.1146/knowable-121118-1

- Endoplasmic-reticulum-rough-and-smooth, softcell-e-learning, learning-resources: bscb.org, Retrieved 2 June, 2019

- Lüllmznn-rauch r (2005). "history and morphology of lysosome". In zaftig p (ed.). Lysosomes (online-ausg. 1 ed.). Georgetown, tex.: landes bioscience/eurekah.com. Pp. 1–16. Isbn 978-0-387-28957-1

- The-endomembrane-system, tour-of-organelles, structure-of-a-cell, biology, science: khanacademy.org, Retrieved 22 May, 2019

- Lodish h, berk a, matsudaira p, kaiser ca, krieger m, scott mp, zipursky sl, darnell j (2004). Molecular cell biology (5th ed.). New york: wh freeman. Isbn 978-0-7167-2672-2

- Organelles-meaning-373368: thoughtco.com, Retrieved 23 February, 2019

- Mounkes lc, stewart cl (june 2004). "aging and nuclear organization: lamins and progeria". Current opinion in cell biology. 16 (3): 322–7. Doi:10.1016/j.ceb.2004.03.009. Pmid 15145358

- Ribosomes: microscopemaster.com, Retrieved 15 July, 2019

- Campbell na, williamson b, heyden rj (2006). Biology: exploring life. Boston, massachusetts: pearson prentice hall. Isbn 978-0-13-250882-7

- Centrosome: biologydictionary.net, Retrieved 16 January, 2019

- Wiemerslage l, lee d (march 2016). "quantification of mitochondrial morphology in neurites of dopaminergic neurons using multiple parameters". Journal of neuroscience methods. 262: 56–65

- Vacuole: biologydictionary.net, Retrieved 5 February, 2019

6

Cellular Activities

Fundamental biological activities in cells include cellular reproduction, cell adhesion, cell signalling and cellular respiration. Some of the methods of cellular reproduction are meiosis, mitosis and fission. All these diverse biological activities of cells have been carefully analyzed in this chapter.

CELLULAR METABOLISM

The collection of chemical reactions in your body is usually referred to as your metabolism. This process is the sum of all chemical changes that take place within the cells in your body. During digestion, for example, cellular metabolism is what releases energy from nutrients. Cellular metabolism sustains life and allows cells to grow, develop, repair damage, and respond to environmental changes.

Cellular metabolism can break down organic matter, a process known as catabolism. Cellular metabolism can also produce substances, a process referred to as anabolism. To provide a more graspable example, breaking down food so the nutrients can be utilized is a catabolic reaction. The production of proteins from amino acids is an example of an anabolic reaction. In general, breaking down releases energy and building up consumes energy. Amino acids, carbohydrates, and lipids (often called fats) are vital for life. Metabolic reactions either produce these molecules during the construction of cells and tissue or digest them and use them as a source of energy.

Catabolic Cellular Metabolism

Catabolic metabolism breaks down complex organic molecules into more simple molecules. These exergonic reactions are characterized by the release of energy. Catabolism reduces protein, fat, and carbohydrates into amino acids, fatty acids, and simple sugars, respectively. The energy released from catabolic reactions drives anabolic reactions. It's a process that has three stages:

1. Breakdown of complex molecules into their basic building blocks.

2. Breakdown of the basic building blocks into even more simple metabolic intermediates.

3. "Combustion" of the acetyl groups of acetyl-coenzymeA by the citric acid cycle and oxidative phosphorylation to produce CO_2 and H_2O. In other words, energy is released.

Anabolic Cellular Metabolism

Whereas catabolic metabolism breaks down molecules into their constituents, anabolic metabolism combines simple substances into more complex substances. When your cells combine amino acids into proteins to produce cells or tissues, that's anabolism. Anabolic reactions are endergonic reactions, which means they use more energy than they produce.

Although catabolism and anabolism occur independently of each other, they are inextricably linked. Without cellular metabolism, the body's cells wouldn't be able to break down or synthesize the compounds needed for energy, growth, function, and healing.

Role of Cells

Your body is comprised of an incalculable number of cells, and many different types of cells at that. Brain cells are different from blood cells, which are different from bone cells, which are different from skin cells, and so on. Every cell in your body is uniquely suited for a specific purpose. Others send electrochemical messages to the brain. Different cells have different features and may even have structural differences depending on their purpose.

Cell Composition

Although cell structure may vary from one to the next, all cells share several basic traits. Organelles are one feature that all cells possess. Organelles are like miniature organs within the cell that perform different specialized functions. For example, mitochondria are a type of organelle responsible for releasing energy. Other examples include the cell nucleus and endoplasmic reticulum.

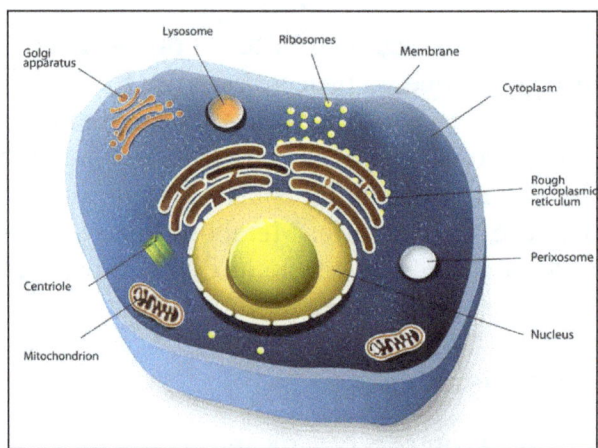

Animal cell.

Cellular Function

Cellular function directly affects how you feel. If your body is operating poorly at the cellular level, it's a sure bet your overall well-being will be affected.

Supporting cellular function in your body really is a "garbage in, garbage out" proposition. If your diet isn't balanced and fails to provide essential nutrients, then your cells aren't being given the fuel they need. If your nutrient intake is incomplete or inconsistent, organic supplements can help fill the gaps.

Vitamin B-12, in particular, is a nutrient that plays a very important role in supporting energy levels and metabolic function. Many people don't get enough vitamin B-12 and, as a result, their energy levels suffer. If you feel chronically fatigued, you may benefit from adding a B-12 supplement to your diet.

Vitamin D also supports normal cellular function. It's difficult to get enough vitamin D in your diet but with adequate skin exposure to sunlight (UVB rays), your body can produce vitamin D. Regular exercise can also boost your health and well-being in many ways; improving cellular function is one of them. It's important to stay hydrated and consume plenty of clean, purified water. Simply put, hydrated cells function better.

Cellular Respiration

Cellular respiration is a set of metabolic reactions and processes that take place in the cells of organisms to convert biochemical energy from nutrients into adenosine triphosphate (ATP), and then release waste products. The reactions involved in respiration are catabolic reactions, which break large molecules into smaller ones, releasing energy in the process, as weak so-called "high-energy" bonds are replaced by stronger bonds in the products. Respiration is one of the key ways a cell releases chemical energy to fuel cellular activity. Cellular respiration is considered an exothermic redox reaction which releases heat. The overall reaction occurs in a series of biochemical steps, most of which are redox reactions themselves. Although cellular respiration is technically a combustion reaction, it clearly does not resemble one when it occurs in a living cell because of the slow release of energy from the series of reactions.

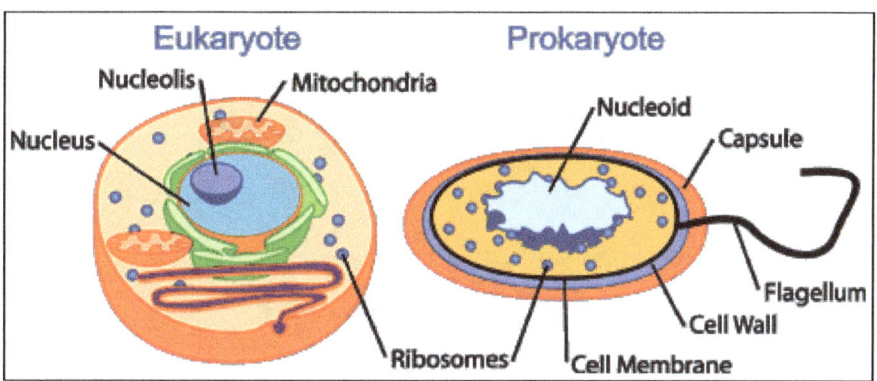

Typical eukaryotic cell.

Nutrients that are commonly used by animal and plant cells in respiration include sugar, amino acids and fatty acids, and the most common oxidizing agent (electron acceptor) is molecular oxygen (O_2). The chemical energy stored in ATP (its third phosphate group is weakly bonded to the rest of the molecule and is cheaply broken allowing stronger bonds to form, thereby transferring energy for use by the cell) can then be used to drive processes requiring energy, including biosynthesis, locomotion or transportation of molecules across cell membranes.

Aerobic Respiration

Aerobic respiration requires oxygen (O_2) in order to create ATP. Although carbohydrates, fats, and proteins are consumed as reactants, it is the preferred method of pyruvate breakdown in glycolysis and requires that pyruvate enter the mitochondria in order to be fully oxidized by the Krebs cycle. The products of this process are carbon dioxide and water, but the energy transferred is used to break bonds in ADP as the third phosphate group is added to form ATP (adenosine triphosphate), by substrate-level phosphorylation, NADH and $FADH_2$:

$$Simplified\ reaction: \quad C_6H_{12}O_6\ (s) + 6\ O_2\ (g) \rightarrow 6\ CO_2\ (g) + 6\ H_2O\ (l) + heat$$
$$\Delta G = -2880\ kJ\ per\ mol\ of\ C_6H_{12}O_6$$

The negative ΔG indicates that the reaction can occur spontaneously.

The potential of NADH and $FADH_2$ is converted to more ATP through an electron transport chain with oxygen as the "terminal electron acceptor". Most of the ATP produced by aerobic cellular respiration is made by oxidative phosphorylation. This works by the energy released in the consumption of pyruvate being used to create a chemiosmotic potential by pumping protons across a membrane. This potential is then used to drive ATP synthase and produce ATP from ADP and a phosphate group. Biology textbooks often state that 38 ATP molecules can be made per oxidised glucose molecule during cellular respiration (2 from glycolysis, 2 from the Krebs cycle, and about 34 from the electron transport system). However, this maximum yield is never quite reached because of losses due to leaky membranes as well as the cost of moving pyruvate and ADP into the mitochondrial matrix, and current estimates range around 29 to 30 ATP per glucose.

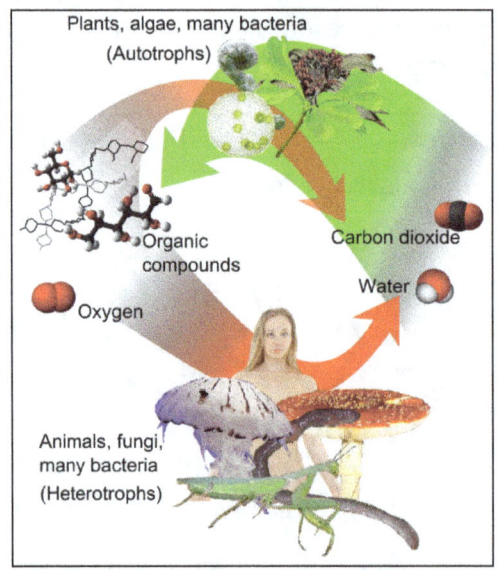

Aerobic respiration (red arrows) is the main means by which both fungi and animals utilize chemical energy in the form of organic compounds that were previously created through photosynthesis (green arrow).

Aerobic metabolism is up to 15 times more efficient than anaerobic metabolism (which yields 2 molecules ATP per 1 molecule glucose). However some anaerobic organisms, such as methanogens are able to continue with anaerobic respiration, yielding more ATP by using other inorganic molecules (not oxygen) as final electron acceptors in the electron transport chain. They share the initial pathway of glycolysis but aerobic metabolism continues with the Krebs cycle and oxidative phosphorylation. The post-glycolytic reactions take place in the mitochondria in eukaryotic cells, and in the cytoplasm in prokaryotic cells.

Glycolysis

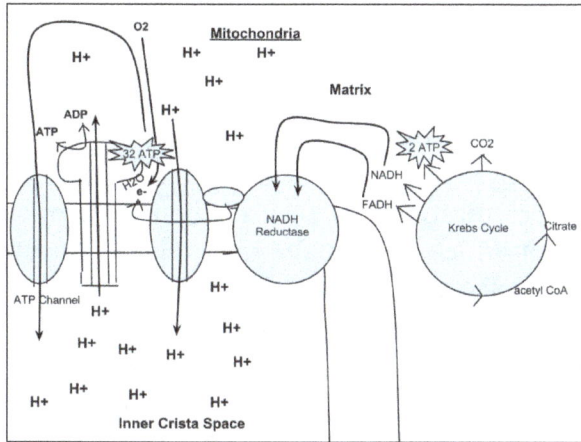

Out of the cytoplasm it goes into the Krebs cycle with the acetyl CoA.

It then mixes with CO_2 and makes 2 ATP, NADH, and FADH. From there the NADH and FADH go into the NADH reductase, which produces the enzyme. The NADH pulls the enzyme's electrons to send through the electron transport chain. The electron transport chain pulls H^+ ions through the chain. From the electron transport chain, the released hydrogen ions make ADP for an end result of 32 ATP. O_2 attracts itself to the left over electron to make water. Lastly, ATP leaves through the ATP channel and out of the mitochondria.

Glycolysis is a metabolic pathway that takes place in the cytosol of cells in all living organisms. This pathway can function with or without the presence of oxygen. In humans, aerobic conditions produce pyruvate and anaerobic conditions produce lactate. In aerobic conditions, the process converts one molecule of glucose into two molecules of pyruvate (pyruvic acid), generating energy in the form of two net molecules of ATP. Four molecules of ATP per glucose are actually produced, however, two are consumed as part of the preparatory phase. The initial phosphorylation of glucose is required to increase the reactivity (decrease its stability) in order for the molecule to be cleaved into two pyruvate molecules by the enzyme aldolase. During the pay-off phase of glycolysis, four phosphate groups are transferred to ADP by substrate-level phosphorylation to make four ATP, and two NADH are produced when the pyruvate are oxidized. The overall reaction can be expressed this way:

$$\text{Glucose} + 2\text{ NAD}^+ + 2\text{ P}_i + 2\text{ ADP} \rightarrow 2\text{ pyruvate} + 2\text{ NADH} + 2\text{ ATP} + 2\text{ H}^+ + 2\text{ H}_2\text{O} + \text{heat}$$

Starting with glucose, 1 ATP is used to donate a phosphate to glucose to produce glucose 6-phosphate. Glycogen can be converted into glucose 6-phosphate as well with the help of glycogen phosphorylase. During energy metabolism, glucose 6-phosphate becomes fructose 6-phosphate. An additional ATP is used to phosphorylate fructose 6-phosphate into fructose 1,6-disphosphate by the help of phosphofructokinase. Fructose 1,6-diphosphate then splits into two phosphorylated molecules with three carbon chains which later degrades into pyruvate. Glycolysis can be literally translated as "sugar splitting".

Oxidative Decarboxylation of Pyruvate

Pyruvate is oxidized to acetyl-CoA and CO_2 by the pyruvate dehydrogenase complex (PDC). The PDC contains multiple copies of three enzymes and is located in the mitochondria of eukaryotic cells and in the cytosol of prokaryotes. In the conversion of pyruvate to acetyl-CoA, one molecule of NADH and one molecule of CO_2 is formed.

Citric Acid Cycle

This is also called the *Krebs cycle* or the *tricarboxylic acid cycle*. When oxygen is present, acetyl-CoA is produced from the pyruvate molecules created from glycolysis. Once acetyl-CoA is formed, aerobic or anaerobic respiration can occur. When oxygen is present, the mitochondria will undergo aerobic respiration which leads to the Krebs cycle. However, if oxygen is not present, fermentation of the pyruvate molecule will occur. In the presence of oxygen, when acetyl-CoA is produced, the molecule then enters the citric acid cycle (Krebs cycle) inside the mitochondrial matrix, and is oxidized to CO_2 while at the same time reducing NAD to NADH. NADH can be used by the electron transport chain to create further ATP as part of oxidative phosphorylation. To fully oxidize the equivalent of one glucose molecule, two acetyl-CoA must be metabolized by the Krebs cycle. Two waste products, H_2O and CO_2, are created during this cycle.

The citric acid cycle is an 8-step process involving 18 different enzymes and co-enzymes. During the cycle, acetyl-CoA (2 carbons) + oxaloacetate (4 carbons) yields citrate (6 carbons), which is rearranged to a more reactive form called isocitrate (6 carbons). Isocitrate is modified to become α-ketoglutarate (5 carbons), succinyl-CoA, succinate, fumarate, malate, and, finally, oxaloacetate.

The net gain of high-energy compounds from one cycle is 3 NADH, 1 $FADH_2$, and 1 GTP; the GTP may subsequently be used to produce ATP. Thus, the total yield from 1 glucose molecule (2 pyruvate molecules) is 6 NADH, 2 $FADH_2$, and 2 ATP.

Oxidative Phosphorylation

In eukaryotes, oxidative phosphorylation occurs in the mitochondrial cristae. It comprises the electron transport chain that establishes a proton gradient (chemiosmotic potential) across the boundary of inner membrane by oxidizing the NADH produced from the Krebs cycle. ATP is synthesized by the ATP synthase enzyme when the chemiosmotic gradient is used to drive the phosphorylation of ADP. The electrons are finally transferred to exogenous oxygen and, with the addition of two protons, water is formed.

Efficiency of ATP Production

The table below describes the reactions involved when one glucose molecule is fully oxidized into carbon dioxide. It is assumed that all the reduced coenzymes are oxidized by the electron transport chain and used for oxidative phosphorylation.

Step	coenzyme yield	ATP yield	Source of ATP
Glycolysis preparatory phase		−2	Phosphorylation of glucose and fructose 6-phosphate uses two ATP from the cytoplasm.
Glycolysis pay-off phase		4	Substrate-level phosphorylation.
	2 NADH	3 or 5	Oxidative phosphorylation : Each NADH produces net 1.5 ATP (instead of usual 2.5) due to NADH transport over the mitochondrial membrane.

Oxidative decarboxylation of pyruvate	2 NADH	5	Oxidative phosphorylation.
Krebs cycle		2	Substrate-level phosphorylation.
	6 NADH	15	Oxidative phosphorylation.
	2 FADH2	3	Oxidative phosphorylation.
Total yield		30 or 32 ATP	From the complete oxidation of one glucose molecule to carbon dioxide and oxidation of all the reduced coenzymes.

Although there is a theoretical yield of 38 ATP molecules per glucose during cellular respiration, such conditions are generally not realized because of losses such as the cost of moving pyruvate (from glycolysis), phosphate, and ADP (substrates for ATP synthesis) into the mitochondria. All are actively transported using carriers that utilize the stored energy in the proton electrochemical gradient.

- Pyruvate is taken up by a specific, low Km transporter to bring it into the mitochondrial matrix for oxidation by the pyruvate dehydrogenase complex.

- The phosphate carrier (PiC) mediates the electroneutral exchange (antiport) of phosphate ($H_2PO_4^-$; P_i) for OH^- or symport of phosphate and protons (H^+) across the inner membrane, and the driving force for moving phosphate ions into the mitochondria is the proton motive force.

- The ATP-ADP translocase (also called adenine nucleotide translocase, ANT) is an antiporter and exchanges ADP and ATP across the inner membrane. The driving force is due to the ATP (−4) having a more negative charge than the ADP (−3), and thus it dissipates some of the electrical component of the proton electrochemical gradient.

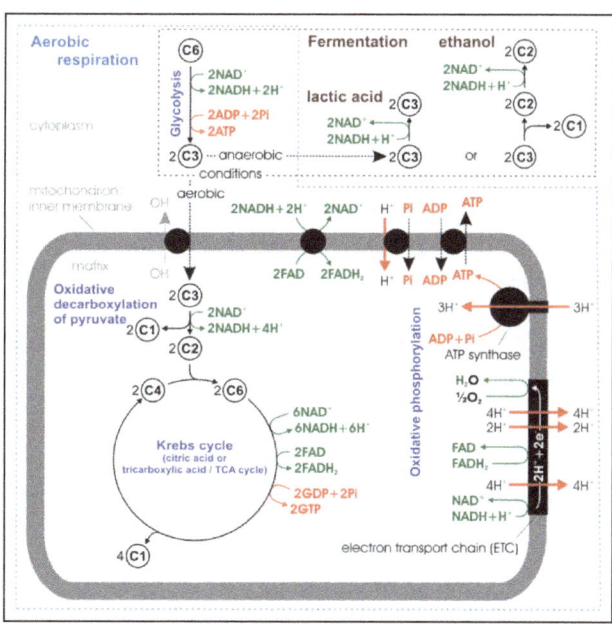

Stoichiometry of aerobic respiration and most known fermentation types in eucaryotic cell. Numbers in circles indicate counts of carbon atoms in molecules, C6 is glucose $C_6H_{12}O_6$, C1 carbon dioxide CO_2. Mitochondrial outer membrane is omitted.

The outcome of these transport processes using the proton electrochemical gradient is that more than 3 H^+ are needed to make 1 ATP. Obviously this reduces the theoretical efficiency of the whole process and the likely maximum is closer to 28–30 ATP molecules. In practice the efficiency may be even lower because the inner membrane of the mitochondria is slightly leaky to protons. Other factors may also dissipate the proton gradient creating an apparently leaky mitochondria. An uncoupling protein known as thermogenin is expressed in some cell types and is a channel that can transport protons. When this protein is active in the inner membrane it short circuits the coupling between the electron transport chain and ATP synthesis. The potential energy from the proton gradient is not used to make ATP but generates heat. This is particularly important in brown fat thermogenesis of newborn and hibernating mammals.

According to some of newer sources the ATP yield during aerobic respiration is not 36–38, but only about 30–32 ATP molecules/1 molecule of glucose, because:

- ATP : NADH+H$^+$ and ATP : FADH$_2$ ratios during the oxidative phosphorylation appear to be not 3 and 2, but 2.5 and 1.5 respectively. Unlike in the substrate-level phosphorylation, the stoichiometry here is difficult to establish.

  ○ ATP synthase produces 1 ATP/3 H$^+$. However the exchange of matrix ATP for cytosolic ADP and Pi (antiport with OH$^-$ or symport with H$^+$) mediated by ATP–ADP translocase and phosphate carrier consumes 1 H$^+$/1 ATP as a result of regeneration of the transmembrane potential changed during this transfer, so the net ratio is 1 ATP : 4 H$^+$.

  ○ The mitochondrial electron transport chain proton pump transfers across the inner membrane 10 H$^+$/1 NADH+H$^+$ (4 + 2 + 4) or 6 H$^+$/1 FADH$_2$ (2 + 4).

So the final stoichiometry is:

　　1 NADH+H$^+$: 10 H$^+$: 10/4 ATP = 1 NADH+H$^+$: 2.5 ATP

　　1 FADH$_2$: 6 H$^+$: 6/4 ATP = 1 FADH$_2$: 1.5 ATP

- ATP : NADH+H$^+$ coming from glycolysis ratio during the oxidative phosphorylation is

  ○ 1.5, as for FADH$_2$, if hydrogen atoms (2H$^+$+2e$^-$) are transferred from cytosolic NADH+H$^+$ to mitochondrial FAD by the glycerol phosphate shuttle located in the inner mitochondrial membrane.

  ○ 2.5 in case of malate-aspartate shuttle transferring hydrogen atoms from cytosolic NADH+H$^+$ to mitochondrial NAD$^+$

So finally we have, per molecule of glucose:

- Substrate-level phosphorylation: 2 ATP from glycolysis + 2 ATP (directly GTP) from Krebs cycle.

- Oxidative phosphorylation:

  ○ 2 NADH+H$^+$ from glycolysis: 2 × 1.5 ATP (if glycerol phosphate shuttle transfers hydrogen atoms) or 2 × 2.5 ATP (malate-aspartate shuttle).

  ○ 2 NADH+H$^+$ from the oxidative decarboxylation of pyruvate and 6 from Krebs cycle: 8 × 2.5 ATP.

○ 2 $FADH_2$ from the Krebs cycle: 2×1.5 ATP.

Altogether this gives $4 + 3$ (or 5) $+ 20 + 3 = 30$ (or 32) ATP per molecule of glucose.

These figures may still require further tweaking as new structural details become available. The above value of 3 H+/ATP for the synthase assumes that the synthase translocates 9 protons, and produces 3 ATP, per rotation. The number of protons depends on the number of c subunits in the Fo c-ring, and it is now known that this is 10 in yeast Fo and 8 for vertebrates. Including one H+ for the transport reactions, this means that synthesis of one ATP requires $1+10/3=4.33$ protons in yeast and $1+8/3 = 3.67$ in vertebrates. This would imply that in human mitochondria the 10 protons from oxidizing NADH would produce 2.72 ATP (instead of 2.5) and the 6 protons from oxidizing succinate or ubiquinol would produce 1.64 ATP (instead of 1.5). This is consistent with experimental results within the margin of error.

The total ATP yield in ethanol or lactic acid fermentation is only 2 molecules coming from glycolysis, because pyruvate is not transferred to the mitochondrion and finally oxidized to the carbon dioxide (CO_2), but reduced to ethanol or lactic acid in the cytoplasm.

Fermentation

Without oxygen, pyruvate (pyruvic acid) is not metabolized by cellular respiration but undergoes a process of fermentation. The pyruvate is not transported into the mitochondrion, but remains in the cytoplasm, where it is converted to waste products that may be removed from the cell. This serves the purpose of oxidizing the electron carriers so that they can perform glycolysis again and removing the excess pyruvate. Fermentation oxidizes NADH to NAD+ so it can be re-used in glycolysis. In the absence of oxygen, fermentation prevents the buildup of NADH in the cytoplasm and provides NAD+ for glycolysis. This waste product varies depending on the organism. In skeletal muscles, the waste product is lactic acid. This type of fermentation is called lactic acid fermentation. In strenuous exercise, when energy demands exceed energy supply, the respiratory chain cannot process all of the hydrogen atoms joined by NADH. During anaerobic glycolysis, NAD+ regenerates when pairs of hydrogen combine with pyruvate to form lactate. Lactate formation is catalyzed by lactate dehydrogenase in a reversible reaction. Lactate can also be used as an indirect precursor for liver glycogen. During recovery, when oxygen becomes available, NAD+ attaches to hydrogen from lactate to form ATP. In yeast, the waste products are ethanol and carbon dioxide. This type of fermentation is known as alcoholic or ethanol fermentation. The ATP generated in this process is made by substrate-level phosphorylation, which does not require oxygen.

Fermentation is less efficient at using the energy from glucose: only 2 ATP are produced per glucose, compared to the 38 ATP per glucose nominally produced by aerobic respiration. This is because the waste products of fermentation still contain chemical potential energy that can be released by oxidation. Ethanol, for example, can be burned in an internal combustion engine like gasoline. Glycolytic ATP, however, is created more quickly. For prokaryotes to continue a rapid growth rate when they are shifted from an aerobic environment to an anaerobic environment, they must increase the rate of the glycolytic reactions. For multicellular organisms, during short bursts of strenuous activity, muscle cells use fermentation to supplement the ATP production from the slower aerobic respiration, so fermentation may be used by a cell even before the oxygen levels are depleted, as is the case in sports that do not require athletes to pace themselves, such as sprinting.

Anaerobic Respiration

Cellular respiration is the process by which biological fuels are oxidised in the presence of an inorganic electron acceptor (such as oxygen) to produce large amounts of energy, to drive the bulk production of ATP.

Anaerobic respiration is used by some microorganisms in which neither oxygen (aerobic respiration) nor pyruvate derivatives (fermentation) is the final electron acceptor. Rather, an inorganic acceptor such as sulfate or nitrate is used. Such organisms are typically found in unusual places such as underwater caves or near hydrothermal vents at the bottom of the ocean.

Photosynthesis

Photosynthesis is the process by which green plants and certain other organisms transform light energy into chemical energy. During photosynthesis in green plants, light energy is captured and used to convert water, carbon dioxide, and minerals into oxygen and energy-rich organic compounds.

It would be impossible to overestimate the importance of photosynthesis in the maintenance of life on Earth. If photosynthesis ceased, there would soon be little food or other organic matter on Earth. Most organisms would disappear, and in time Earth's atmosphere would become nearly devoid of gaseous oxygen. The only organisms able to exist under such conditions would be the chemosynthetic bacteria, which can utilize the chemical energy of certain inorganic compounds and thus are not dependent on the conversion of light energy.

Energy produced by photosynthesis carried out by plants millions of years ago is responsible for the fossil fuels (i.e., coal, oil, and gas) that power industrial society. In past ages, green plants and small organisms that fed on plants increased faster than they were consumed, and their remains were deposited in Earth's crust by sedimentation and other geological processes. There, protected from oxidation, these organic remains were slowly converted to fossil fuels. These fuels not only provide much of the energy used in factories, homes, and transportation but also serve as the raw material for plastics and other synthetic products. Unfortunately, modern civilization is using up in a few centuries the excess of photosynthetic production accumulated over millions of years. Consequently, the carbon dioxide that has been removed from the air to make carbohydrates in photosynthesis over millions of years is being returned at an incredibly rapid rate. The carbon dioxide concentration in Earth's atmosphere is rising the fastest it ever has in Earth's history, and this phenomenon is expected to have major implications on Earth's climate.

Requirements for food, materials, and energy in a world where human population is rapidly growing have created a need to increase both the amount of photosynthesis and the efficiency of converting photosynthetic output into products useful to people. One response to those needs—the so-called Green Revolution, begun in the mid-20th century—achieved enormous improvements in agricultural yield through the use of chemical fertilizers, pest and plant-disease control, plant breeding, and mechanized tilling, harvesting, and crop processing. This effort limited severe famines to a few areas of the world despite rapid population growth, but it did not eliminate widespread malnutrition. Moreover, beginning in the early 1990s, the rate at which yields of major crops increased began to decline. This was especially true for rice in Asia. Rising costs associated with sustaining high rates of agricultural production, which required ever-increasing inputs of fertilizers and pesticides and constant development of new plant varieties, also became problematic for farmers in many countries.

A second agricultural revolution, based on plant genetic engineering, was forecast to lead to increases in plant productivity and thereby partially alleviate malnutrition. Since the 1970s, molecular biologists have possessed the means to alter a plant's genetic material (deoxyribonucleic acid, or DNA) with the aim of achieving improvements in disease and drought resistance, product yield and quality, frost hardiness, and other desirable properties. However, such traits are inherently complex, and the process of making changes to crop plants through genetic engineering has turned out to be more complicated than anticipated. In the future such genetic engineering may result in improvements in the process of photosynthesis, but by the first decades of the 21st century, it had yet to demonstrate that it could dramatically increase crop yields.

Another intriguing area in the study of photosynthesis has been the discovery that certain animals are able to convert light energy into chemical energy. The emerald green sea slug (Elysia chlorotica), for example, acquires genes and chloroplasts from Vaucheria litorea, an alga it consumes, giving it a limited ability to produce chlorophyll. When enough chloroplasts are assimilated, the slug may forgo the ingestion of food. The pea aphid (Acyrthosiphon pisum) can harness light to manufacture the energy-rich compound adenosine triphosphate (ATP); this ability has been linked to the aphid's manufacture of carotenoid pigments.

Overall Reaction of Photosynthesis

In chemical terms, photosynthesis is a light-energized oxidation–reduction process. (Oxidation refers to the removal of electrons from a molecule; reduction refers to the gain of electrons by a molecule.) In plant photosynthesis, the energy of light is used to drive the oxidation of water (H_2O), producing oxygen gas (O_2), hydrogen ions (H^+), and electrons. Most of the removed electrons and hydrogen ions ultimately are transferred to carbon dioxide (CO_2), which is reduced to organic products. Other electrons and hydrogen ions are used to reduce nitrate and sulfate to amino and sulfhydryl groups in amino acids, which are the building blocks of proteins. In most green cells, carbohydrates—especially starch and the sugar sucrose—are the major direct organic products of photosynthesis. The overall reaction in which carbohydrates—represented by the general formula (CH_2O)—are formed during plant photosynthesis can be indicated by the following equation:

$$CO_2 + 2\,H_2O \xrightarrow[\text{green plants}]{\text{light}} (CH_2O) + O_2 + H_2O$$

This equation is merely a summary statement, for the process of photosynthesis actually involves numerous reactions catalyzed by enzymes (organic catalysts). These reactions occur in two stages: the "light" stage, consisting of photochemical (i.e., light-capturing) reactions; and the "dark" stage, comprising chemical reactions controlled by enzymes. During the first stage, the energy of light is absorbed and used to drive a series of electron transfers, resulting in the synthesis of ATP and the electron-donor-reduced nicotine adenine dinucleotide phosphate (NADPH). During the dark stage, the ATP and NADPH formed in the light-capturing reactions are used to reduce carbon dioxide to organic carbon compounds. This assimilation of inorganic carbon into organic compounds is called carbon fixation.

During the 20th century, comparisons between photosynthetic processes in green plants and in certain photosynthetic sulfur bacteria provided important information about the photosynthetic mechanism. Sulfur bacteria use hydrogen sulfide (H_2S) as a source of hydrogen atoms and produce sulfur instead of oxygen during photosynthesis. The overall reaction is:

$$CO_2 + 2\,H_2S \xrightarrow[\substack{\text{sulfur}\\\text{bacteria}}]{\text{light}} (CH_2O) + S_2 + H_2O$$

In the 1930s Dutch biologist Cornelis van Niel recognized that the utilization of carbon dioxide to form organic compounds was similar in the two types of photosynthetic organisms. Suggesting that differences existed in the light-dependent stage and in the nature of the compounds used as a source of hydrogen atoms, he proposed that hydrogen was transferred from hydrogen sulfide (in bacteria) or water (in green plants) to an unknown acceptor (called A), which was reduced to H_2A. During the dark reactions, which are similar in both bacteria and green plants, the reduced acceptor (H_2A) reacted with carbon dioxide (CO_2) to form carbohydrate (CH_2O) and to oxidize the unknown acceptor to A. This putative reaction can be represented as:

$$CO_2 + 2\,H_2A \xrightarrow{\text{light}} (CH_2O) + 2\,A + H_2O$$

Van Niel's proposal was important because the popular (but incorrect) theory had been that oxygen was removed from carbon dioxide (rather than hydrogen from water, releasing oxygen) and that carbon then combined with water to form carbohydrate (rather than the hydrogen from water combining with CO_2 to form CH_2O).

By 1940 chemists were using heavy isotopes to follow the reactions of photosynthesis. Water marked with an isotope of oxygen (^{18}O) was used in early experiments. Plants that photosynthesized in the presence of water containing $H_2^{18}O$ produced oxygen gas containing ^{18}O; those that photosynthesized in the presence of normal water produced normal oxygen gas. These results provided definitive support for van Niel's theory that the oxygen gas produced during photosynthesis is derived from water.

Basic Products of Photosynthesis

As has been stated, carbohydrates are the most-important direct organic product of photosynthesis in the majority of green plants. The formation of a simple carbohydrate, glucose, is indicated by a chemical equation,

$$6\,CO_2 + 12\,H_2O \xrightarrow[\text{green plants}]{\text{light}} C_6H_{12}O_6 + 6\,O_2 + 6\,H_2O$$

Little free glucose is produced in plants; instead, glucose units are linked to form starch or are joined with fructose, another sugar, to form sucrose.

Not only carbohydrates, as was once thought, but also amino acids, proteins, lipids (or fats), pigments, and other organic components of green tissues are synthesized during photosynthesis.

Minerals supply the elements (e.g., nitrogen, N; phosphorus, P; sulfur, S) required to form these compounds. Chemical bonds are broken between oxygen (O) and carbon (C), hydrogen (H), nitrogen, and sulfur, and new bonds are formed in products that include gaseous oxygen (O_2) and organic compounds. More energy is required to break the bonds between oxygen and other elements (e.g., in water, nitrate, and sulfate) than is released when new bonds form in the products. This difference in bond energy accounts for a large part of the light energy stored as chemical energy in the organic products formed during photosynthesis. Additional energy is stored in making complex molecules from simple ones.

Evolution of the Process

Although life and the quality of the atmosphere today depend on photosynthesis, it is likely that green plants evolved long after the first living cells. When Earth was young, electrical storms and solar radiation probably provided the energy for the synthesis of complex molecules from abundant simpler ones, such as water, ammonia, and methane. The first living cells probably evolved from these complex molecules. For example, the accidental joining (condensation) of the amino acid glycine and the fatty acid acetate may have formed complex organic molecules known as porphyrins. These molecules, in turn, may have evolved further into coloured molecules called pigments—e.g., chlorophylls of green plants, bacteriochlorophyll of photosynthetic bacteria, hemin (the red pigment of blood), and cytochromes, a group of pigment molecules essential in both photosynthesis and cellular respiration.

Primitive coloured cells then had to evolve mechanisms for using the light energy absorbed by their pigments. At first, the energy may have been used immediately to initiate reactions useful to the cell. As the process for utilization of light energy continued to evolve, however, a larger part of the absorbed light energy probably was stored as chemical energy, to be used to maintain life. Green plants, with their ability to use light energy to convert carbon dioxide and water to carbohydrates and oxygen, are the culmination of this evolutionary process.

He first oxygenic (oxygen-producing) cells probably were the blue-green algae (cyanobacteria), which appeared about two billion to three billion years ago. These microscopic organisms are believed to have greatly increased the oxygen content of the atmosphere, making possible the development of aerobic (oxygen-using) organisms. Cyanophytes are prokaryotic cells; that is, they contain no distinct membrane-enclosed subcellular particles (organelles), such as nuclei and chloroplasts. Green plants, by contrast, are composed of eukaryotic cells, in which the photosynthetic apparatus is contained within membrane-bound chloroplasts. The complete genome sequences of cyanobacteria and higher plants provide evidence that the first photosynthetic eukaryotes were likely the red algae that developed when nonphotosynthetic eukaryotic cells engulfed cyanobacteria. Within the host cells, these cyanobacteria evolved into chloroplasts.

There are a number of photosynthetic bacteria that are not oxygenic. The evolutionary pathway that led to these bacteria diverged from the one that resulted in oxygenic organisms. In addition to the absence of oxygen production, nonoxygenic photosynthesis differs from oxygenic photosynthesis in two other ways: light of longer wavelengths is absorbed and used by pigments called bacteriochlorophylls, and reduced compounds other than water (such as hydrogen sulfide or organic molecules) provide the electrons needed for the reduction of carbon dioxide.

Factors that Influence the Rate of Photosynthesis

The rate of photosynthesis is defined in terms of the rate of oxygen production either per unit mass (or area) of green plant tissues or per unit weight of total chlorophyll. The amount of light, the carbon dioxide supply, temperature, water supply, and the availability of minerals are the most important environmental factors that affect the rate of photosynthesis in land plants. The rate of photosynthesis is also determined by the plant species and its physiological state—e.g., its health, its maturity, and whether it is in flower.

Light Intensity and Temperature

As has been mentioned, the complex mechanism of photosynthesis includes a photochemical, or light-harvesting, stage and an enzymatic, or carbon-assimilating, stage that involves chemical reactions. These stages can be distinguished by studying the rates of photosynthesis at various degrees of light saturation (i.e., intensity) and at different temperatures. Over a range of moderate temperatures and at low to medium light intensities (relative to the normal range of the plant species), the rate of photosynthesis increases as the intensity increases and is relatively independent of temperature. As the light intensity increases to higher levels, however, the rate becomes saturated; light "saturation" is achieved at a specific light intensity, dependent on species and growing conditions. In the light-dependent range before saturation, therefore, the rate of photosynthesis is determined by the rates of photochemical steps. At high light intensities, some of the chemical reactions of the dark stage become rate-limiting. In many land plants, a process called photorespiration occurs, and its influence upon photosynthesis increases with rising temperatures. More specifically, photorespiration competes with photosynthesis and limits further increases in the rate of photosynthesis, especially if the supply of water is limited.

Carbon Dioxide

Included among the rate-limiting steps of the dark stage of photosynthesis are the chemical reactions by which organic compounds are formed by using carbon dioxide as a carbon source. The rates of these reactions can be increased somewhat by increasing the carbon dioxide concentration. Since the middle of the 19th century, the level of carbon dioxide in the atmosphere has been rising because of the extensive combustion of fossil fuels, cement production, and land-use changes associated with deforestation. The atmospheric level of carbon dioxide climbed from about 0.028 percent in 1860 to 0.032 percent by 1958 (when improved measurements began) and to 0.040 percent by 2016. This increase in carbon dioxide directly increases plant photosynthesis, but the size of the increase depends on the species and physiological condition of the plant. Furthermore, most scientists maintain that increasing levels of atmospheric carbon dioxide affect climate, increasing global temperatures and changing rainfall patterns. Such changes will also affect photosynthesis rates.

Water

For land plants, water availability can function as a limiting factor in photosynthesis and plant growth. Besides the requirement for a small amount of water in the photosynthetic reaction itself, large amounts of water are transpired from the leaves; that is, water evaporates from the leaves to the atmosphere via the stomata. Stomata are small openings through the leaf epidermis, or outer skin; they permit the entry of carbon dioxide but inevitably also allow the exit of water vapour. The stomata open and close according to the physiological needs of the leaf. In hot and arid climates the stomata may close to conserve water, but this closure limits the entry of carbon dioxide and hence the rate of photosynthesis. The decreased transpiration means there is less cooling of the leaves and hence leaf temperatures rise. The decreased carbon dioxide concentration inside the leaves and the increased leaf temperatures favour the wasteful process of photorespiration. If the level of carbon dioxide in the atmosphere increases, more carbon dioxide could enter through a smaller opening of the stomata, so more photosynthesis could occur with a given supply of water.

Minerals

Several minerals are required for healthy plant growth and for maximum rates of photosynthesis. Nitrogen, sulfate, phosphate, iron, magnesium, calcium, and potassium are required in substantial amounts for the synthesis of amino acids, proteins, coenzymes, deoxyribonucleic acid (DNA) and ribonucleic acid (RNA), chlorophyll and other pigments, and other essential plant constituents. Smaller amounts of such elements as manganese, copper, and chloride are required in photosynthesis. Some other trace elements are needed for various nonphotosynthetic functions in plants.

Internal Factors

Each plant species is adapted to a range of environmental factors. Within this normal range of conditions, complex regulatory mechanisms in the plant's cells adjust the activities of enzymes (i.e., organic catalysts). These adjustments maintain a balance in the overall photosynthetic process and control it in accordance with the needs of the whole plant. With a given plant species, for example, doubling the carbon dioxide level might cause a temporary increase of nearly twofold in the rate of photosynthesis; a few hours or days later, however, the rate might fall to the original level because photosynthesis produced more sucrose than the rest of the plant could use. By contrast, another plant species provided with such carbon dioxide enrichment might be able to use more sucrose, because it had more carbon-demanding organs, and would continue to photosynthesize and to grow faster throughout most of its life cycle.

Energy Efficiency of Photosynthesis

The energy efficiency of photosynthesis is the ratio of the energy stored to the energy of light absorbed. The chemical energy stored is the difference between that contained in gaseous oxygen and organic compound products and the energy of water, carbon dioxide, and other reactants. The amount of energy stored can only be estimated because many products are formed, and these vary with the plant species and environmental conditions. If the equation for glucose formation given earlier is used to approximate the actual storage process, the production of one mole (i.e., 6.02×10^{23} molecules; abbreviated N) of oxygen and one-sixth mole of glucose results in the storage of about 117 kilocalories (kcal) of chemical energy. This amount must then be compared with the energy of light absorbed to produce one mole of oxygen in order to calculate the efficiency of photosynthesis.

Light can be described as a wave of particles known as photons; these are units of energy, or light quanta. The quantity N photons is called an einstein. The energy of light varies inversely with the length of the photon waves; that is, the shorter the wavelength, the greater the energy content.

The energy (e) of a photon is given by the equation e = hc/λ, where c is the velocity of light, h is Planck's constant, and λ is the light wavelength. The energy (E) of an einstein is E = Ne = Nhc/λ = 28,600/λ, when E is in kilocalories and λ is given in nanometres (nm; 1 nm = 10^{-9} metres). An einstein of red light with a wavelength of 680 nm has an energy of about 42 kcal. Blue light has a shorter wavelength and therefore more energy than red light. Regardless of whether the light is blue or red, however, the same number of einsteins are required for photosynthesis per mole of oxygen formed. The part of the solar spectrum used by plants has an estimated mean wavelength of 570 nm; therefore, the energy of light used during photosynthesis is approximately 28,600/570, or 50 kcal per einstein.

In order to compute the amount of light energy involved in photosynthesis, one other value is needed: the number of einsteins absorbed per mole of oxygen evolved. This is called the quantum requirement. The minimum quantum requirement for photosynthesis under optimal conditions is about nine. Thus, the energy used is 9 × 50, or 450 kcal per mole of oxygen evolved. Therefore, the estimated maximum energy efficiency of photosynthesis is the energy stored per mole of oxygen evolved, 117 kcal, divided by 450—that is, 117/450, or 26 percent.

The actual percentage of solar energy stored by plants is much less than the maximum energy efficiency of photosynthesis. An agricultural crop in which the biomass (total dry weight) stores as much as 1 percent of total solar energy received on an annual areawide basis is exceptional, although a few cases of higher yields (perhaps as much as 3.5 percent in sugarcane) have been reported. There are several reasons for this difference between the predicted maximum efficiency of photosynthesis and the actual energy stored in biomass. First, more than half of the incident sunlight is composed of wavelengths too long to be absorbed, and some of the remainder is reflected or lost to the leaves. Consequently, plants can at best absorb only about 34 percent of the incident sunlight. Second, plants must carry out a variety of physiological processes in such nonphotosynthetic tissues as roots and stems; these processes, as well as cellular respiration in all parts of the plant, use up stored energy. Third, rates of photosynthesis in bright sunlight sometimes exceed the needs of the plants, resulting in the formation of excess sugars and starch. When this happens, the regulatory mechanisms of the plant slow down the process of photosynthesis, allowing more absorbed sunlight to go unused. Fourth, in many plants, energy is wasted by the process of photorespiration. Finally, the growing season may last only a few months of the year; sunlight received during other seasons is not used. Furthermore, it should be noted that if only agricultural products (e.g., seeds, fruits, and tubers, rather than total biomass) are considered as the end product of the energy-conversion process of photosynthesis, the efficiency falls even further.

Chloroplasts, the Photosynthetic Units of Green Plants

The process of plant photosynthesis takes place entirely within the chloroplasts. Detailed studies of the role of these organelles date from the work of British biochemist Robert Hill. About 1940 Hill discovered that green particles obtained from broken cells could produce oxygen from water in the presence of light and a chemical compound, such as ferric oxalate, able to serve as an electron acceptor. This process is known as the Hill reaction. During the 1950s Daniel Arnon and other American biochemists prepared plant cell fragments in which not only the Hill reaction but also the synthesis of the energy-storage compound ATP occurred. In addition, the coenzyme NADP was used as the final acceptor of electrons, replacing the nonphysiological electron acceptors used by Hill. His procedures were refined further so that small individual pieces of isolated chloroplast membranes, or lamellae, could perform the Hill reaction. These small pieces of lamellae were then fragmented into pieces so small that they performed only the light reactions of the photosynthetic process. It is now possible also to isolate the entire chloroplast so that it can carry out the complete process of photosynthesis, from light absorption, oxygen formation, and the reduction of carbon dioxide to the formation of glucose and other products.

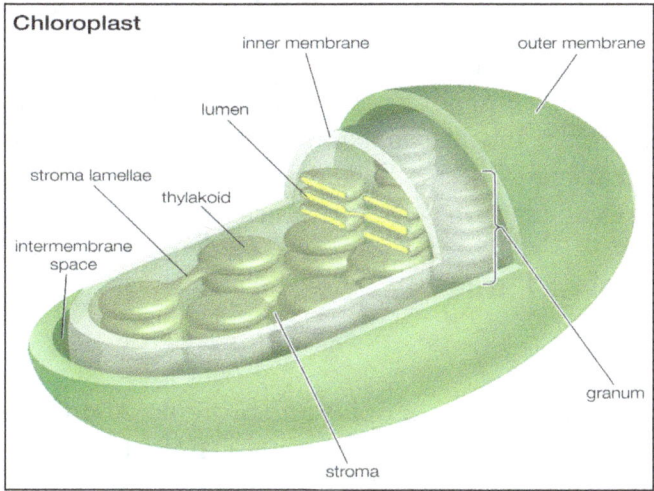

Chloroplast structure.

The internal (thylakoid) membrane vesicles are organized into stacks, which reside in a matrix known as the stroma. All the chlorophyll in the chloroplast is contained in the membranes of the thylakoid vesicles.

Structural Features

The intricate structural organization of the photosynthetic apparatus is essential for the efficient performance of the complex process of photosynthesis. The chloroplast is enclosed in a double outer membrane, and its size approximates a spheroid about 2,500 nm thick and 5,000 nm long. Some single-celled algae have one chloroplast that occupies more than half the cell volume. Leaf cells of higher plants contain many chloroplasts, each approximately the size of the one in some algal cells.

When thin sections of a chloroplast are examined under the electron microscope, several features are apparent. Chief among these are the intricate internal membranes (i.e., the lamellae) and the stroma, a colourless matrix in which the lamellae are embedded. Also visible are starch granules, which appear as dense bodies.

The stroma is basically a solution of enzymes and small molecules. The dark reactions occur in the stroma, the soluble enzymes of which catalyze the conversion of carbon dioxide and minerals to carbohydrates and other organic compounds. The capacity for carbon fixation and reduction is lost if the outer membrane of the chloroplast is broken, allowing the stroma enzymes to leak out.

A single lamella, which contains all the photosynthetic pigments, is approximately 10–15 nm thick. The lamellae exist in more-or-less flat sheets, a few of which extend through much of the length of the chloroplast. Examination of cross sections of lamellae under the electron microscope shows that their edges are joined to form closed hollow disks that are called thylakoids ("saclike"). The chloroplasts of most higher plants have regions, called grana, in which the thylakoids are very tightly stacked. When viewed by electron microscopy at an oblique angle, the grana appear as stacks of disks. When viewed in cross section, it is apparent that some thylakoids extend from one grana through the stroma into other grana. The thin aqueous spaces inside the thylakoids are believed to be connected with each other via these stroma thylakoids. These thylakoid spaces are isolated from the stroma spaces by the relatively impermeable lamellae.

The light reactions occur exclusively in the thylakoids. The complex structural organization of lamellae is required for proper thylakoid function; intact thylakoids are necessary for the formation of ATP. Thylakoids that have been broken down to smaller units can no longer form ATP, even when the conversion of light into chemical energy occurs during electron transport in these units. Such lamellar fragments can carry out the Hill reaction, with the transfer of electrons from water to $NADP^+$.

Chemical Composition of Lamellae

Lipids

Lamellae consist of about equal amounts of lipids and proteins. About one-fourth of the lipid portion of the lamellae consists of pigments and coenzymes; the remainder consists of various lipids, including polar compounds such as phospholipids and galactolipids. These polar lipid molecules have "head" groups that attract water (i.e., are hydrophilic) and fatty acid "tails" that are oil soluble and repel water (i.e., are hydrophobic). When polar lipids are placed in an aqueous environment, they can line up with the fatty acid tails side by side. A second layer of phospholipids forms tail-to-tail with the first, establishing a lipid bilayer in which the hydrophilic heads are in contact with the aqueous solution on each side of the bilayer. Sandwiched between the heads are the hydrophobic tails, creating a hydrophobic environment from which water is excluded. This lipid bilayer is an essential feature of all biological membranes. The hydrophobic parts of proteins and lipid-soluble cofactors and pigments are dissolved or embedded in the lipid bilayer. Lamellar membranes can function as electrical insulating material and permit a charge, or potential difference, to develop across the membrane. Such a charge can be a source of chemical or electrical energy.

Approximately one-fifth of the lamellar lipids are chlorophyll molecules; one type, chlorophyll a, is more abundant than the second type, chlorophyll b. The chlorophyll molecules are specifically bound to small protein molecules. Most of these chlorophyll-proteins are "light-harvesting" pigments. These absorb light and pass its energy on to special chlorophyll a molecules that are directly involved in the conversion of light energy to chemical energy. When one of these special chlorophyll a molecules is excited by light energy (as described later), it gives up an electron. There are two types of these special chlorophyll a molecules: one, called P_{680}, has an absorption spectrum that peaks at 684 nm; the other, called P_{700}, shows an absorption peak at 700 nm.

Although chlorophylls are the main light-absorbing molecules in green plants, there are other pigments such as carotenes and carotenoids (which are responsible for the yellow-orange colour of carrots). Carotenes can also absorb light and may supplement chlorophyll as the light-absorbing molecules in some plant cells. The light energy absorbed by carotenes must be passed to chlorophyll before conversion to chemical energy can occur. Carotenoids are part of a cycle that renders excess energy beyond the level of light saturation harmless, effectively serving as "lightning rods" in the process.

Proteins

Many of the lamellar proteins are components of the chlorophyll–protein complexes. Other proteins include enzymes and protein-containing coenzymes. Enzymes are required as organic catalysts for specific reactions within the lamellae. Protein coenzymes, also called cofactors, include important electron carrier molecules called cytochromes, which are iron-containing pigments with the pigment portions attached to protein molecules. During electron transfer, an electron is accepted by an iron atom in the pigment portion of a cytochrome molecule, which thus is reduced; then the electron is transferred to the iron atom in the next cytochrome carrier in the electron transfer chain, thus oxidizing the first cytochrome and reducing the next one in the chain.

In addition to the metal atoms found in the pigment portions of cytochrome molecules, metal atoms also are found in other protein molecules of the lamellae. In proteins with a total molecular weight of 900,000 (based on the weight of hydrogen as one), there are 2 atoms of manganese, 10 atoms of iron, and 6 atoms of copper. These metal atoms are required for the catalytic activity of some of the enzymes important in photosynthesis. The manganese atoms are involved in water-splitting and oxygen formation. Both copper- and iron-containing proteins function in electron transport between water and the final electron-acceptor molecule of the light stage of photosynthesis, an iron-containing protein called ferredoxin. Ferredoxin is a soluble component in the chloroplasts. In its reduced form, it gives electrons directly to the systems that reduce nitrate and sulfate and via NADPH to the system that reduces carbon dioxide. A copper-containing protein called plastocyanin (PC) carries electrons at one point in the electron transport chain. PC molecules are water soluble and can move through the inner space of the thylakoids, carrying electrons from one place to another.

Quinones

Small molecules called plastoquinones are found in substantial numbers in the lamellae. Like the cytochromes, quinones have important roles in carrying electrons between the components of the light reactions. Since they are lipid soluble, they can diffuse through the membrane. They can carry one or two electrons, and, in their reduced form (with added electrons), they carry hydrogen atoms that can be released as hydrogen ions when the added electrons are passed on, for example, to a cytochrome.

Process of Photosynthesis: The Light Reactions

Light Absorption and Energy Transfer

The light energy absorbed by a chlorophyll molecule excites some electrons within the structure of the molecule to higher energy levels, or excited states. Light of shorter wavelength (such as blue) has more energy than light of longer wavelength (such as red), so absorption of blue light creates an excited state of higher energy. A molecule raised to this higher energy state quickly gives up the "extra" energy as heat and falls to its lowest excited state. This lowest excited state is similar to that of a molecule that has just absorbed the longest wavelength light capable of exciting it. In the case of chlorophyll a, this lowest excited state corresponds to that of a molecule that has absorbed red light of about 680 nm.

The return of a chlorophyll a molecule from its lowest excited state to its original low-energy state (ground state) requires the release of the extra energy of the excited state. This can occur in one of several ways. In photosynthesis, most of this energy is conserved as chemical energy by the transfer of an electron from a special chlorophyll a molecule (P_{680} or P_{700}) to an electron acceptor. When this electron transfer is blocked by inhibitors, such as the herbicide dichlorophenylmethylurea (DCMU), or by low temperature, the energy can be released as red light. Such reemission of light is called fluorescence. The examination of fluorescence from photosynthetic material in which electron transfer has been blocked has proved to be a valuable tool for scientists studying the light reactions.

Pathway of Electrons

The general features of a widely accepted mechanism for photoelectron transfer, in which two light reactions (light reaction I and light reaction II) occur during the transfer of electrons from water to carbon dioxide, were proposed by Robert Hill and Fay Bendall in 1960. This mechanism is based on the relative potential (in volts) of various cofactors of the electron-transfer chain to be oxidized or reduced. Molecules that in their oxidized form have the strongest affinity for electrons (i.e., are strong oxidizing agents) have a low relative potential. In contrast, molecules that in their oxidized form are difficult to reduce have a high relative potential once they have accepted electrons. The molecules with a low relative potential are considered to be strong oxidizing agents, and those with a high relative potential are considered to be strong reducing agents.

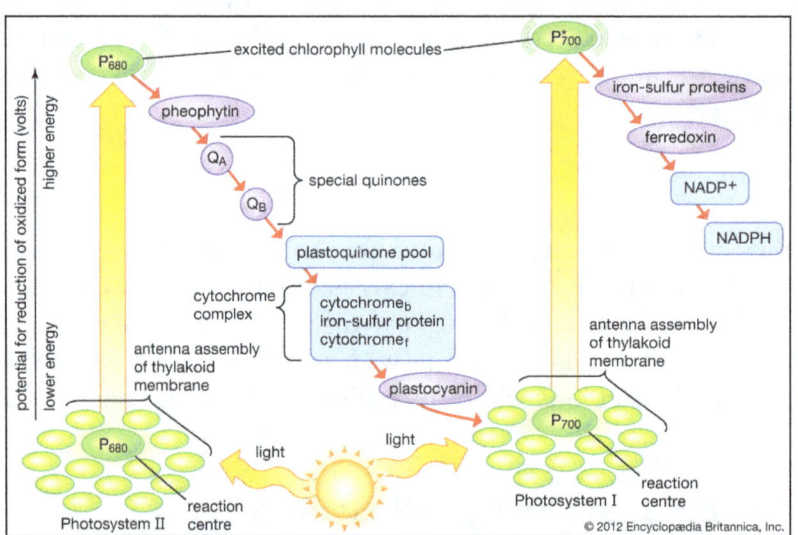

Flow of electrons during the light reaction stage of photosynthesisArrows pointing upward represent light reactions that increase the chemical potential; arrows slanting downward represent flow of electrons via carriers in the membrane.

In diagrams that describe the light reaction stage of photosynthesis, the actual photochemical steps are typically represented by two vertical arrows. These arrows signify that the special pigments P_{680} and P_{700} receive light energy from the light-harvesting chlorophyll-protein molecules and are raised in energy from their ground state to excited states. In their excited state, these pigments are extremely strong reducing agents that quickly transfer electrons to the first acceptor. These first acceptors also are strong reducing agents and rapidly pass electrons to more stable carriers. In light reaction II, the first acceptor may be pheophytin, which is a molecule similar to chlorophyll that also has a strong reducing potential and quickly transfers electrons to the next acceptor. Special quinones are next in the series. These molecules are similar to plastoquinone; they receive electrons from pheophytin and pass them to the intermediate electron carriers, which include the plastoquinone pool and the cytochromes b and f associated in a complex with an iron-sulfur protein.

In light reaction I, electrons are passed on to iron-sulfur proteins in the lamellar membrane, after which the electrons flow to ferredoxin, a small water-soluble iron-sulfur protein. When $NADP^+$ and a suitable enzyme are present, two ferredoxin molecules, carrying one electron each, transfer two electrons to $NADP^+$, which picks up a proton (i.e., a hydrogen ion) and becomes NADPH.

Each time a P_{680} or P_{700} molecule gives up an electron, it returns to its ground (unexcited) state, but with a positive charge due to the loss of the electron. These positively charged ions are extremely strong oxidizing agents that remove an electron from a suitable donor. The P_{680}^+ of light reaction II is capable of taking electrons from water in the presence of appropriate catalysts. There is good evidence that two or more manganese atoms complexed with protein are involved in this catalysis, taking four electrons from two water molecules (with release of four hydrogen ions). The manganese-protein complex gives up these electrons one at a time via an unidentified carrier to P_{680}^+, reducing it to P_{680}. When manganese is selectively removed by chemical treatment, the thylakoids lose the capacity to oxidize water, but all other parts of the electron pathway remain intact.

In light reaction I, P_{700}^+ recovers electrons from plastocyanin, which in turn receives them from intermediate carriers, including the plastoquinone pool and cytochrome b and cytochrome f molecules. The pool of intermediate carriers may receive electrons from water via light reaction II and the quinones. Transfer of electrons from water to ferredoxin via the two light reactions and intermediate carriers is called noncyclic electron flow. Alternatively, electrons may be transferred only by light reaction I, in which case they are recycled from ferredoxin back to the intermediate carriers. This process is called cyclic electron flow.

Evidence of two Light Reactions

Many lines of evidence support the concept of electron flow via two light reactions. An early study by American biochemist Robert Emerson employed the algae Chlorella, which was illuminated with red light alone, with blue light alone, and with red and blue light at the same time. Oxygen evolution was measured in each case. It was substantial with blue light alone but not with red light alone. With both red and blue light together, the amount of oxygen evolved far exceeded the sum of that seen with blue and red light alone. These experimental data pointed to the existence of two types of light reactions that, when operating in tandem, would yield the highest rate of oxygen evolution. It is now known that light reaction I can use light of a slightly longer wavelength than red ($\lambda = 680$ nm), while light reaction II requires light with a wavelength of 680 nm or shorter.

Since those early studies, the two light reactions have been separated in many ways, including separation of the membrane particles in which each reaction occurs. Lamellae can be disrupted mechanically into fragments that absorb light energy and break the bonds of water molecules (i.e., oxidize water) to produce oxygen, hydrogen ions, and electrons. These electrons can be transferred to ferredoxin, the final electron acceptor of the light stage. No transfer of electrons from water to ferredoxin occurs if the herbicide DCMU is present. The subsequent addition of certain reduced dyes (i.e., electron donors) restores the light reduction of $NADP^+$ but without oxygen production, suggesting that light reaction I but not light reaction II is functioning. It is now known that DCMU blocks the transfer of electrons between the first quinone and the plastoquinone pool in light reaction II.

When treated with certain detergents, lamellae can be broken down into smaller particles capable of carrying out single light reactions. One type of particle can absorb light energy, oxidize water, and produce oxygen (light reaction II), but a special dye molecule must be supplied to accept the electrons. In the presence of electron donors, such as a reduced dye, a second type of lamellar particle can absorb light and transfer electrons from the electron donor to ferredoxin (light reaction I).

Photosystems I and II

The structural and photochemical properties of the minimum particles capable of performing light reactions I and II have received much study. Treatment of lamellar fragments with neutral detergents releases these particles, designated photosystem I and photosystem II, respectively. Subsequent harsher treatment (with charged detergents) and separation of the individual polypeptides with electrophoretic techniques have helped identify the components of the photosystems. Each photosystem consists of a light-harvesting complex and a core complex. Each core complex contains a reaction centre with the pigment (either P_{700} or P_{680}) that can be photochemically oxidized, together with electron acceptors and electron donors. In addition, the core complex has some 40 to 60 chlorophyll molecules bound to proteins. In addition to the light absorbed by the chlorophyll molecules in the core complex, the reaction centres receive a major part of their excitation from the pigments of the light-harvesting complex.

Quantum Requirements

The quantum requirements of the individual light reactions of photosynthesis are defined as the number of light photons absorbed for the transfer of one electron. The quantum requirement for each light reaction has been found to be approximately one photon. The total number of quanta required, therefore, to transfer the four electrons that result in the formation of one molecule of oxygen via the two light reactions should be four times two, or eight. It appears, however, that additional light is absorbed and used to form ATP by a cyclic photophosphorylation pathway. (The cyclic photophosphorylation pathway is an ATP-forming process in which the excited electron returns to the reaction centre.) The actual quantum requirement, therefore, probably is 9 to 10.

Process of Photosynthesis: The Conversion of Light Energy to ATP

The electron transfers of the light reactions provide the energy for the synthesis of two compounds vital to the dark reactions: NADPH and ATP.

ATP is formed by the addition of a phosphate group to a molecule of adenosine diphosphate (ADP)—or to state it in chemical terms, by the phosphorylation of ADP. This reaction requires a substantial input of energy, much of which is captured in the bond that links the added phosphate group to ADP. Because light energy powers this reaction in the chloroplasts, the production of ATP during photosynthesis is referred to as photophosphorylation, as opposed to oxidative phosphorylation in the electron-transport chain in the mitochondrion.

Unlike the production of NADPH, the photophosphorylation of ADP occurs in conjunction with both cyclic and noncyclic electron flow. In fact, researchers speculate that the sole purpose of cyclic electron flow may be for photophosphorylation, since this process involves no net transfer of electrons to reducing agents. The relative amounts of cyclic and noncyclic flow may be adjusted in accordance with changing physiological needs for ATP and reduced ferredoxin and NADPH in chloroplasts. In contrast to electron transfer in light reactions I and II, which can occur in membrane fragments, intact thylakoids are required for efficient photophosphorylation. This requirement stems from the special nature of the mechanism linking photophosphorylation to electron flow in the lamellae.

The theory relating the formation of ATP to electron flow in the membranes of both chloroplasts and mitochondria (the organelles responsible for ATP formation during cellular respiration) was first proposed by English biochemist Peter Dennis Mitchell, who received the 1978 Nobel Prize for Chemistry. This chemiosmotic theory has been somewhat modified to fit later experimental facts. The general features are now widely accepted. A central feature is the formation of a hydrogen ion (proton) concentration gradient and an electrical charge across intact lamellae. The potential energy stored by the proton gradient and electrical charge is then used to drive the energetically unfavourable conversion of ADP and inorganic phosphate (P_i) to ATP and water.

The manganese-protein complex associated with light reaction II is exposed to the interior of the thylakoid. Consequently, the oxidation of water during light reaction II leads to release of hydrogen ions (protons) into the inner thylakoid space. Furthermore, it is likely that photoreaction II entails the transfer of electrons across the lamella toward its outer face, so that when plastoquinone molecules are reduced, they can receive protons from the outside of the thylakoid. When these reduced plastoquinone molecules are oxidized, giving up electrons to the cytochrome-iron-sulfur complex, protons are released inside the thylakoid. Because the lamella is impermeable to them, the release of protons inside the thylakoid by oxidation of both water and plastoquinone leads to a higher concentration of protons inside the thylakoid than outside it. In other words, a proton gradient is established across the lamella. Since protons are positively charged, the movement of protons across the thylakoid lamella during both light reactions results in the establishment of an electrical charge across the lamella.

An enzyme complex located partly in and on the lamellae catalyzes the reaction in which ATP is formed from ADP and inorganic phosphate. The reverse of this reaction is catalyzed by an enzyme called ATP-ase; hence, the enzyme complex is sometimes called an ATP-ase complex. It is also called the coupling factor. It consists of hydrophilic polypeptides (F_1), which project from the outer surface of the lamellae, and hydrophobic polypeptides (F_o), which are embedded inside the lamellae. F_o forms a channel that permits protons to flow through the lamellar membrane to F_1. The enzymes in F_1 then catalyze ATP formation, using both the proton supply and the lamellar transmembrane charge.

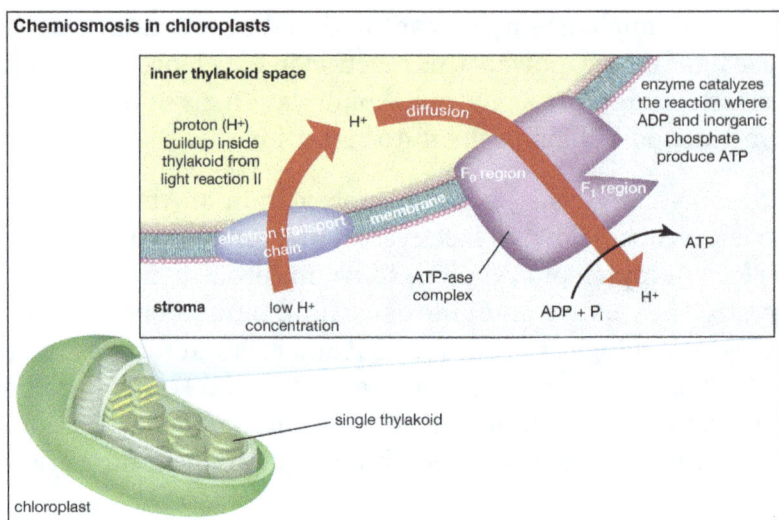

Chemiosmosis in chloroplasts

Chemiosmosis in chloroplasts that results in the donation of a proton for the production of adenosine triphosphate (ATP) in plants

In summary, the use of light energy for ATP formation occurs indirectly: a proton gradient and electrical charge—built up in or across the lamellae as a consequence of electron flow in the light reactions—provide the energy to drive the synthesis of ATP from ADP and P_i.

Process of Photosynthesis: Carbon Fixation and Reduction

The assimilation of carbon into organic compounds is the result of a complex series of enzymatically regulated chemical reactions—the dark reactions. This term is something of a misnomer, for these reactions can take place in either light or darkness. Furthermore, some of the enzymes involved in the so-called dark reactions become inactive in prolonged darkness; however, they are activated when the leaves that contain them are exposed to light.

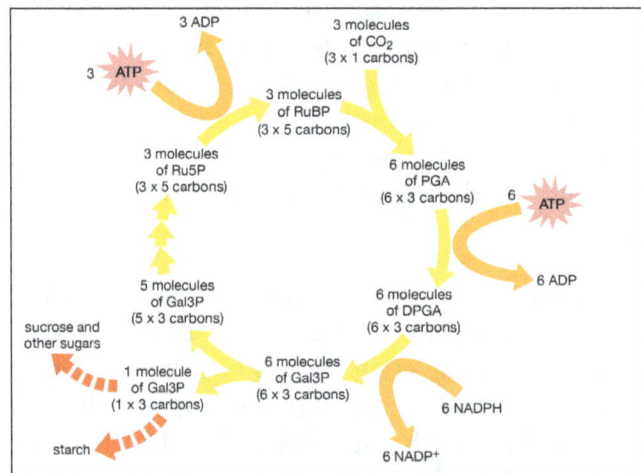

C_3 carbon fixation pathway

Pathway of carbon dioxide fixation and reduction in photosynthesis, the reductive pentose phosphate cycle. The diagram represents one complete turn of the cycle, with the net production of one molecule of Gal3P. The nine molecules of ATP and six molecules of NADPH come from the light reactions.

Elucidation of the Carbon Pathway

Radioactive isotopes of carbon (^{14}C) and phosphorus (^{32}P) have been valuable in identifying the intermediate compounds formed during carbon assimilation. A photosynthesizing plant does not strongly discriminate between the most abundant natural carbon isotope (^{12}C) and ^{14}C. During photosynthesis in the presence of $^{14}CO_2$, the compounds formed become labeled with the radioisotope. During very short exposures, only the first intermediates in the carbon-fixing pathway become labeled. Early investigations showed that some radioactive products were formed even when the light was turned off and the $^{14}CO_2$ was added just afterward in the dark, confirming the nature of the carbon fixation as a "dark" reaction.

American biochemist Melvin Calvin, a Nobel Prize recipient for his work on the carbon-reduction cycle, allowed green plants to photosynthesize in the presence of radioactive carbon dioxide for a few seconds under various experimental conditions. Products that became labeled with radioactive carbon during Calvin's experiments included a three-carbon compound called 3-phosphoglycerate (abbreviated PGA), sugar phosphates, amino acids, sucrose, and carboxylic acids. When photosynthesis was stopped after two seconds, the principal radioactive product was PGA, which therefore was identified as the first stable compound formed during carbon dioxide fixation in green plants. PGA is a three-carbon compound, and the mode of photosynthesis is thus referred to as C_3. In the two other known pathways, C_4 and CAM (crassulacean acid metabolism), the C_3 pathway follows the fixation of CO_2 into oxaloacetate, a four-carbon acid, and its reduction to malate. PGA is formed from 2-carboxy-3-keto-D-arabinitol 1,5-bisphosphate, which is a highly unstable six-carbon compound formed from the carboxylation of ribulose-1,5-bisphosphate, a five-carbon compound.

Further studies with ^{14}C as well as with inorganic phosphate labeled with ^{32}P led to the mapping of the carbon fixation and reduction pathway called the reductive pentose phosphate (RPP) cycle, or the Calvin-Benson cycle. An additional pathway for carbon transport in certain plants was later discovered in other laboratories. All the steps in these pathways can be carried out in the laboratory by isolated enzymes in the dark. Several steps require the ATP or NADPH generated by the light reactions. In addition, some of the enzymes are fully active only when conditions simulate those in green cells exposed to light. In living plants, these enzymes are active during photosynthesis but not in the dark.

Calvin-benson Cycle

The Calvin-Benson cycle, in which carbon is fixed, reduced, and utilized, involves the formation of intermediate sugar phosphates in a cyclic sequence. One complete cycle incorporates three molecules of carbon dioxide and produces one molecule of the three-carbon compound glyceraldehyde-3-phosphate (Gal3P). This three-carbon sugar phosphate usually is either exported from the chloroplasts or converted to starch inside the chloroplast.

ATP and NADPH formed during the light reactions are utilized for key steps in this pathway and provide the energy and reducing equivalents (i.e., electrons) to drive the sequence in the direction shown. For each molecule of carbon dioxide that is fixed, two molecules of NADPH and three molecules of ATP from the light reactions are required. The overall reaction can be represented as follows:

$$9\,ATP + 6\,NADPH + 3\,CO_2 \rightarrow Gal\,3\,P + 6\,NADP^+\,9\,ADP + 8\,P\,i$$

The cycle is composed of four stages: (1) carboxylation, (2) reduction, (3) isomerization/condensation/dismutation, and (4) phosphorylation.

Carboxylation

The initial incorporation of carbon dioxide, which is catalyzed by the enzyme ribulose 1,5-bisphosphate carboxylase (Rubisco), proceeds by the addition of carbon dioxide to the five-carbon compound ribulose 1,5-bisphosphate (RuBP) and the splitting of the resulting six-carbon compound into two molecules of PGA. This reaction occurs three times during each complete turn of the cycle; thus, six molecules of PGA are produced.

Reduction

The six molecules of PGA are first phosphorylated with ATP by the enzyme PGA-kinase, yielding six molecules of 1,3-diphosphoglycerate (DPGA). These molecules are subsequently reduced with NADPH and the enzyme glyceraldehyde-3-phosphate dehydrogenase to give six molecules of Gal3P. These reactions are the reverse of two steps of the process glycolysis in cellular respiration.

Isomerization/Condensation/Dismutation

For each complete Calvin-Benson cycle, one of the Gal3P molecules, with its three carbon atoms, is the net product and may be transferred out of the chloroplast or converted to starch inside the chloroplast. For the cycle to regenerate, the other five Gal3P molecules (with a total of 15 carbon atoms) must be converted back to three molecules of five-carbon RuBP. The conversion of Gal3P to RuBP begins with a complex series of enzymatically regulated reactions that lead to the synthesis of the five-carbon compound ribulose-5-phosphate (Ru5P).

Phosphorylation

The three molecules of Ru5P are converted to the carboxylation substrate, RuBP, by the enzyme phosphoribulokinase, using ATP. This reaction, shown below, completes the cycle:

$$3\,Ru\,5\,P + 3\,ATP \rightarrow 3\,RuBP + 3\,ADP$$

Regulation of the Cycle

Photosynthesis cannot occur at night, but the respiratory process of glycolysis—which uses some of the same reactions as the Calvin-Benson cycle, except in the reverse—does take place. Thus, some steps in this cycle would be wasteful if allowed to occur in the dark, because they would counteract the reactions of glycolysis. For this reason, some enzymes of the Calvin-Benson cycle are "turned off" (i.e., become inactive) in the dark.

Even in the presence of light, changes in physiological conditions frequently necessitate adjustments in the relative rates of reactions of the Calvin-Benson cycle, so that enzymes for some reactions change in their catalytic activity. These alterations in enzyme activity typically are brought about by changes in levels of such chloroplast components as reduced ferredoxin, acids, and soluble components (e.g., P_i and magnesium ions).

Products of carbon Reduction

The most important use of Gal3P is its export from the chloroplasts to the cytosol of green cells, where it is used for biosynthesis of products needed by the plant. In land plants, a principal product is sucrose, which is translocated from the green cells of the leaves to other parts of the plant. Other key products include the carbon skeletons of certain primary amino acids, such as alanine, glutamate, and aspartate. To complete the synthesis of these compounds, amino groups are added to the appropriate carbon skeletons made from Gal3P. Sulfur amino acids such as cysteine are formed by adding sulfhydryl groups and amino groups. Other biosynthesis pathways lead from Gal3P to lipids, pigments, and most of the constituents of green cells.

Starch synthesis and accumulation in the chloroplasts occur particularly when photosynthetic carbon fixation exceeds the needs of the plant. Under such circumstances, sugar phosphates accumulate in the cytosol, binding cytosolic P_i. The export of Gal3P from the chloroplasts is tied to a one-for-one exchange of P_i for Gal3P, so less cytosolic Pi results in decreased export of Gal3P and decreased P_i in the chloroplast. These changes trigger alterations in the activities of regulated enzymes, leading in turn to increased starch synthesis. This starch can be broken down at night and used as a source of reduced carbon and energy for the physiological needs of the plant. Too much starch in the chloroplasts leads to diminished rates of photosynthesis, however. In addition, high levels of sugars in the cytosol lead to the suppression of the normal activities of the genes involved in photosynthesis. Thus, under what would seem to be the ideal photosynthetic conditions of a bright warm day, many plants in fact have-slower-than expected rates of photosynthesis.

Photorespiration

Under conditions of high light intensity, hot weather, and water limitation, the productivity of the Calvin-Benson cycle is limited in many plants by the occurrence of photorespiration. This process converts sugar phosphates back to carbon dioxide; it is initiated by the oxygenation of RuBP (i.e., the combination of gaseous oxygen [O_2] with RuBP). This oxygenation reaction yields only one molecule of PGA and one molecule of a two-carbon acid, phosphoglycolate, which is subsequently converted in part to carbon dioxide. The reaction of oxygen with RuBP is in direct competition with the carboxylation reaction (CO_2 + RuBP) that initiates the Calvin-Benson cycle and is, in fact, catalyzed by the same protein, ribulose 1,5-bisphosphate carboxylase. The relative concentrations of oxygen and carbon dioxide within the chloroplasts as well as leaf temperature determine whether oxygenation or carboxylation is favoured. The concentration of oxygen inside the chloroplasts may be higher than atmospheric (20 percent) because of photosynthetic oxygen evolution, whereas the internal carbon dioxide concentration may be lower than atmospheric (0.039 percent) because of photosynthetic uptake. Any increase in the internal carbon dioxide pressure tends to help the carboxylation reaction compete more effectively with oxygenation.

Carbon Fixation in C_4 Plants

Certain plants—including the important crops sugarcane and corn (maize), as well as other diverse species that are thought to have expanded their geographic ranges into tropical areas—have developed a special mechanism of carbon fixation that largely prevents photorespiration. The leaves of these plants have special anatomy and biochemistry. In particular, photosynthetic functions are divided between mesophyll and bundle-sheath leaf cells. The carbon-fixation pathway begins in the mesophyll cells, where carbon dioxide is converted into bicarbonate, which is then added to the three-carbon acid phosphoenolpyruvate (PEP) by an enzyme called phosphoenolpyruvate carboxylase. The product of this reaction is the four-carbon acid oxaloacetate, which is reduced to malate, another four-carbon acid, in one form of the C_4 pathway. Malate then is transported to bundle-sheath cells, which are located near the vascular system of the leaf. There, malate enters the chloroplasts and is oxidized and decarboxylated (i.e., loses CO_2) by malic enzyme. This yields high concentrations of carbon dioxide, which is fed into the Calvin-Benson cycle of the bundle sheath cells, and pyruvate, a three-carbon acid that is translocated back to the mesophyll cells. In the mesophyll chloroplasts, the enzyme pyruvate orthophosphate dikinase (PPDK) uses ATP and Pi to convert pyruvate back to PEP, completing the C_4 cycle. There are several variations of this pathway in different species. For example, the amino acids aspartate and alanine can substitute for malate and pyruvate in some species.

The C_4 pathway acts as a mechanism to build up high concentrations of carbon dioxide in the chloroplasts of the bundle sheath cells. The resulting higher level of internal carbon dioxide in these chloroplasts serves to increase the ratio of carboxylation to oxygenation, thus minimizing photorespiration. Although the plant must expend extra energy to drive this mechanism, the energy loss is more than compensated by the near elimination of photorespiration under conditions where it would otherwise occur. Sugarcane and certain other plants that employ this pathway have the highest annual yields of biomass of all species. In cool climates, where photorespiration is insignificant, C_4 plants are rare. Carbon dioxide is also used efficiently in carbohydrate synthesis in the bundle sheath.

PEP carboxylase, which is located in the mesophyll cells, is an essential enzyme in C_4 plants. In hot and dry environments, carbon dioxide concentrations inside the leaf fall when the plant closes or partially closes its stomata to reduce water loss from the leaves. Under these conditions, photorespiration is likely to occur in plants that use Rubisco as the primary carboxylating enzyme, since Rubisco adds oxygen to RuBP when carbon dioxide concentrations are low. PEP carboxylase, however, does not use oxygen as a substrate, and it has a greater affinity for carbon dioxide than Rubisco does. Thus, it has the ability to fix carbon dioxide in reduced carbon dioxide conditions, such as when the stomata on the leaves are only partially open. As a consequence, at similar rates of photosynthesis, C_4 plants lose less water when compared with C_3 plants. This explains why C_4 plants are favoured in dry and warm environments.

Carbon Fixation via Crassulacean Acid Metabolism (CAM)

In addition to C_3 and C_4 species, there are many succulent plants that make use of a third photosynthetic pathway: crassulacean acid metabolism (CAM). This pathway is named after the Crassulaceae, a family in which many species display this type of metabolism, but it also occurs commonly in other families, such as the Cactaceae, the Euphorbiaceae, the Orchidaceae, and the

Bromeliaceae. CAM species number more than 20,000 and span 34 families. Almost all CAM plants are angiosperms; however, quillworts and ferns also use the CAM pathway. In addition, some scientists note that CAM might be used by Welwitschia, a gymnosperm. CAM plants are often characterized by their succulence, but this quality is not pronounced in epiphytes that use the CAM pathway.

CAM plants are known for their capacity to fix carbon dioxide at night, using PEP carboxylase as the primary carboxylating enzyme and the accumulation of malate (which is made by the enzyme malate dehydrogenase) in the large vacuoles of their cells. Deacidification occurs during the day, when carbon dioxide is released from malate and fixed in the Calvin-Benson cycle, using Rubisco. During daylight hours, the stomata are closed to prevent water loss. The stomata are open at night when the air is cooler and more humid, and this setting allows the leaves of the plant to assimilate carbon dioxide. Since their stomata are closed during the day, CAM plants require considerably less water than both C_3 and C_4 plants that fix the same amount of carbon dioxide in photosynthesis.

The productivity of most CAM plants is fairly low, however. This is not an inherent trait of CAM species, because some cultivated CAM plants (e.g., Agave mapisaga and A. salmiana) can achieve a high aboveground productivity. In fact, some cultivated species that are irrigated, fertilized, and carefully pruned are highly productive. For example, prickly pear (Opuntia ficus-indica) and its thornless variety, O. amyclea, produce 4.6 kg per square metre (0.9 pound per square foot) of new growth per year. Such productivity is among the highest of any plant species. Thus, the rates of photosynthesis of CAM plants may be as high as those of C_3 plants, if morphologically similar plants adapted to the similar habitats are compared.

The unusual capacity of CAM plants to fix carbon dioxide into organic acids in the dark, causing nocturnal acidification, with deacidification occurring during the day, has been known to science since the 19th century. (There is evidence, however, that the Romans noticed the difference between the morning acid taste of some of the house plants they cultivated.) On the other hand, the C_4 pathway was discovered during the middle of the 20th century. A full appreciation of CAM as a photosynthetic pathway was greatly stimulated by analogies with C_4 species.

Differences in Carbon Fixation Pathways

A comparison of the differences between the various carbon pathways is provided in the table.

pathway	carbon-assimilation process	first stable intermediate product	stomate activity	Photorespiration	plant types using this pathway
C_3	Calvin-Benson cycle only.	phosphoglycerate (PGA), a three-carbon acid.	open during the day, closed at night.	not suppressed	plants living in colder, wetter environments characterized by low-to-medium light intensities.
C_4	adds CO_2 to phosphoenolpyruvate (PEP) to form oxaloacetate first; the Calvin-Benson cycle follows.	oxaloacetate, a four-carbon acid, which is later reduced to malate.	open during the day, closed at night.	suppressed	plants living in warmer, drier environments characterized by high light intensity.

| CAM* | adds CO_2 to phosphoe-nolpyruvate (PEP) to form oxaloacetate first; the Calvin-Benson cycle follows. | oxaloacetate, a four-carbon acid, which is later re-duced to malate and stored in vacuoles. | open at night, closed during the day. | suppressed | succulents (members of Crassulaceae), which occur in warmer, drier environments charac-terized by high light intensity. |

Molecular Biology of Photosynthesis

Oxygenic photosynthesis occurs in a certain type of prokaryotic cells called cyanobacteria and eu-karyotic plant cells (algae and higher plants). In eukaryotic plant cells, which contain chloroplasts and a nucleus, the genetic information needed for the reproduction of the photosynthetic appara-tus is contained partly in the chloroplast chromosome and partly in chromosomes of the nucleus. For example, the carboxylation enzyme ribulose 1,5-bisphosphate carboxylase is a large protein molecule comprising a complex of eight large polypeptide subunits and eight small polypeptide subunits. The gene for the large subunits is located in the chloroplast chromosome, whereas the gene for the small subunits is in the nucleus. Transcription of the DNA of the nuclear gene yields messenger RNA (mRNA) that encodes the information for the synthesis of the small polypeptides. During this synthesis, which occurs on the cytosolic ribosomes, some extra amino acid residues are added to form a recognition leader on the end of the polypeptide chain. This leader is recog-nized by special receptor sites on the outer chloroplast membrane; these receptor sites then allow the polypeptide to penetrate the membrane and enter the chloroplast. The leader is removed, and the small subunits combine with the large subunits, which have been synthesized on chloroplast ribosomes according to mRNA transcribed from the chloroplast DNA. The expression of nuclear genes that code for proteins needed in the chloroplasts appears to be under control of events in the chloroplasts in some cases; for example, the synthesis of some nuclear-encoded chloroplast enzymes may occur only when light is absorbed by chloroplasts.

CELLULAR REPRODUCTION

One of the most defining characteristics of the living condition is the ability to reproduce. All living cells and organisms reproduce, producing offspring like themselves, and pass on the hereditary information contained in their DNA molecules. The processes of reproduction, while varied and complex, depends upon the ability of individual cells to replicate. All cells arise from preexisting cells by some mechanism of cell reproduction or division.

Cell reproduction is often divided into two major types: asexual and sexual cell reproduction. Typ-ically in asexual reproduction, a single cell gives rise to a genetic duplicate of the parental cell, without any genetic contribution from another individual, while sexual cell reproduction involves the genetic recombination of two cells.

In procaryotes, asexual cell division often proceeds by a process of fission. Reproduction occurs when a parental cell replicates its bacterial DNA producing a complete and faithful copy of its chromosome. Growth of the bacterial cell to an appropriate size seems to induce division by binary fission. New plasma membranes and wall material are laid down constricting the cell into two pieces. The splitting of the cytoplasmic domains and the two DNA molecules into nearly equal halves results in two daughter cells.

Budding, which is another method of asexual reproduction, occurs in most yeast, hydra, and in some filamentous fungi. In this process, a small cytoplasmic swelling protrudes (a bud) develops on the surface of either the yeast cell or the hypha, with its cytoplasm being continuous with that of the parent cell. The parental cell nucleus then divides and one of the daughter nuclei migrates into the bud. Continuous synthesis of cytoplasm and repeated nuclear divisions results in many buds over the cells surface. Buds are pinched off and behave as spores, germinating and forming a new hypha, all genetically identical to the parent.

In higher organisms (most eucaryotes) asexual cell reproduction involves an elaborate duplication of the chromosomes followed by their separation in a nuclear division called mitosis. Mitosis is often followed by cytokinesis, a division of the cytoplasm. In the hard-walled cells of higher plants, a medial cell plate forms and divides the parental cell into two compartments. In animal cells, which do not have a hard cell wall, a membrane furrow, made of a microfilament contractile ring that constricts as a camera diaphragm, pinching the cell in two daughter cells.

Cells that reproduce sexually are characterized by meiosis, the nuclear division process by which sex cells (gametes) are formed. Every chromosome of a somatic cell occurs in a pair (diploid). During meiosis these diploid pairs of chromosomes duplicate and are separated so that each meiotic sex cell has only one chromosome (haploid) of each pair. Two successive meiotic divisions result in the production of haploid sperm and egg cells, each with one-half of the amount of the parental DNA.

During the life cycle of sexually reproducing organisms, fertilization results in the fusion of haploid gametes (sperm and egg) producing the zygote. Dividing by asexual cell reproduction, the zygote undergoes cellular differentiation, whereby cells become structurally, functionally, and biochemically distinct from each other.

Cell Cycle

The cell cycle can be thought of as the life cycle of a cell. In other words, it is the series of growth and development steps a cell undergoes between its "birth"—formation by the division of a mother cell—and reproduction—division to make two new daughter cells.

Stages of the Cell Cycle

To divide, a cell must complete several important tasks: it must grow, copy its genetic material (DNA), and physically split into two daughter cells. Cells perform these tasks in an organized, predictable series of steps that make up the cell cycle. The cell cycle is a cycle, rather than a linear pathway, because at the end of each go-round, the two daughter cells can start the exact same process over again from the beginning.

In eukaryotic cells, or cells with a nucleus, the stages of the cell cycle are divided into two major phases: interphase and the mitotic (M) phase.

- During interphase, the cell grows and makes a copy of its DNA.

- During the mitotic (M) phase, the cell separates its DNA into two sets and divides its cytoplasm, forming two new cells.

Interphase

Let's enter the cell cycle just as a cell forms, by division of its mother cell. What must this newborn cell do next if it wants to go on and divide itself? Preparation for division happens in three steps:

- G_1 Phase. During G_1 phase, also called the first gap phase, the cell grows physically larger, copies organelles, and makes the molecular building blocks it will need in later steps. [Do cells always grow before they divide?]

- S phase. In S phase, the cell synthesizes a complete copy of the DNA in its nucleus. It also duplicates a microtubule-organizing structure called the centrosome. The centrosomes help separate DNA during M phase.

- G_2 phase. During the second gap Phase or G_2 Phase the cell grows more, makes proteins and organelles, and begins to reorganize its contents in preparation for mitosis. G_2 phase ends when mitosis begins.

- The G_1, S, and G_2 phases together are known as interphase. The prefix intermeans between, reflecting that interphase takes place between one mitotic (M) phase and the next.

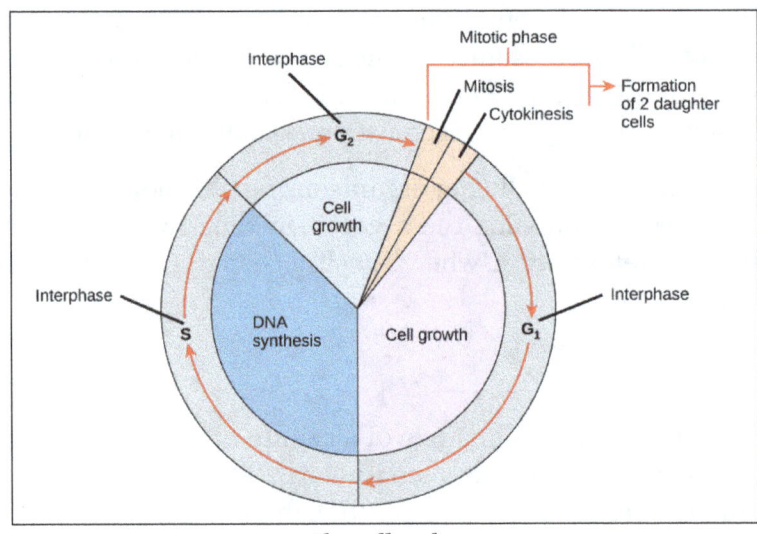

The cell cycle.

M Phase

During the mitotic (M) phase, the cell divides its copied DNA and cytoplasm to make two new cells. M phase involves two distinct division-related processes: mitosis and cytokinesis.

In mitosis, the nuclear DNA of the cell condenses into visible chromosomes and is pulled apart by the mitotic spindle, a specialized structure made out of microtubules. Mitosis takes place in four stages: prophase (sometimes divided into early prophase and prometaphase), metaphase, anaphase, and telophase. You can learn more about these stages in the video on mitosis.

In cytokinesis, the cytoplasm of the cell is split in two, making two new cells. Cytokinesis usually begins just as mitosis is ending, with a little overlap. Importantly, cytokinesis takes place differently in animal and plant cells.

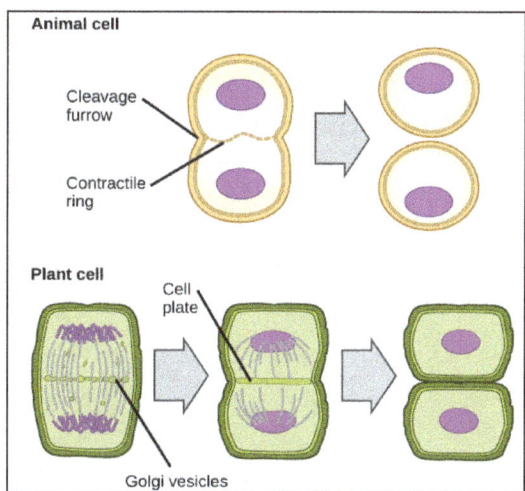

- In animals, cell division occurs when a band of cytoskeletal fibers called the contractile ring contracts inward and pinches the cell in two, a process called contractile cytokinesis. The indentation produced as the ring contracts inward is called the cleavage furrow. Animal cells can be pinched in two because they're relatively soft and squishy.

- Plant cells are much stiffer than animal cells; they're surrounded by a rigid cell wall and have high internal pressure. Because of this, plant cells divide in two by building a new structure down the middle of the cell. This structure, known as the cell plate, is made up of plasma membrane and cell wall components delivered in vesicles, and it partitions the cell in two.

Cell Cycle Exit and G_0

What happens to the two daughter cells produced in one round of the cell cycle? This depends on what type of cells they are. Some types of cells divide rapidly, and in these cases, the daughter cells may immediately undergo another round of cell division. For instance, many cell types in an early embryo divide rapidly, and so do cells in a tumor.

Other types of cells divide slowly or not at all. These cells may exit the G_1 phase and enter a resting state called G_0 phase. In G_0, a cell is not actively preparing to divide, it's just doing its job. For instance, it might conduct signals as a neuron (like the one in the drawing below) or store carbohydrates as a liver cell. G_0 is a permanent state for some cells, while others may re-start division if they get the right signals.

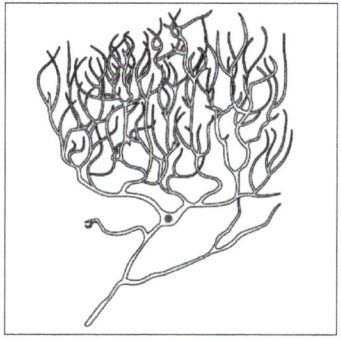

Neurons and glial cells

Duration of the Cell Cycle

Different cells take different lengths of time to complete the cell cycle. A typical human cell might take about 24 hours to divide, but fast-cycling mammalian cells, like the ones that line the intestine, can complete a cycle every 9-10 hours when they're grown in culture.

Different types of cells also split their time between cell cycle phases in different ways. In early frog embryos, for example, cells spend almost no time in G_1 and G_2 and instead rapidly cycle between S and M phases—resulting in the division of one big cell, the zygote, into many smaller cells.

Meiosis

Meiosis is a special type of cell division that reduces the chromosome number by half, creating four haploid cells, each genetically distinct from the parent cell that gave rise to them. This process occurs in all sexually reproducing single-celled and multicellular eukaryotes, including animals, plants, and fungi. Errors in meiosis resulting in aneuploidy are the leading known cause of miscarriage and the most frequent genetic cause of developmental disabilities.

In meiosis, DNA replication is followed by two rounds of cell division to produce four daughter cells, each with half the number of chromosomes as the original parent cell. The two meiotic divisions are known as Meiosis I and Meiosis II. Before meiosis begins, during S phase of the cell cycle, the DNA of each chromosome is replicated so that it consists of two identical sister chromatids, which remain held together through sister chromatid cohesion. This S-phase can be referred to as "pre-meiotic S-phase" or "meiotic S-phase". Immediately following DNA replication, meiotic cells enter a prolonged G2-like stage known as meiotic prophase. During this time, homologous chromosomes pair with each other and undergo genetic recombination, a programmed process in which DNA is cut and then repaired, which allows them to exchange some of their genetic information. A subset of recombination events results in crossovers, which create physical links known as chiasmata (singular: chiasma, for the Greek letter Chi (X)) between the homologous chromosomes. In most organisms, these links are essential to direct each pair of homologous chromosomes to segregate away from each other during Meiosis I, resulting in two haploid cells that have half the number of chromosomes as the parent cell. During Meiosis II, the cohesion between sister chromatids is released and they segregate from one another, as during mitosis. In some cases all four of the meiotic products form gametes such as sperm, spores, or pollen. In female animals, three of the four meiotic products are typically eliminated by extrusion into polar bodies, and only one cell develops to produce an ovum. Because the number of chromosomes is halved during meiosis, gametes can fuse (i.e. fertilization) to form a diploid zygote that contains two copies of each chromosome, one from each parent. Thus, alternating cycles of meiosis and fertilization enable sexual reproduction, with successive generations maintaining the same number of chromosomes. For example, diploid human cells contain 23 pairs of chromosomes including 1 pair of sex chromosomes (46 total), half of maternal origin and half of paternal origin. Meiosis produces haploid gametes (ova or sperm) that contain one set of 23 chromosomes. When two gametes (an egg and a sperm) fuse, the resulting zygote is once again diploid, with the mother and father each contributing 23 chromosomes. This same pattern, but not the same number of chromosomes, occurs in all organisms that utilize meiosis.

Although the process of meiosis is related to the more general cell division process of mitosis, it differs in two important respects:

Recombination	Meiosis	Shuffles the genes between the two chromosomes in each pair (one received from each parent), producing recombinant chromosomes with unique genetic combinations in every gamete.
	Mitosis	Occurs only if needed to repair DNA damage; usually occurs between identical sister chromatids and does not result in genetic changes.
Chromosome number (ploidy)	Meiosis	Produces four genetically unique cells, each with half the number of chromosomes as in the parent.
	Mitosis	Produces two genetically identical cells, each with the same number of chromosomes as in the parent.

Meiosis begins with a diploid cell, which contains two copies of each chromosome, termed homologs. First, the cell undergoes DNA replication, so each homolog now consists of two identical sister chromatids. Then each set of homologs pair with each other and exchange DNA by homologous recombination leading to physical connections (crossovers) between the homologs. In the first meiotic division, the homologs are segregated to separate daughter cells by the spindle apparatus. The cells then proceed to a second division without an intervening round of DNA replication. The sister chromatids are segregated to separate daughter cells to produce a total of four haploid cells. Female animals employ a slight variation on this pattern and produce one large ovum and two small polar bodies. Because of recombination, an individual chromatid can consist of a new combination of maternal and paternal DNA, resulting in offspring that are genetically distinct from either parent. Furthermore, an individual gamete can include an assortment of maternal, paternal, and recombinant chromatids. This genetic diversity resulting from sexual reproduction contributes to the variation in traits upon which natural selection can act.

Meiosis uses many of the same mechanisms as mitosis, the type of cell division used by eukaryotes to divide one cell into two identical daughter cells. In some plants, fungi, and protists meiosis results in the formation of spores: haploid cells that can divide vegetatively without undergoing fertilization. Some eukaryotes, like bdelloid rotifers, do not have the ability to carry out meiosis and have acquired the ability to reproduce by parthenogenesis.

Meiosis does not occur in archaea or bacteria, which generally reproduce asexually via binary fission. However, a "sexual" process known as horizontal gene transfer involves the transfer of DNA from one bacterium or archaeon to another and recombination of these DNA molecules of different parental origin.

Phases

Meiosis is divided into meiosis I and meiosis II which are further divided into Karyokinesis I and Cytokinesis I and Karyokinesis II and Cytokinesis II respectively. The preparatory steps that lead up to meiosis are identical in pattern and name to interphase of the mitotic cell cycle. Interphase is divided into three phases:

- Growth 1 (G_1) phase: In this very active phase, the cell synthesizes its vast array of proteins, including the enzymes and structural proteins it will need for growth. In G_1, each of the chromosomes consists of a single linear molecule of DNA.

- Synthesis (S) phase: The genetic material is replicated; each of the cell's chromosomes duplicates to become two identical sister chromatids attached at a centromere. This replication does not change the ploidy of the cell since the centromere number remains the same. The identical sister chromatids have not yet condensed into the densely packaged chromosomes visible with the light microscope. This will take place during prophase I in meiosis.

- Growth 2 (G_2) phase: G_2 phase as seen before mitosis is not present in meiosis. Meiotic prophase corresponds most closely to the G_2 phase of the mitotic cell cycle.

Interphase is followed by meiosis I and then meiosis II. Meiosis I separates replicated homologous chromosomes, each still made up of two sister chromatids, into two daughter cells, thus reducing the chromosome number by half. During meiosis II, sister chromatids decouple and the resultant daughter chromosomes are segregated into four daughter cells. For diploid organisms, the daughter cells resulting from meiosis are haploid and contain only one copy of each chromosome. In some species, cells enter a resting phase known as interkinesis between meiosis I and meiosis II.

Meiosis I and II are each divided into prophase, metaphase, anaphase, and telophase stages, similar in purpose to their analogous subphases in the mitotic cell cycle. Therefore, meiosis includes the stages of meiosis I (prophase I, metaphase I, anaphase I, telophase I) and meiosis II (prophase II, metaphase II, anaphase II, telophase II).

Meiosis generates gamete genetic diversity in two ways: (1) Law of Independent Assortment. The independent orientation of homologous chromosome pairs along the metaphase plate during metaphase I & orientation of sister chromatids in metaphase II, this is the subsequent separation of homologs and sister chromatids during anaphase I & II, it allows a random and independent distribution of chromosomes to each daughter cell (and ultimately to gametes); and (2) Crossing Over. The physical exchange of homologous chromosomal regions by homologous recombination during prophase I results in new combinations of DNA within chromosomes.

During meiosis, specific genes are more highly transcribed. In addition to strong meiotic stage-specific expression of mRNA, there are also pervasive translational controls (e.g. selective usage of pre-formed mRNA), regulating the ultimate meiotic stage-specific protein expression of genes during meiosis. Thus, both transcriptional and translational controls determine the broad restructuring of meiotic cells needed to carry out meiosis.

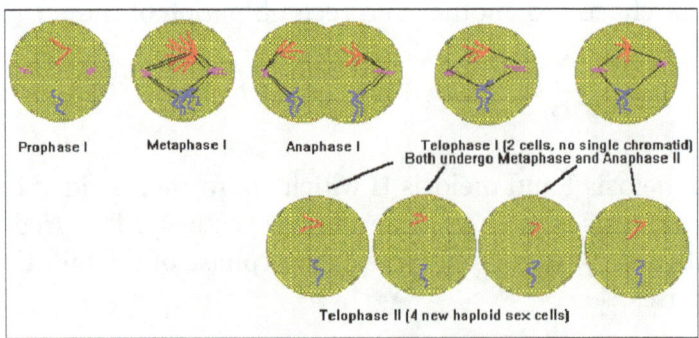

Diagram of the meiotic phases.

Meiosis I

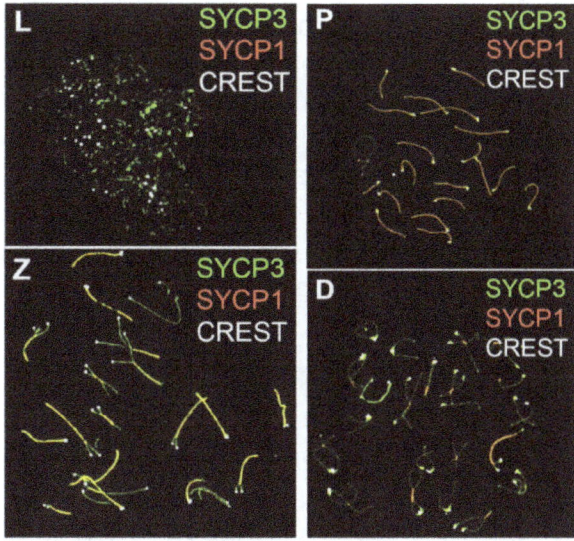

Meiosis Prophase I in mice. L:Leptotene,
Z:Zygotene, P:Pachytene, D:Diplotene.

Meiosis I segregates homologous chromosomes, which are joined as tetrads (2n, 4c), producing two haploid cells (n chromosomes, 23 in humans) which each contain chromatid pairs (1n, 2c). Because the ploidy is reduced from diploid to haploid, meiosis I is referred to as a *reductional division*. Meiosis II is an *equational division* analogous to mitosis, in which the sister chromatids are segregated, creating four haploid daughter cells (1n, 1c).

Prophase I

Prophase I is typically the longest phase of meiosis. During prophase I, homologous chromosomes pair and exchange DNA (homologous recombination). This often results in chromosomal crossover. This process is critical for pairing between homologous chromosomes and hence for accurate segregation of the chromosomes at the first meiosis division. The new combinations of DNA created during crossover are a significant source of genetic variation, and result in new combinations of alleles, which may be beneficial. The paired and replicated chromosomes are called bivalents or tetrads, which have two chromosomes and four chromatids, with one chromosome coming from each parent. The process of pairing the homologous chromosomes is called synapsis. At this stage, non-sister chromatids may cross-over at points called chiasmata (plural; singular chiasma). Prophase I has historically been divided into a series of substages which are named according to the appearance of chromosomes.

Leptotene

The first stage of prophase I is the *leptotene* stage, also known as *leptonema*. In this stage of prophase I, individual chromosomes—each consisting of two sister chromatids—become "individualized" to form visible strands within the nucleus. The two sister chromatids closely associate and are visually indistinguishable from one another. During leptotene, lateral elements of the synaptonemal complex assemble. Leptotene is of very short duration and progressive condensation and coiling of chromosome fibers takes place.

Zygotene

The *zygotene* stage, also known as *zygonema*, occurs as the chromosomes approximately line up with each other into homologous chromosome pairs. In some organisms, this is called the bouquet stage because of the way the telomeres cluster at one end of the nucleus. At this stage, the synapsis (pairing/coming together) of homologous chromosomes takes place, facilitated by assembly of central element of the synaptonemal complex. Pairing is brought about in a zipper-like fashion and may start at the centromere (procentric), at the chromosome ends (proterminal), or at any other portion (intermediate). Individuals of a pair are equal in length and in position of the centromere. Thus pairing is highly specific and exact. The paired chromosomes are called bivalent or tetrad chromosomes.

Pachytene

The *pachytene* stage, also known as *pachynema*. At this point a tetrad of the chromosomes has formed known as a bivalent. This is the stage when homologous recombination, including chromosomal crossover (crossing over), occurs. Nonsister chromatids of homologous chromosomes may exchange segments over regions of homology. Sex chromosomes, however, are not wholly identical, and only exchange information over a small region of homology. At the sites where exchange happens, chiasmata form. The exchange of information between the non-sister chromatids results in a recombination of information; each chromosome has the complete set of information it had before, and there are no gaps formed as a result of the process. Because the chromosomes cannot be distinguished in the synaptonemal complex, the actual act of crossing over is not perceivable through the microscope, and chiasmata are not visible until the next stage.

Diplotene

During the *diplotene* stage, also known as *diplonema,* the synaptonemal complex degrades and homologous chromosomes separate from one another a little. The chromosomes themselves uncoil a bit, allowing some transcription of DNA. However, the homologous chromosomes of each bivalent remain tightly bound at chiasmata, the regions where crossing-over occurred. The chiasmata remain on the chromosomes until they are severed at the transition to anaphase I.

In human fetal oogenesis, all developing oocytes develop to this stage and are arrested in prophase I before birth. This suspended state is referred to as the *dictyotene stage* or dictyate. It lasts until meiosis is resumed to prepare the oocyte for ovulation, which happens at puberty or even later.

Diakinesis

Chromosomes condense further during the *diakinesis* stage, from Greek words meaning "moving through". This is the first point in meiosis where the four parts of the tetrads are actually visible. Sites of crossing over entangle together, effectively overlapping, making chiasmata clearly visible. Other than this observation, the rest of the stage closely resembles prometaphase of mitosis; the nucleoli disappear, the nuclear membrane disintegrates into vesicles, and the meiotic spindle begins to form.

Synchronous Processes

During these stages, two centrosomes, containing a pair of centrioles in animal cells, migrate to the two poles of the cell. These centrosomes, which were duplicated during S-phase, function as microtubule organizing centers nucleating microtubules, which are essentially cellular ropes and poles. The microtubules invade the nuclear region after the nuclear envelope disintegrates, attaching to the chromosomes at the kinetochore. The kinetochore functions as a motor, pulling the chromosome along the attached microtubule toward the originating centrosome, like a train on a track. There are four kinetochores on each tetrad, but the pair of kinetochores on each sister chromatid fuses and functions as a unit during meiosis I.

Microtubules that attach to the kinetochores are known as *kinetochore microtubules*. Other microtubules will interact with microtubules from the opposite centrosome: these are called *nonkinetochore microtubules* or *polar microtubules*. A third type of microtubules, the aster microtubules, radiates from the centrosome into the cytoplasm or contacts components of the membrane skeleton.

Metaphase I

Homologous pairs move together along the metaphase plate: As *kinetochore microtubules* from both centrosomes attach to their respective kinetochores, the paired homologous chromosomes align along an equatorial plane that bisects the spindle, due to continuous counterbalancing forces exerted on the bivalents by the microtubules emanating from the two kinetochores of homologous chromosomes. This attachment is referred to as a bipolar attachment. The physical basis of the independent assortment of chromosomes is the random orientation of each bivalent along the metaphase plate, with respect to the orientation of the other bivalents along the same equatorial line. The protein complex cohesin holds sister chromatids together from the time of their replication until anaphase. In mitosis, the force of kinetochore microtubules pulling in opposite directions creates tension. The cell senses this tension and does not progress with anaphase until all the chromosomes are properly bi-oriented. In meiosis, establishing tension requires at least one crossover per chromosome pair in addition to cohesin between sister chromatids.

Anaphase I

Kinetochore microtubules shorten, pulling homologous chromosomes (which each consist of a pair of sister chromatids) to opposite poles. Nonkinetochore microtubules lengthen, pushing the centrosomes farther apart. The cell elongates in preparation for division down the center. Unlike in mitosis, only the cohesin from the chromosome arms is degraded while the cohesin surrounding the centromere remains protected by a protein named Shugoshin (japanese for "guardian spirit"), what prevents chromatids to apart. This allows the sister chromatids to remain together while homologs are segregated.

Telophase I

The first meiotic division effectively ends when the chromosomes arrive at the poles. Each daughter cell now has half the number of chromosomes but each chromosome consists of a pair of chromatids. The microtubules that make up the spindle network disappear, and a new nuclear membrane surrounds each haploid set. The chromosomes uncoil back into chromatin. Cytokinesis, the pinching of the cell membrane in animal cells or the formation of the cell wall in plant cells, occurs, completing the creation of two daughter cells. Sister chromatids remain attached during telophase I.

Cells may enter a period of rest known as interkinesis or interphase II. No DNA replication occurs during this stage.

Meiosis II

Meiosis II is the second meiotic division, and usually involves equational segregation, or separation of sister chromatids. Mechanically, the process is similar to mitosis, though its genetic results are fundamentally different. The end result is production of four haploid cells (n chromosomes, 23 in humans) from the two haploid cells (with n chromosomes, each consisting of two sister chromatids) produced in meiosis I. The four main steps of meiosis II are: prophase II, metaphase II, anaphase II, and telophase II.

In prophase II we see the disappearance of the nucleoli and the nuclear envelope again as well as the shortening and thickening of the chromatids. Centrosomes move to the polar regions and arrange spindle fibers for the second meiotic division.

In metaphase II, the centromeres contain two kinetochores that attach to spindle fibers from the centrosomes at opposite poles. The new equatorial metaphase plate is rotated by 90 degrees when compared to meiosis I, perpendicular to the previous plate.

This is followed by anaphase II, in which the remaining centromeric cohesin, not protected by Shugoshin anymore, is cleaved, allowing the sister chromatids to segregate. The sister chromatids by convention are now called sister chromosomes as they move toward opposing poles.

The process ends with telophase II, which is similar to telophase I, and is marked by decondensation and lengthening of the chromosomes and the disassembly of the spindle. Nuclear envelopes re-form and cleavage or cell plate formation eventually produces a total of four daughter cells, each with a haploid set of chromosomes.

Meiosis is now complete and ends up with four new daughter cells.

Origin and Function

The origin and function of meiosis are fundamental to understanding the evolution of sexual reproduction in eukaryotes. There is no current consensus among biologists on the questions of how sex in eukaryotes arose in evolution, what basic function sexual reproduction serves, and why it is maintained, given the basic two-fold cost of sex. It is clear that it evolved over 1.2 billion years ago, and that almost all species which are descendants of the original sexually reproducing species are still sexual reproducers, including plants, fungi, and animals.

Meiosis is a key event of the sexual cycle in eukaryotes. It is the stage of the life cycle when a cell gives rise to two haploid cells (gametes) each having half as many chromosomes. Two such haploid gametes, arising from different individual organisms, fuse by the process of fertilization, thus completing the sexual cycle.

Meiosis is ubiquitous among eukaryotes. It occurs in single-celled organisms such as yeast, as well as in multicellular organisms, such as humans. Eukaryotes arose from prokaryotes more than 2.2 billion years ago and the earliest eukaryotes were likely single-celled organisms. To understand sex in eukaryotes, it is necessary to understand (1) how meiosis arose in single celled eukaryotes, and (2) the function of meiosis.

Occurrence

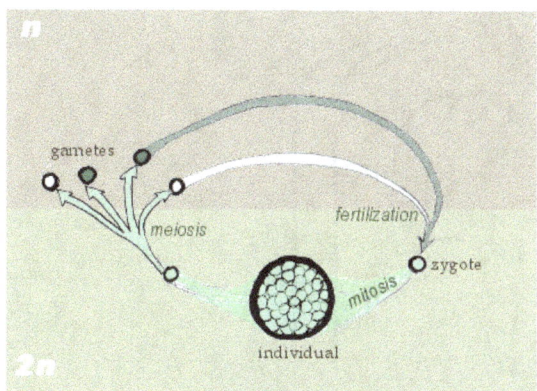 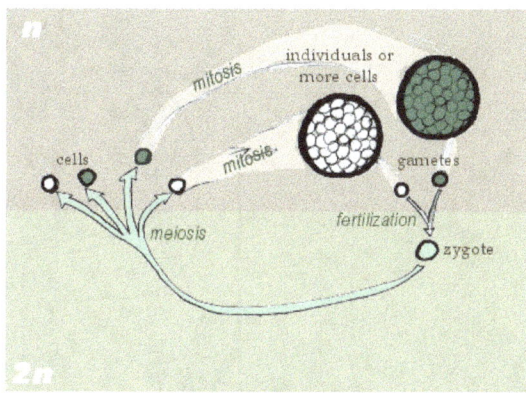

Diplontic life cycle.   Haplontic life cycle.

Meiosis occurs in eukaryotic life cycles involving sexual reproduction, consisting of the constant cyclical process of meiosis and fertilization. This takes place alongside normal mitotic cell division. In multicellular organisms, there is an intermediary step between the diploid and haploid transition where the organism grows. At certain stages of the life cycle, germ cells produce gametes. Somatic cells make up the body of the organism and are not involved in gamete production.

Cycling meiosis and fertilization events produces a series of transitions back and forth between alternating haploid and diploid states. The organism phase of the life cycle can occur either during the diploid state (*diplontic* life cycle), during the haploid state (*haplontic* life cycle), or both (*haplodiplontic* life cycle, in which there are two distinct organism phases, one during the haploid state and the other during the diploid state). In this sense there are three types of life cycles that utilize sexual reproduction, differentiated by the location of the organism phase(s).

In the *diplontic life cycle* (with pre-gametic meiosis), of which humans are a part, the organism is diploid, grown from a diploid cell called the zygote. The organism's diploid germ-line stem cells undergo meiosis to create haploid gametes (the spermatozoa for males and ova for females), which fertilize to form the zygote. The diploid zygote undergoes repeated cellular division by mitosis to grow into the organism.

In the *haplontic life cycle* (with post-zygotic meiosis), the organism is haploid instead, spawned by the proliferation and differentiation of a single haploid cell called the gamete. Two organisms of opposing sex contribute their haploid gametes to form a diploid zygote. The zygote undergoes meiosis immediately, creating four haploid cells. These cells undergo mitosis to create the organism. Many fungi and many protozoa utilize the haplontic life cycle.

Finally, in the *haplodiplontic life cycle* (with sporic or intermediate meiosis), the living organism alternates between haploid and diploid states. Consequently, this cycle is also known as the alternation of generations. The diploid organism's germ-line cells undergo meiosis to produce spores. The spores proliferate by mitosis, growing into a haploid organism. The haploid organism's gamete then combines with another haploid organism's gamete, creating the zygote. The zygote undergoes repeated mitosis and differentiation to become a diploid organism again. The haplodiplontic life cycle can be considered a fusion of the diplontic and haplontic life cycles.

Plants and Animals

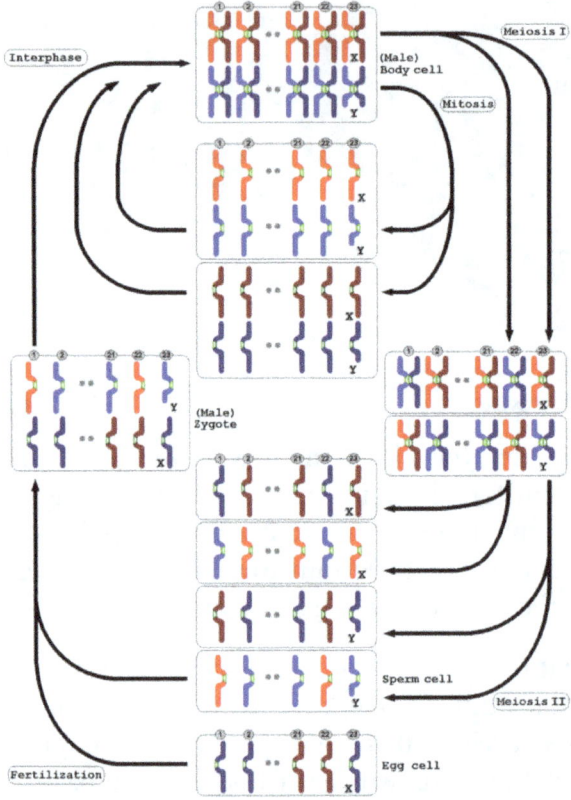

Chromatides' and chromosomes' distribution within
the mitotic and meiotic cycle of a male human cell.

Meiosis occurs in all animals and plants. The end result, the production of gametes with half the number of chromosomes as the parent cell, is the same, but the detailed process is different. In animals, meiosis produces gametes directly. In land plants and some algae, there is an alternation of generations such that meiosis in the diploid sporophyte generation produces haploid spores. These spores multiply by mitosis, developing into the haploid gametophyte generation, which then gives rise to gametes directly (i.e. without further meiosis). In both animals and plants, the final stage is for the gametes to fuse, restoring the original number of chromosomes.

Mammals

In females, meiosis occurs in cells known as oocytes (singular: oocyte). Each primary oocyte divides twice in meiosis, unequally in each case. The first division produces a daughter cell, and a much smaller polar body which may or may not undergo a second division. In meiosis II, division of the daughter cell produces a second polar body, and a single haploid cell, which enlarges to become an ovum. Therefore, in females each primary oocyte that undergoes meiosis results in one mature ovum and one or two polar bodies.

Note that there are pauses during meiosis in females. Maturing oocytes are arrested in prophase I of meiosis I and lie dormant within a protective shell of somatic cells called the follicle. At the beginning of each menstrual cycle, FSH secretion from the anterior pituitary stimulates a few follicles to mature in a process known as folliculogenesis. During this process, the maturing oocytes resume meiosis and continue until metaphase II of meiosis II, where they are again arrested just before ovulation. If these oocytes are fertilized by sperm, they will resume and complete meiosis. During folliculogenesis in humans, usually one follicle becomes dominant while the others undergo atresia. The process of meiosis in females occurs during oogenesis, and differs from the typical meiosis in that it features a long period of meiotic arrest known as the dictyate stage and lacks the assistance of centrosomes.

In males, meiosis occurs during spermatogenesis in the seminiferous tubules of the testicles. Meiosis during spermatogenesis is specific to a type of cell called spermatocytes, which will later mature to become spermatozoa. Meiosis of primordial germ cells happens at the time of puberty, much later than in females. Tissues of the male testis suppress meiosis by degrading retinoic acid, proposed to be a stimulator of meiosis. This is overcome at puberty when cells within seminiferous tubules called Sertoli cells start making their own retinoic acid. Sensitivity to retinoic acid is also adjusted by proteins called nanos and DAZL. Genetic loss-of-function studies on retinoic acid-generating enzymes have shown that retinoic acid is required postnatally to stimulate spermatogonia differentiation which results several days later in spermatocytes undergoing meiosis, however retinoic acid is not required during the time when meiosis initiates.

In female mammals, meiosis begins immediately after primordial germ cells migrate to the ovary in the embryo. Some studies suggest that retinoic acid derived from the primitive kidney (mesonephros) stimulates meiosis in embryonic ovarian oogonia and that tissues of the embryonic male testis suppress meiosis by degrading retinoic acid. However, genetic loss-of-function studies on retinoic acid-generating enzymes have shown that retinoic acid is not required for initiation of either female meiosis which occurs during embryogenesis or male meiosis which initiates postnatally.

Variations

Nondisjunction

The normal separation of chromosomes in meiosis I or sister chromatids in meiosis II is termed *disjunction*. When the segregation is not normal, it is called *nondisjunction*. This results in the production of gametes which have either too many or too few of a particular chromosome, and is a common mechanism for trisomy or monosomy. Nondisjunction can occur in the meiosis I or meiosis II, phases of cellular reproduction, or during mitosis.

Most monosomic and trisomic human embryos are not viable, but some aneuploidies can be tolerated, such as trisomy for the smallest chromosome, chromosome 21. Phenotypes of these aneuploidies range from severe developmental disorders to asymptomatic. Medical conditions include but are not limited to:

- Down syndrome: Trisomy of chromosome 21

- Patau syndrome: Trisomy of chromosome 13

- Edwards syndrome: Trisomy of chromosome 18

- Klinefelter syndrome: Extra X chromosomes in males: i.e. XXY, XXXY, XXXXY, etc.

- Turner syndrome: Lacking of one X chromosome in females: i.e. Xo

- Triple X syndrome: An extra X chromosome in females

- XYY syndrome: An extra Y chromosome in males.

The probability of nondisjunction in human oocytes increases with increasing maternal age, presumably due to loss of cohesin over time.

Other

Alongside with the variations of meiosis related to the moment when meiosis occur in life cycles, resulting in post-zygotic, pre-gametic and intermediate meiosis, the number of nuclear divisions in meiosis is also variable. The majority of eukaryotes have a two-divisional meiosis (though sometimes achiasmatic), but a very rare form, one-divisional meiosis, occurs in some flagellates (parabasalids and oxymonads) from the gut of the wood-feeding cockroach *Cryptocercus*.

Comparison to Mitosis

In order to understand meiosis, a comparison to mitosis is helpful. The table below shows the differences between meiosis and mitosis.

	Meiosis	Mitosis
End result	Normally four cells, each with half the number of chromosomes as the parent.	Two cells, having the same number of chromosomes as the parent.
Function	Production of gametes (sex cells) in sexually reproducing eukaryotes with diplont life cycle.	Cellular reproduction, growth, repair, asexual reproduction.
Where does it happen?	Almost all eukaryotes (animals, plants, fungi, and protists); In gonads, before gametes (in diplontic life cycles); After zygotes (in haplontic); Before spores (in haplodiplontic).	All proliferating cells in all eukaryotes.
Steps	Prophase I, Metaphase I, Anaphase I, Telophase I, Prophase II, Metaphase II, Anaphase II, Telophase II.	Prophase, Prometaphase, Metaphase, Anaphase, Telophase .
Genetically same as parent?	No	Yes
Crossing over happens?	Yes, normally occurs between each pair of homologous chromosomes.	Very rarely
Pairing of homologous chromosomes?	Yes	No
Cytokinesis	Occurs in Telophase I and Telophase II.	Occurs in Telophase .
Centromeres split	Does not occur in Anaphase I, but occurs in Anaphase II	Occurs in Anaphase.

Mitosis

Mitosis is a type of cell division in which one cell (the mother) divides to produce two new cells (the daughters) that are genetically identical to itself. In the context of the cell cycle, mitosis is the part of the division process in which the DNA of the cell's nucleus is split into two equal sets of chromosomes.

The great majority of the cell divisions that happen in your body involve mitosis. During development and growth, mitosis populates an organism's body with cells, and throughout an organism's life, it replaces old, worn-out cells with new ones. For single-celled eukaryotes like yeast, mitotic divisions are actually a form of reproduction, adding new individuals to the population.

In all of these cases, the "goal" of mitosis is to make sure that each daughter cell gets a perfect, full set of chromosomes. Cells with too few or too many chromosomes usually don't function well: they may not survive, or they may even cause cancer. So, when cells undergo mitosis, they don't just divide their DNA at random and toss it into piles for the two daughter cells. Instead, they split up their duplicated chromosomes in a carefully organized series of steps.

Phases of Mitosis

Mitosis consists of four basic phases: prophase, metaphase, anaphase, and telophase. Some textbooks list five, breaking prophase into an early phase (called prophase) and a late phase (called prometaphase). These phases occur in strict sequential order, and cytokinesis - the process of dividing the cell contents to make two new cells - starts in anaphase or telophase.

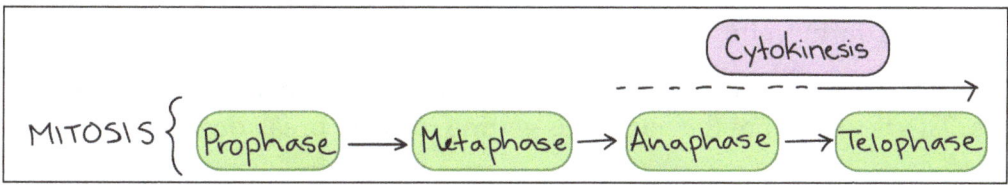

You can remember the order of the phases with the famous mnemonic: Please Pee on the MAT. But don't get too hung up on names – what's most important to understand is what's happening at each stage, and why it's important for the division of the chromosomes.

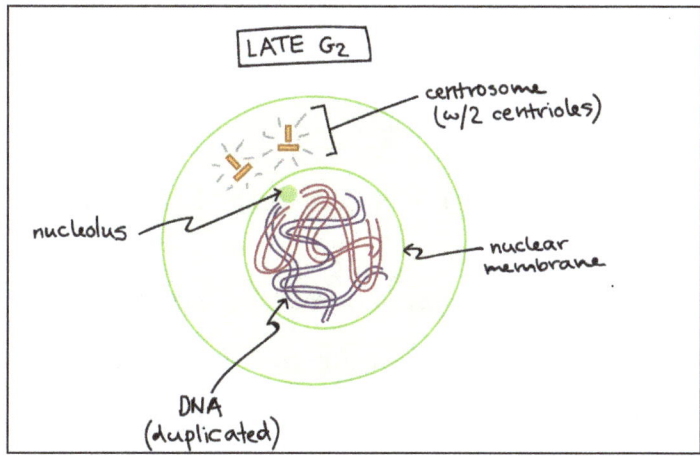

A cell right before it begins mitosis. This cell is in interphase (late G_2 phase) and has already copied its DNA, so the chromosomes in the nucleus each consist of two connected copies, called sister chromatids. You can't see the chromosomes very clearly at this point, because they are still in their long, stringy, decondensed form.

This animal cell has also made a copy of its centrosome, an organelle that will play a key role in orchestrating mitosis, so there are two centrosomes. (Plant cells generally don't have centrosomes with centrioles, but have a different type of microtubule organizing center that plays a similar role.)

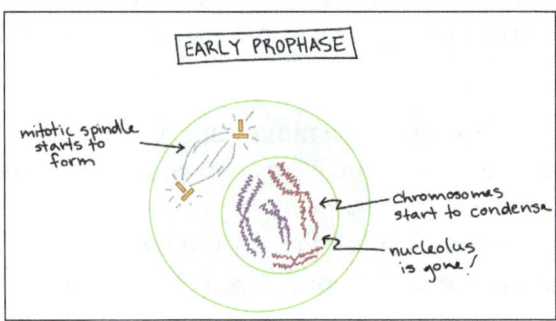

In early prophase, the cell starts to break down some structures and build others up, setting the stage for division of the chromosomes.

- The chromosomes start to condense (making them easier to pull apart later on).

- The mitotic spindle begins to form. The spindle is a structure made of microtubules, strong fibers that are part of the cell's "skeleton." Its job is to organize the chromosomes and move them around during mitosis. The spindle grows between the centrosomes as they move apart.

- The nucleolus (or nucleoli, plural), a part of the nucleus where ribosomes are made, disappears. This is a sign that the nucleus is getting ready to break down.

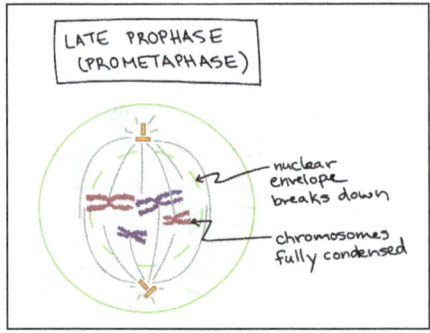

In late prophase (sometimes also called prometaphase), the mitotic spindle begins to capture and organize the chromosomes.

- The chromosomes finish condensing, so they are very compact.

- The nuclear envelope breaks down, releasing the chromosomes.

- The mitotic spindle grows more, and some of the microtubules start to "capture" chromosomes.

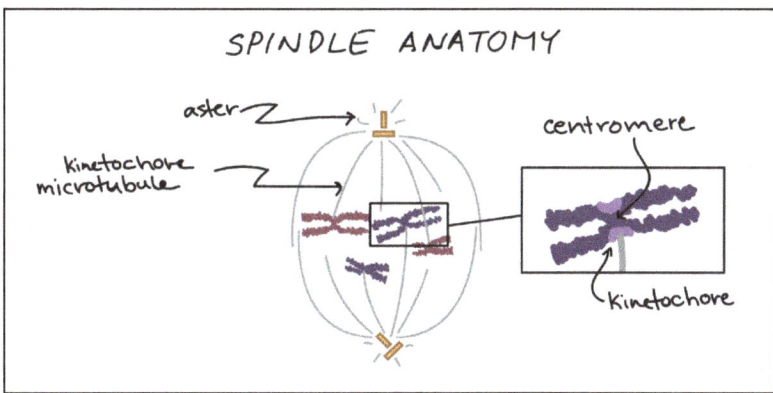

Microtubules can bind to chromosomes at the kinetochore, a patch of protein found on the centromere of each sister chromatid. (Centromeres are the regions of DNA where the sister chromatids are most tightly connected.)

Microtubules that bind a chromosome are called kinetochore microtubules. Microtubules that don't bind to kinetochores can grab on to microtubules from the opposite pole, stabilizing the spindle. More microtubules extend from each centrosome towards the edge of the cell, forming a structure called the aster.

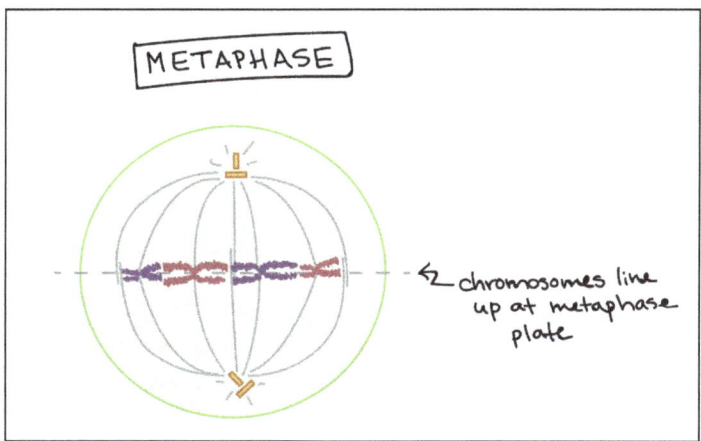

In metaphase, the spindle has captured all the chromosomes and lined them up at the middle of the cell, ready to divide.

- All the chromosomes align at the metaphase plate (not a physical structure, just a term for the plane where the chromosomes line up).

- At this stage, the two kinetochores of each chromosome should be attached to microtubules from opposite spindle poles.

Before proceeding to anaphase, the cell will check to make sure that all the chromosomes are at the metaphase plate with their kinetochores correctly attached to microtubules. This is called the spindle checkpoint and helps ensure that the sister chromatids will split evenly between the two daughter cells when they separate in the next step. If a chromosome is not properly aligned or attached, the cell will halt division until the problem is fixed.

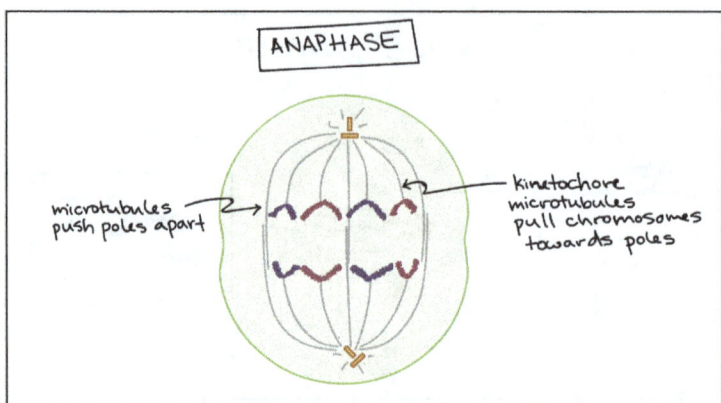

In anaphase, the sister chromatids separate from each other and are pulled towards opposite ends of the cell.

- The protein "glue" that holds the sister chromatids together is broken down, allowing them to separate. Each is now its own chromosome. The chromosomes of each pair are pulled towards opposite ends of the cell.

- Microtubules not attached to chromosomes elongate and push apart, separating the poles and making the cell longer.

All of these processes are driven by motor proteins, molecular machines that can "walk" along microtubule tracks and carry a cargo. In mitosis, motor proteins carry chromosomes or other microtubules as they walk.

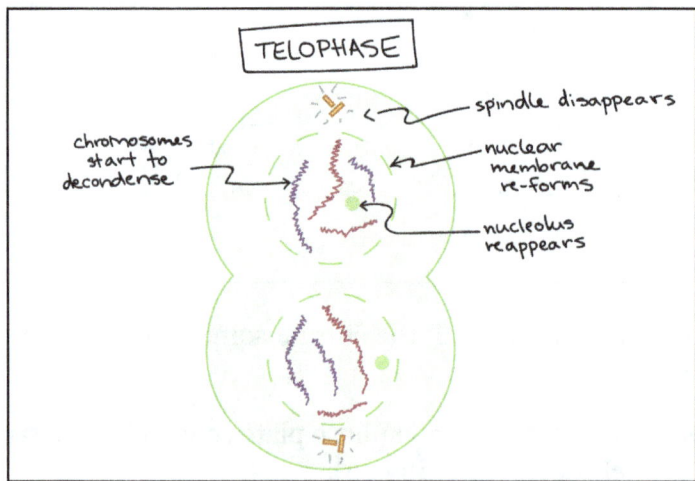

In telophase, the cell is nearly done dividing, and it starts to re-establish its normal structures as cytokinesis (division of the cell contents) takes place.

- The mitotic spindle is broken down into its building blocks.

- Two new nuclei form, one for each set of chromosomes. Nuclear membranes and nucleoli reappear.

- The chromosomes begin to decondense and return to their "stringy" form.

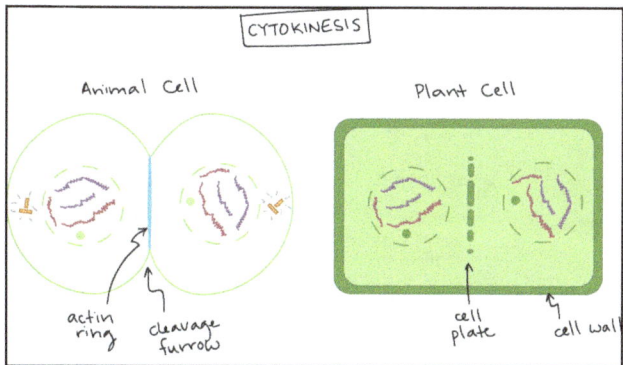

Cytokinesis, the division of the cytoplasm to form two new cells, overlaps with the final stages of mitosis. It may start in either anaphase or telophase, depending on the cell, and finishes shortly after telophase.

In animal cells, cytokinesis is contractile, pinching the cell in two like a coin purse with a drawstring. The "drawstring" is a band of filaments made of a protein called actin, and the pinch crease is known as the cleavage furrow. Plant cells can't be divided like this because they have a cell wall and are too stiff. Instead, a structure called the cell plate forms down the middle of the cell, splitting it into two daughter cells separated by a new wall.

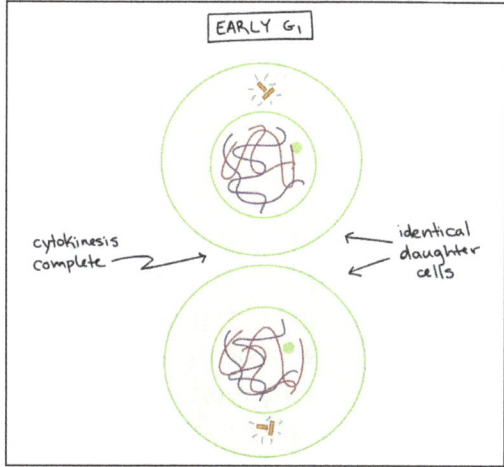

When cytokinesis finishes, we end up with two new cells, each with a complete set of chromosomes identical to those of the mother cell. The daughter cells can now begin their own cellular "lives," and – depending on what they decide to be when they grow up – may undergo mitosis themselves, repeating the cycle.

Fission

Fission, in biology, is the division of a single entity into two or more parts and the regeneration of those parts to separate entities resembling the original. The object experiencing fission is usually a cell, but the term may also refer to how organisms, bodies, populations, or species split into discrete parts. The fission may be *binary fission*, in which a single organism produces two parts, or *multiple fission*, in which a single entity produces multiple parts.

Binary Fission

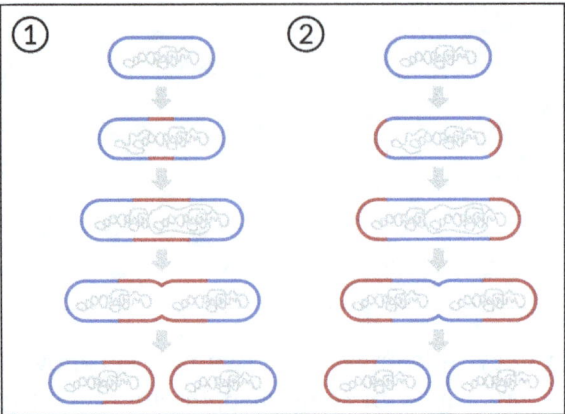

Schematic diagram of cellular growth (elongation) and binary fission of bacilli. Blue and red lines indicate old and newly generated bacterial cell wall, respectively. (1) growth at the centre of the bacterial body. e.g. *Bacillus subtilis*, *E. coli*, and others. (2) apical growth. e.g. *Corynebacterium diphtheriae*.

Organisms in the domains of Archaea and Bacteria reproduce with binary fission. This form of asexual reproduction and cell division is also used by some organelles within eukaryotic organisms (e.g., mitochondria). Binary fission results in the reproduction of a living prokaryotic cell (or organelle) by dividing the cell into two parts, each with the potential to grow to the size of the original.

Fission of Prokaryotes

The single DNA molecule first replicates, then attaches each copy to a different part of the cell membrane. When the cell begins to pull apart, the replicated and original chromosomes are separated. The consequence of this asexual method of reproduction is that all the cells are genetically identical, meaning that they have the same genetic material (barring random mutations). Unlike the processes of mitosis and meiosis used by eukaryotic cells, binary fission takes place without the formation of a spindle apparatus on the cell. Like in mitosis (and unlike in meiosis), the parental identity is preserved.

Process of Bacterial Fission

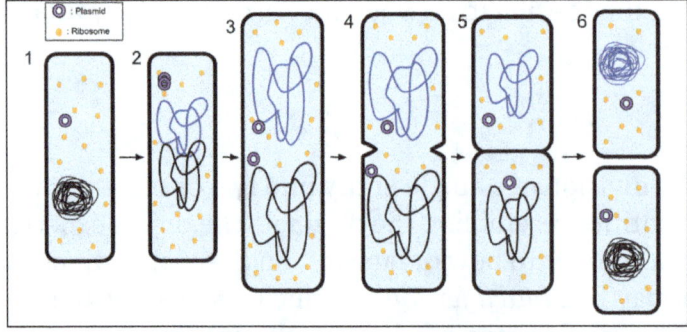

Binary fission in a prokaryote

The process of binary fission in bacteria involves the following steps. First, the cell's DNA is replicated. The replicated DNA copies then move to opposite poles of the cell in an energy-dependent process. The cell lengthens. Then, the equatorial plane of the cell constricts and separates the plasma membrane such that each new cell has exactly the same genetic material.

More specifically, the following steps occur:

1. The bacterium before binary fission is when the DNA is tightly coiled.

2. The DNA of the bacterium has uncoiled and duplicated.

3. The DNA is pulled to the separate poles of the bacterium as it increases the size to prepare for splitting.

4. The growth of a new cell wall begins to separate the bacterium.

5. The new cell wall fully develops, resulting in the complete split of the bacterium.

6. The new daughter cells have tightly coiled DNA rods, ribosomes, and plasmids; these are now brand-new organisms.

Speed of Bacterial Fission

Binary fission is generally rapid though its speed varies between species. For *E. coli*, cells typically divide about every 20 minutes at 37 °C. Because the new cells will, in turn, undergo binary fission on their own, the time binary fission requires is also the time the bacterial culture requires to double in the number of cells it contains. This time period can, therefore, be referred to as the doubling time. Some species other than *E. coli* may have faster or slower doubling times: some strains of *Mycobacterium tuberculosis* may have doubling times of nearly 100 hours. Bacterial growth is limited by factors including nutrient availability and available space, so binary fission occurs at much lower rates in bacterial cultures once they enter the stationary phase of growth.

Fission of Organelles

Some organelles in eukaryotic cells reproduce using binary fission. Mitochondrial fission occurs frequently within the cell, even when the cell is not actively undergoing mitosis, and is necessary to regulate the cell's metabolism.

Types of Binary Fission

Binary fission in organisms can occur in four ways, irregular,longitudinal, transverse, oblique .i.e. left oblique & right oblique

1. Irregular: In this fission, cytokinesis may take place along any plane but it is always perpendicular to the plane of karyokinesis. e.g. amoeba.

2. Longitudinal: Here cytokinesis takes place along the longitudinal axis. e.g. in flagellates like *Euglena.*

3. Transverse: Here cytokinesis takes place along the transverse axis. e.g. in ciliate protozoans like *Paramecium*.

4. Oblique: In this type of binary fission cytokinesis occurs obliquely. e.g. *Ceratium*.

Multiple Fission

Fission of Protists

Multiple fission at the cellular level occurs in many protists, e.g. sporozoans and algae. The nucleus of the parent cell divides several times by amitosis, producing several nuclei. The cytoplasm then separates, creating multiple daughter cells.

Some parasitic, single-celled organisms undergo a multiple fission-like process to produce numerous daughter cells from a single parent cell. Isolates of the human parasite *Blastocystis hominis* were observed to begin such a process within 4 to 6 days. Cells of the fish parasite *Trypanosoma borreli* have also been observed participating in both binary and multiple fission.

Fission of Apicomplexans

In the apicomplexans, a phylum of parasitic protists, multiple fission, or schizogony, is manifested either as merogony, sporogony or gametogony. Merogony results in merozoites, which are multiple daughter cells, that originate within the same cell membrane, sporogony results in sporozoites, and gametogony results in microgametes.

Fission of Green Algae

Green algae can divide into more than two daughter cells. The exact number of daughter cells depends on the species of algae and is an effect of temperature and light.

Multiple Fission of Bacteria

Most species of bacteria primarily undergo binary reproduction. Some species and groups of bacteria may undergo multiple fission as well, sometimes beginning or ending with the production of spores. The species *Metabacterium polyspora*, a symbiont of guinea pigs, has been found to produce multiple endospores in each division. Some species of cyanobacteria have also been found to reproduce through multiple fission.

Plasmotomy

Some protozoans reproduce by yet another mechanism of fission called as plasmotomy. In this type of fission, a multinucleate adult parent undergoes cytokinesis to form two multinucleate (or coenocytic) daughter cells. The daughter cells so produced undergo karyokinesis, further. *Opalina* and *Pelomyxa* reproduce in this way.

Clonal Fragmentation

Fragmentation in multicellular or colonial organisms is a form of asexual reproduction or cloning where an organism is split into fragments. Each of these fragments develop into mature, fully grown individuals that are clones of the original organism. In echinoderms, this method of reproduction is usually known as fissiparity.

Population Fission

Any splitting of a single population of individuals into discrete parts may be considered fission. A population may undergo fission for a variety of reasons, including migration or geographic isolation. Because the fission leads to genetic variance in the newly isolated, smaller populations, population fission is a precursor to speciation.

CELL SIGNALING

Cells typically communicate using chemical signals. These chemical signals, which are proteins or other molecules produced by a sending cell, are often secreted from the cell and released into the extracellular space. There, they can float – like messages in a bottle – over to neighboring cells.

Not all cells can "hear" a particular chemical message. In order to detect a signal (that is, to be a target cell), a neighbor cell must have the right receptor for that signal. When a signaling molecule binds to its receptor, it alters the shape or activity of the receptor, triggering a change inside of the cell. Signaling molecules are often called ligands, a general term for molecules that bind specifically to other molecules (such as receptors).

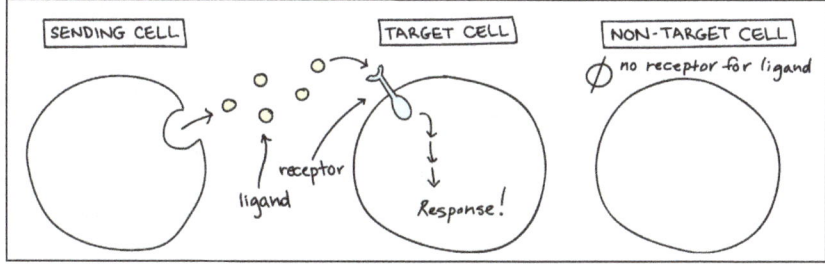

The message carried by a ligand is often relayed through a chain of chemical messengers inside the cell. Ultimately, it leads to a change in the cell, such as alteration in the activity of a gene or even the induction of a whole process, such as cell division. Thus, the original intercellular (between-cells) signal is converted into an intracellular (within-cell) signal that triggers a response.

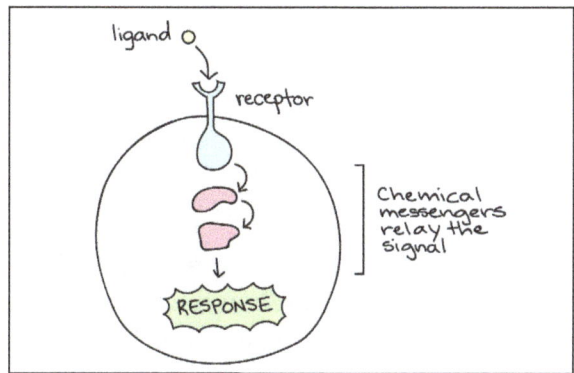

Forms of Signaling

Cell-cell signaling involves the transmission of a signal from a sending cell to a receiving cell. However, not all sending and receiving cells are next-door neighbors, nor do all cell pairs exchange signals in the same way.

There are four basic categories of chemical signaling found in multicellular organisms: paracrine signaling, autocrine signaling, endocrine signaling, and signaling by direct contact. The main difference between the different categories of signaling is the distance that the signal travels through the organism to reach the target cell.

Paracrine Signaling

Often, cells that are near one another communicate through the release of chemical messengers (ligands that can diffuse through the space between the cells). This type of signaling, in which cells communicate over relatively short distances, is known as paracrine signaling.

Paracrine signaling allows cells to locally coordinate activities with their neighbors. Although they're used in many different tissues and contexts, paracrine signals are especially important during development, when they allow one group of cells to tell a neighboring group of cells what cellular identity to take on.

Synaptic Signaling

One unique example of paracrine signaling is synaptic signaling, in which nerve cells transmit signals. This process is named for the synapse, the junction between two nerve cells where signal transmission occurs.

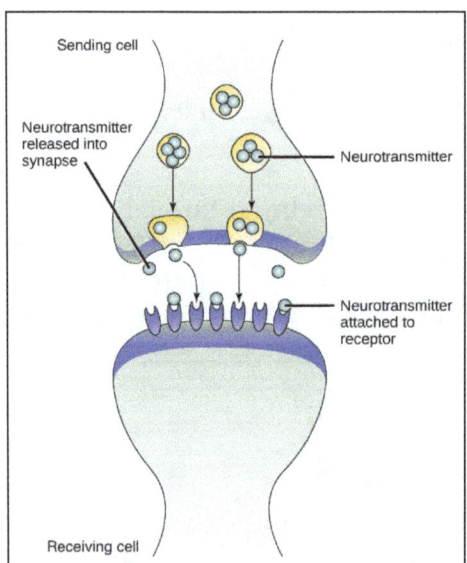

When the sending neuron fires, an electrical impulse moves rapidly through the cell, traveling down a long, fiber-like extension called an axon. When the impulse reaches the synapse, it triggers the release of ligands called neurotransmitters, which quickly cross the small gap between the nerve cells. When the neurotransmitters arrive at the receiving cell, they bind to receptors and cause a chemical change inside of the cell (often, opening ion channels and changing the electrical potential across the membrane).

The neurotransmitters that are released into the chemical synapse are quickly degraded or taken back up by the sending cell. This "resets" the system so they synapse is prepared to respond quickly to the next signal.

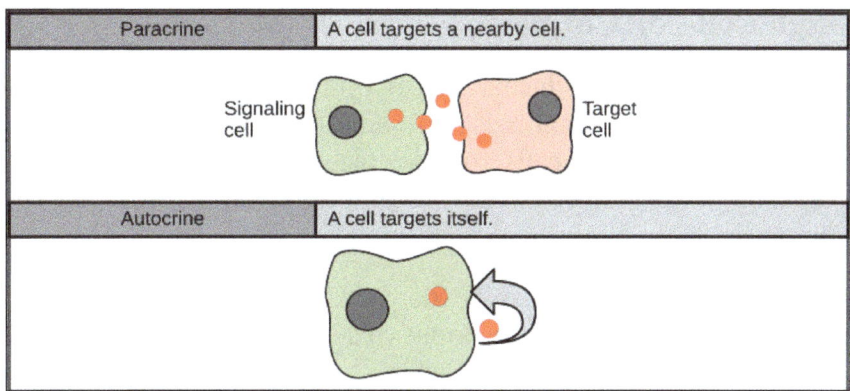

Autocrine Signaling

In autocrine signaling, a cell signals to itself, releasing a ligand that binds to receptors on its own surface (or, depending on the type of signal, to receptors inside of the cell). This may seem like an odd thing for a cell to do, but autocrine signaling plays an important role in many processes.

For instance, autocrine signaling is important during development, helping cells take on and re-inforce their correct identities. From a medical standpoint, autocrine signaling is important in cancer and is thought to play a key role in metastasis (the spread of cancer from its original site to other parts of the body). In many cases, a signal may have both autocrine and paracrine effects, binding to the sending cell as well as other similar cells in the area.

Endocrine Signaling

When cells need to transmit signals over long distances, they often use the circulatory system as a distribution network for the messages they send. In long-distance endocrine signaling, signals are produced by specialized cells and released into the bloodstream, which carries them to target cells in distant parts of the body. Signals that are produced in one part of the body and travel through the circulation to reach far-away targets are known as hormones.

In humans, endocrine glands that release hormones include the thyroid, the hypothalamus, and the pituitary, as well as the gonads (testes and ovaries) and the pancreas. Each endocrine gland releases one or more types of hormones, many of which are master regulators of development and physiology.

For example, the pituitary releases growth hormone (GH), which promotes growth, particularly of the skeleton and cartilage. Like most hormones, GH affects many different types of cells through-out the body. However, cartilage cells provide one example of how GH functions: it binds to recep-tors on the surface of these cells and encourages them to divide.

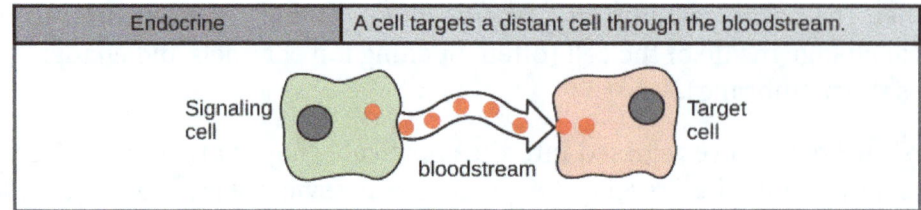

Signaling through Cell-cell Contact

Gap junctions in animals and plasmodesmata in plants are tiny channels that directly connect neighboring cells. These water-filled channels allow small signaling molecules, called intracellular mediators, to diffuse between the two cells. Small molecules, such as calcium ions(Ca^{2+}) are able to move between cells, but large molecules like proteins and DNA cannot fit through the channels without special assistance.

The transfer of signaling molecules transmits the current state of one cell to its neighbor. This allows a group of cells to coordinate their response to a signal that only one of them may have received. In plants, there are plasmodesmata between almost all cells, making the entire plant into one giant network.

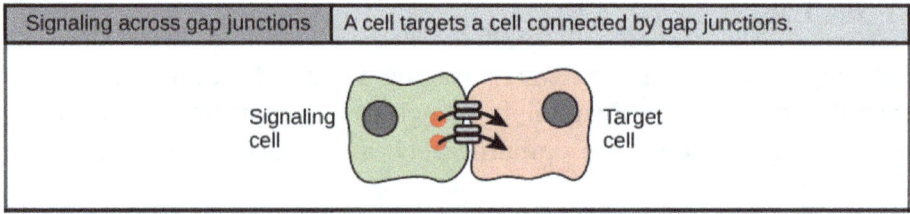

In another form of direct signaling, two cells may bind to one another because they carry complementary proteins on their surfaces. When the proteins bind to one another, this interaction changes the shape of one or both proteins, transmitting a signal. This kind of signaling is especially important in the immune system, where immune cells use cell-surface markers to recognize "self" cells (the body's own cells) and cells infected by pathogens.

CELL ADHESION

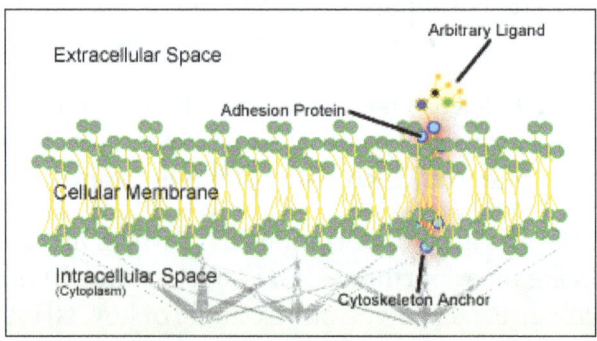

Schematic of cell adhesion

Cell adhesion is the process by which cells interact and attach to neighbouring cells through specialised molecules of the cell surface. This process can occur either through direct contact between cell surfaces or indirect interaction, where cells attach to surrounding extracellular matrix, a gel-like structure containing molecules released by cells into spaces between them. Cells adhesion occurs from the interactions between cell-adhesion molecules (CAMs), transmembrane proteins located on the cell surface. Cell adhesion link cells in different ways and can be involved in signal transduction for cells to detect and respond to changes in the surroundings. Other cellular processes regulated by cell adhesion include cell migration and tissue development in multicellular organisms. Alterations in cell adhesion can disrupt important cellular processes and lead to a variety of diseases, including cancer and arthritis. Cell adhesion is also essential for infectious organisms, such as bacteria or viruses, to cause diseases.

General Mechanism

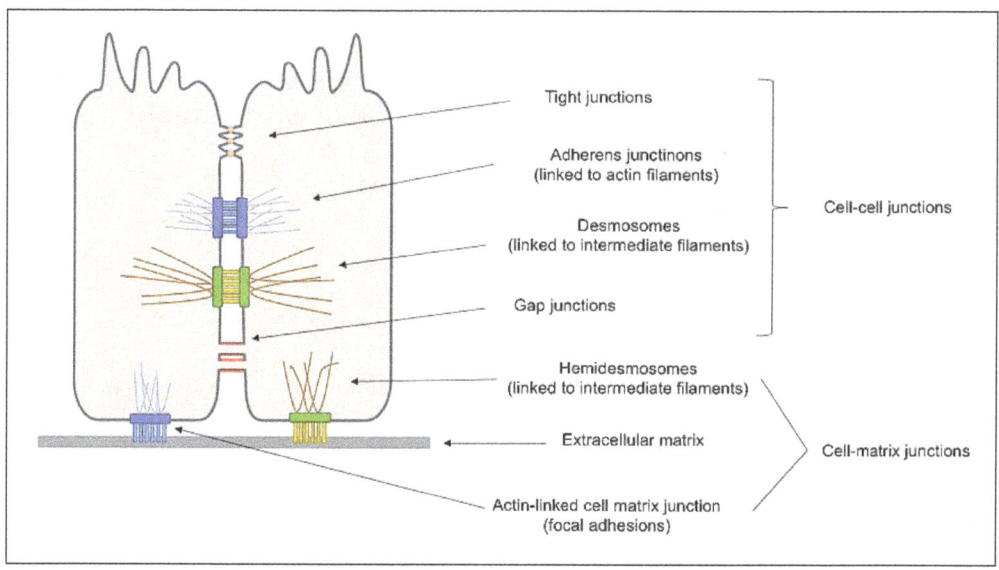

Overview diagram of different types of cell junctions present in epithelial cells,
including cell–cell junctions and cell–matrix junctions.

CAMs are classified into four major families: integrins, immunoglobulin (Ig) superfamily, cadherins, and selectins. Each of these adhesion molecules has a different function and recognizes different ligands. Cadherins and immunoglobulins are homophilic CAMs, as they directly bind to the same type of CAMs on another cell, while integrins and selectins are heterophilic CAMs that bind to different types of CAMs. Defects in cell adhesion are usually attributable to defects in expression of CAMs.

In multicellular organisms, bindings between CAMs allow cells to adhere to one another and creates structures called cell junctions. According to their functions, the cell junctions can be classified as:

- Anchoring junctions (adherens junctions, desmosomes and hemidesmosomes), which maintain cells together and strengthens contact between cells.

- Occluding junctions (tight junctions), which seal gaps between cells through cell–cell contact, making an impermeable barrier for diffusion

- Channel-forming junctions (gap junctions), which links cytoplasm of adjacent cells allowing transport of molecules to occur between cells

- Signal-relaying junctions, which can be synapses in the nervous system

Alternatively, cell junctions can be categorised into two main types according to what interacts with the cell: cell–cell junctions, mainly mediated by cadherins, and cell–matrix junctions, mainly mediated by integrins.

Cell–cell Junctions

Cell–cell junctions can occur in different forms. In anchoring junctions between cells such as adherens junctions and desmosomes, the main CAMs present are the cadherins. This family of CAMs are membrane proteins that mediate cell–cell adhesion through its extracellular domains and require extracellular Ca^{2+} ions to function correctly. Cadherins forms homophilic attachment between themselves, which results in cells of a similar type sticking together and can lead to selective cell adhesion, allowing vertebrate cells to assemble into organised tissues. Cadherins are essential for cell–cell adhesion and cell signalling in multicellular animals and can be separated into two types: classical cadherins and non-classical cadherins.

Adherens Junctions

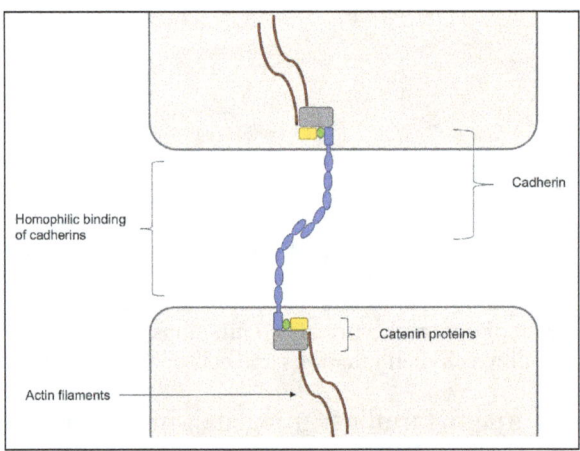

Adheren junction showing homophilic binding between
cadherins and how catenin links it to actin filaments

Adherens junctions mainly function to maintain shape of tissues and hold cells together. In adherens junctions, classical cadherins between neighbouring cells interact through their extracellular domains. Classical cadherins share a conserved calcium-sensitive region in their extracellular domains. When this region comes into contact with Ca^{2+} ions, extracellular domains of cadherins change from the inactive flexible conformation to a more rigid conformation to undergo homophilic binding. Intracellular domains of classical cadherins are also highly conserved as they bind to proteins called catenins forming catenin-cadherin complexes, which links classical cadherins to actin filaments. This association with actin filaments is essential for adherens junctions to stabilises cell–cell adhesion. Interactions with actin filaments can also promote clustering of cadherins, involved in assembly of adherens junctions, as cadherin clusters promotes actin filaments polymerisation which in turn assembles adherens junctions by binding to cadherin–catenin complexes forming at the junction.

Desmosomes

Desmosomes are structurally similar to adherens junctions but composed of different components. Instead of classical cadherins, non-classical cadherins such as desmogleins and desmocollins act as adhesion molecules and they are linked to intermediate filaments instead of actin filaments. No catenin is present in desmosomes as intracellular domains of desmosomal cadherins interact with desmosomal plaque proteins, which form the thick cytoplasmic plaques in desmosomes and link cadherins to intermediate filaments. Desmosomes provides strength and resistance to mechanical stress by unloading forces onto the flexible but resilient intermediate filaments, something that cannot occur with the rigid actin filaments. This makes desmosomes important in tissues that encounter high levels of mechanical stress, such as heart muscle and epithelia, and explains why it appears frequently in these types of tissues.

Tight Junctions

Tight junctions are normally present in epithelial and endothelial tissues, where they seal gaps and regulate paracellular transport of solutes and extracellular fluids in these tissues that function as barriers. Tight junction is formed by transmembrane proteins, including claudins, occludins and tricellulins, that bind closely to each other on adjacent membranes in a homophilic manner. Similar to anchoring junctions, intracellular domains of these tight junction proteins are bound with scaffold proteins that keep these proteins in clusters and link them to actin filaments in order to maintain structure of the tight junction. Claudins, essential for formation of tight junctions, form paracellular pores which allow selective passage of specific ions across tight junctions making the barrier selectively permeable.

Gap Junctions

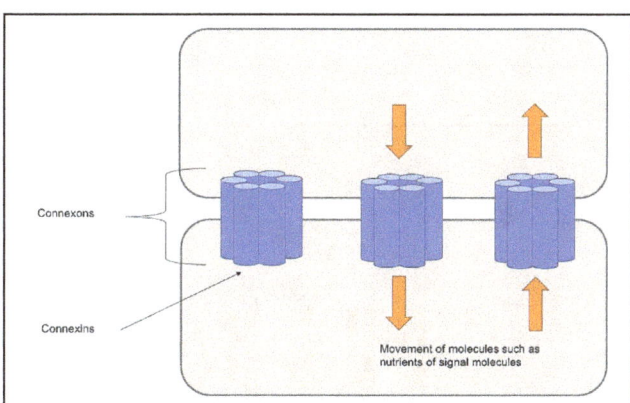

Gap junctions showing connexons and connexins

Gap junctions are composed of channels called connexons, which consist of transmembrane proteins called connexins clustered in groups of six. Connexons from adjacent cells form continuous channels when they come into contact and align with each other. These channels allow transport of ions and small molecules between cytoplasm of two adjacent cells, apart from holding cells together and provide structural stability like anchoring junctions or tight junctions. Gap junction channels are selectively permeable to specific ions depending on which connexins form the connexons, which allows gap junctions to be involved in cell signalling by regulating the transfer of molecules involved in signalling cascades. Channels can respond to many different stimuli and are regulated dynamically either by rapid mechanisms, such as voltage gating, or by slow mechanism, such as altering numbers of channels present in gap junctions.

Adhesion Mediated by Selectins

Selectins are a family of specialised CAMs involved in transient cell–cell adhesion occurring in the circulatory system. They mainly mediate the movement of white blood cells (leukocytes) in the bloodstream by allowing the white blood cells to "roll" on endothelial cells through reversible bindings of selections. Selectins undergo heterophilic bindings, as its extracellular domain binds to carbohydrates on adjacent cells instead of other selectins, while it also require Ca^{2+} ions to function, same as cadherins. cell–cell adhesion of leukocytes to endothelial cells is important for immune responses as leukocytes can travel to sites of infection or injury through this mechanism. At these sites, integrins on the rolling white blood cells are activated and bind firmly to the local endothelial cells, allowing the leukocytes to stop migrating and move across the endothelial barrier.

Adhesion Mediated by Members of the Immunoglobulin Superfamily

The immunoglobulin superfamily (IgSF) is one of the largest superfamily of proteins in the body and it contains many diverse CAMs involved in different functions. These transmembrane proteins have one or more immunoglobulin-like domains in their extracellular domains and undergo calcium-independent binding with ligands on adjacent cells. Some IgSF CAMs, such as neural cell adhesion molecules (NCAMs), can perform homophilic binding while others, such as intercellular cell adhesion molecules (ICAMs) or vascular cell adhesion molecules (VCAMs) undergo heterophilic binding with molecules like carbohydrates or integrins. Both ICAMs and VCAMs are expressed on vascular endothelial cells and they interact with integrins on the leukocytes to assist leukocyte attachment and its movement across the endothelial barrier.

Cell–matrix Junctions

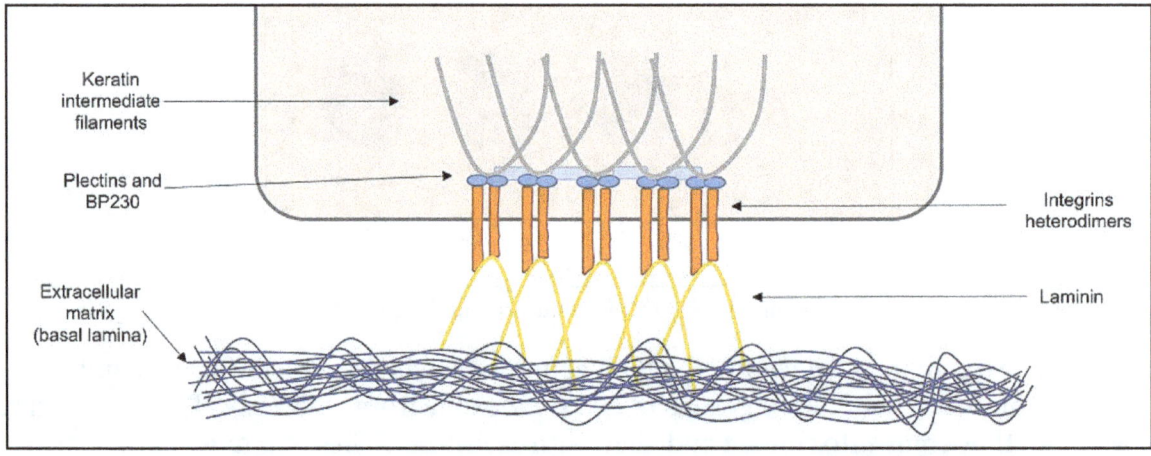

Hemidesmosomes diagram showing interaction between integrins and laminin, including how integrins are linked to keratin intermediate filaments

Cells creates extracellular matrix by releasing molecules into its surrounding extracellular space. Cells have specific CAMs that will bind to molecules in the extracellular matrix and link the matrix to the intracellular cytoskeleton. Extracellular matrix can act as a support when organising cells into tissues and can also be involved in cell signalling by activating intracellular pathways when bound to the CAMs. Cell–matrix junctions are mainly mediated by integrins, which also clusters like cadherins to form firm adhesions. Integrins are transmembrane heterodimers formed by different α and β subunits, both subunits with different domain structures. Integrins can signal in both directions: inside-out signalling, intracellular signals modifying the intracellular domains, can regulate affinity of integrins for their ligands, while outside-in signalling, extracellular ligands binding to extracellular domains, can induce conformational changes in integrins and initiate signalling cascades. Extracellular domains of integrins can bind to different ligands through heterophilic binding while intracellular domains can either be linked to intermediate filaments, forming hemidesmosomes, or to actin filaments, forming focal adhesions.

Hemidesmosomes

In hemidesmosomes, integrins attach to extracellular matrix proteins called laminins in the basal lamina, which is the extracellular matrix secreted by epithelial cells. Integrins link extracellular matrix to keratin intermediate filaments, which interacts with intracellular domain of integrins via adapter proteins such as plectins and BP230. Hemidesmosomes are important in maintaining structural stability of epithelial cells by anchoring them together indirectly through the extracellular matrix.

Focal Adhesions

In focal adhesions, integrins attach fibronectins, a component in the extracellular matrix, to actin filaments inside cells. Adapter proteins, such as talins, vinculins, α-actinins and filamins, form a complex at the intracellular domain of integrins and bind to actin filaments. This multi-protein complex linking integrins to actin filaments is important for assembly of signalling complexes that act as signals for cell growth and cell motility.

Other Organisms

Eukaryotes

Plants cells adhere closely to each other and are connected through plasmodesmata, channels that cross the plant cell walls and connect cytoplasms of adjacent plant cells. Molecules that are either nutrients or signals required for growth are transported, either passively or selectively, between plant cells through plasmodesmata.

Protozoans express multiple adhesion molecules with different specificities that bind to carbohydrates located on surfaces of their host cells. cell–cell adhesion is key for pathogenic protozoans to attach en enter their host cells. An example of a pathogenic protozoan is the malarial parasite (*Plasmodium falciparum*), which uses one adhesion molecule called the circumsporozoite protein to bind to liver cells, and another adhesion molecule called the merozoite surface protein to bind red blood cells.

Pathogenic fungi use adhesion molecules present on its cell wall to attach, either through protein-protein or protein-carbohydrate interactions, to host cells or fibronectins in the extracellular matrix.

Prokaryotes

Prokaryotes have adhesion molecules on their cell surface termed bacterial adhesins, apart from using its pili (fimbriae) and flagella for cell adhesion. Adhesins can recognise a variety of ligands present on the host cell surfaces and also components in the extracellular matrix. These molecules also control host specificity and regulate tropism (tissue- or cell-specific interactions) through their interaction with their ligands.

Viruses

Viruses also have adhesion molecules required for viral binding to host cells. For example, influenza virus has a hemagglutinin on its surface that is required for recognition of the sugar sialic acid on host cell surface molecules. HIV has an adhesion molecule termed gp120 that binds to its ligand CD4, which is expressed on lymphocytes. Viruses can also target components of cell junctions to enter host cells, which is what happens when the hepatitis C virus targets occludins and claudins in tight junctions to enter liver cells.

Clinical Implications

Dysfunction of cell adhesion occurs during cancer metastasis. Loss of cell–cell adhesion in metastatic tumour cells allows them to escape their site of origin and spread through the circulatory system. One example of CAMs deregulated in cancer are cadherins, which are inactivated either by genetic mutations or by other oncogenic signalling molecules, allowing cancer cells to migrate and be more invasive. Other CAMs, like selectins and integrins, can facilitate metastasis by mediating cell–cell interactions between migrating metastatic tumour cells in the circulatory system with endothelial cells of other distant tissues. Due to the link between CAMs and cancer metastasis, these molecules could be potential therapeutic targets for cancer treatment.

There are also other human genetic diseases caused by an inability to express specific adhesion molecules. An example is leukocyte adhesion deficiency-I (LAD-I), where expression of the β_2 integrin subunit is reduced or lost. This leads to reduced expression of β_2 integrin heterodimers, which are required for leukocytes to firmly attach to the endothelial wall at sites of inflammation in order to fight infections. Leukocytes from LAD-I patients are unable to adhere to endothelial cells and patients exhibit serious episodes of infection that can be life-threatening.

An autoimmune disease called pemphigus is also caused by loss of cell adhesion, as it results from autoantibodies targeting a person's own desmosomal cadherins which leads to epidermal cells detaching from each other and causes skin blistering.

Pathogenic microorganisms, including bacteria, viruses and protozoans, have to first adhere to host cells in order to infect and cause diseases. Anti-adhesion therapy can be used to prevent infection by targeting adhesion molecules either on the pathogen or on the host cell. Apart from altering the production of adhesion molecules, competitive inhibitors that bind to adhesion molecules to prevent binding between cells can also be used, acting as anti-adhesive agents.

References

- What-is-cellular-metabolism, natural-health: globalhealingcenter.com, Retrieved 13 May, 2019

- Rich, p. R. (2003). "the molecular machinery of keilin's respiratory chain". Biochemical society transactions. 31 (pt 6): 1095–1105. Doi:10.1042/bst0311095. Pmid 14641005

- Photosynthesis, science: britannica.com, Retrieved 3 May, 2019

- P.hinkle (2005). "p/o ratios of mitochondrial oxidative phosphorylation". Biochimica et biophysica acta. 1706: 1–11. Doi:10.1016/j.bbabio.2004.09.004. Pmid 15620362

- Reproduction-text., 255mitos, cmallery: miami.edu, Retrieved 8 June, 2019

- Bernstein h, bernstein c (2010). "evolutionary origin of recombination during meiosis". Bioscience. 60 (7): 498–505. Doi:10.1525/bio.2010.60.7.5

- Cell-cycle-phases, mitosis, cellular-molecular-biology, biology, science: khanacademy.org, Retrieved 5 March, 2019

- Lodé t (june 2011). "sex is not a solution for reproduction: the libertine bubble theory". Bioessays. 33 (6): 419–22. Doi:10.1002/bies.201000125. Pmid 21472739

- Phases-of-mitosis, mitosis, cellular-molecular-biology, biology, science: khanacademy.org, Retrieved 12 July, 2019

- Carlson, b. M. (2007). Principals of regenerative biology. Elsevier academic press. P. 379. Isbn 978-0-12-369439-3

- Introduction-to-cell-signaling, mechanisms-of-cell-signaling, cell-signaling, biology, science: khanacademy.org, Retrieved 1 August, 2019

- Alberts, bruce; johnson, alexander; lewis, julian; morgan, david; raff, martin; roberts, keith; walter, peter (2014). Molecular biology of the cell (6th ed.). Garland science. Isbn 9780815344322

Permissions

Index

www.ingramcontent.com/pod-product-compliance
Lightning Source LLC
Chambersburg PA
CBHW080409190526
45161CB00003B/182